DOCUMENTS
IN
ENGLISH ECONOMIC
HISTORY

Documents in English Economic History

Edited by

B.W. CLAPP, H.E.S. FISHER
and A.R.J. JUŘICA

England since 1760

Edited by
B.W. Clapp

LONDON

G. Bell & Sons Ltd

1976

India
Orient Longman Ltd.
Calcutta, Bombay, Madras and New Delhi

Canada
Clarke, Irwin & Co. Ltd, Toronto

Australia
Edward Arnold (Australia) Pty Ltd, Port Melbourne, Vic.

New Zealand
Book Reps (New Zealand) Ltd, 46 Lake Road, Northcote, Auckland

East Africa
J. E. Budds, P.O. Box 44 536, Nairobi

West Africa
Thos, Nelson (Nigeria) Ltd, P.O. Box 336, Apapa, Lagos

South and Central Africa
Book Promotions (Pty), Ltd, 311 Sanlam Centre,
Main Road, Wynberg, Cape Province

ISBN 0 7135 1871 5

Printed in Hungary

Preface

It is now sixty years since A. E. Bland, P. A. Brown and R. H. Tawney compiled their well-known documentary account of English economic history since the Norman Conquest, *English Economic History: Select Documents* (Bell, 1914). The present volume provides students, sixth-formers and their teachers with a documentary account of the past two hundred years of English economic and social history. A companion volume is in preparation under the editorship of Dr H. E. S. Fisher and Mr A. R. J. Juřica, covering the period from the Norman Conquest to the Industrial Revolution. Taken together the two volumes will provide the first full-scale collection of documents illustrating the economic history of England since 'Bland, Brown and Tawney' was published. While their aim is a compilation that more closely reflects modern conceptions of what English economic history is about, the editors of these volumes are also anxious to preserve some continuity. With the agreement of the publishers, Messrs. G. Bell and Sons, and of the literary executors of Bland, Brown and Tawney they have incorporated into the new work a few documents from the old. Again following Bland, Brown and Tawney, the editors of this collection have been sparing of comment on the documents.

An extended collection of documents illustrating English economic history has several jobs to do. The most important, no doubt, is to familiarise students with a representative if small sample of the primary materials from which history comes to be written. A collection of documents on English economic history should, secondly, illustrate the themes that would be treated in the ideal textbook that one would wish to write. Economic history as it is written in England is far from the ideal. Despite Adam Smith's reminder that 'consumption is the sole end and purpose of all production', historians have attended but little to consumption, or to distribution either for that matter, but have turned instead to more easily worked fields of research, particularly manufacture, landholding and foreign trade. The economic history of England as written is too often a conventional account of exporting industries, agriculture, and overseas commerce, with little attempt to convey a credible picture even of the occupations of the people, let alone of their way of life, diet, social customs, or aspirations. The editors of these volumes have aimed to present to the reader a more rounded and realistic view of the past. The discipline of economic history like economic life itself does not advance steadily on a broad front. There are growing points that often reflect contemporary pre-occupations—in the 1930s with economic fluctuations, in the 1960s with growth—that provoke much research, some enlightenment and a little academic controversy. A general collection of documents cannot hope to marshall the evidence that would settle these thorny problems, but it does incidentally and can deliberately include material that has a bearing on such points of debate. Some documents, therefore, may have a double interest as reflecting past economic life and as providing a piece of evidence to support or rebut a fashionable interpretation of the past.†

Finally, the editors hope that their collection may have a little value as a guide to sources. Like earlier collections of documents this one relies, though not exclusively, on printed sources, and makes no claim to be considered as a handbook for research workers, whose concern is mostly with manuscript material. Within its limitations, however, the present collection takes a more liberal view of what constitutes a piece of historical evidence than its predecessor, Bland, Brown and Tawney. Historians need no longer apologise as Macaulay did for citing humble sources. Ballads, local newspapers, the autobiography of a painter, an excerpt from *The Gondoliers* are cited here as well as graver works such as Parliamentary Papers and statutes of the realm. Historians now realise that there is no virtue in official sources as such. As we live in a statistical age it is the historian's duty to look for representative or typical evidence. Whether he finds it in official documents or in ephemeral verse is immaterial. In practice any compilation is likely to rely substantially on the sources favoured by historians down the ages—official publications, topographical works, the literature of protest. Provided that the compiler remains alert to the dangers there need be no harm in this addiction. As Sir John Clapham remarked, '[the bluebooks], carefully read, tell us almost as much about normal as about diseased social tissue'. The evidence submitted to the Select Committee on Import Duties, 1840, for

---

† For example, document 8 in Chapter 1, section B. This document may be read both as a contemporary comment on the rise in interest rates, 1896–1914, and as an anticipation of W. W. Rostow's theories on the evolution of the British economy in the nineteenth century.

example, has useful information about the coffeehouses of London. The Richmond Commission on the depression of agriculture, 1881, illustrates the economics of ship-owning. And the royal commission on coal supplies, 1903, has evidence from Parsons about the progress of the steam turbine.[†] One of the pleasures of documentary history lies in such unexpected discoveries.

The editor gratefully acknowledges the help he has received from his colleagues in the preparation of this volume. Professor W. E. Minchinton gave freely of his advice; Mr. M. A. Havinden put his expert knowledge of agricultural history at the editor's disposal. The late A. J. Rogerson, research assistant in the department of economic history of the University of Exeter, did much of the preliminary work in the selection of documents. Above all, Dr. H. E. S. Fisher has taken a close interest in the progress of this volume from early planning to proof-reading; his prudence has saved the editor from many a slip. For the faults and errors that remain the editor alone is responsible.

March 1975   B. W. C

---

† Chapter 9, section B, 4; Chapter 4, 13; Chapter 3, section B, 6.

Acknowledgments

A book of this kind could not be put together without the consent of many authors, executors and publishers to the reproduction of copyright material. This consent has been readily given and the editor wishes to acknowledge his gratitude for the generosity with which his troublesome requests have been met. If he has inadvertently included copyright material without making due acknowledgment he offers his sincere apologies.

The following have kindly consented to the reproduction of copyright material:

The Comptroller, Her Majesty's Stationery Office, Professor G. C. Allen, *British industries and their organisation*, (Longman); British Railways Board, *The reshaping of British railways;* Lord Franks, *Central planning and control in war and peace;* Tom Harrisson, Director Mass-Observation Archive, University of Sussex, *Britain revisited;* Professor R. A. Humphreys and the Council of the Royal Historical Society, *British consular reports on the trade and politics of Latin America;* the Liberal Party, *Britain's industrial future;* Philip Mair, executor of W. H., Lord Beveridge, *Unemployment: a problem of industry;* A. Plummer, *Witney blanket industry;* J. B. Priestley, *English journey;* H. Pollins, 'Railway auditing—a report of 1867'; The Observer, *Rethinking our future;* Allen and Unwin Ltd., publishers of R. D. Best, *Brass chandelier*, and of R. H. Tawney, *The radical tradition;* Edward Arnold Ltd., publishers of Lady Hugh Bell, *At the works;* The Business Archives Council, publishers of T. S. Ashton, *Letters of a West African trader;* Cambridge University Press and the Pilgrim Trust, *Men without work;* the Council of the Bibliographical Society, London, publishers of E. Howe, *The London Compositor;* Doubleday and Company, Inc., A. de Tocqueville, *Journeys to England and Ireland*, ed. J. P. Mayer; the Estate of H. Rider Haggard and A. P. Watt and Son, *Rural England;* Fabian Society, publishers of

S. Webb, *The decline in the birthrate;* the Longman Group, publishers of C. R. Fay, *Huskisson and his age,* and of A. Loveday, *Britain and world trade;* Macmillan, publishers of C. Wright and C. E. Fayle, *A history of Lloyd's,* of B. S. Rowntree, *Poverty: a study of town life,* and of J. B. Orr, *Food, health and income;* Manchester University Press and Chetham Society, publishers of F. Collier, *The family economy of the working classes, 1784–1833,* and of C. S. Davies, *The agricultural history of Cheshire, 1750–1850;* John Murray Ltd., publishers of H. Withers, *Poverty and waste,* and of A. D. Hall, *Pilgrimage of British farming;* Thomas Nelson and Sons Ltd, publishers of Edward Gibbon, *Memoirs of my life,* ed. G. A. Bonnard; the Royal Statistical Society, publishers of Sir George Paish, 'Great Britain's investments in other lands'; and the George Sturt Memorial Fund and Cambridge University Press, publishers of G. Sturt, *The wheelwright's shop.*

Contents

2 AGRICULTURE

3 THE EVOLUTION OF INDUSTRY

4 TRANSPORT AND INTERNAL TRADE

5 FINANCE

ABBREVIATIONS AND SYMBOLS

B.P.P.   British Parliamentary Papers

...   A small deletion has been made.

✳ ✳ ✳   A substantial passage has been omitted.

The place of publication of books quoted is London unless otherwise stated.

[ ]   Matter added by the editor

qq   Questions

†   Editor's footnote

The numbering of footnotes in quoted passages is that in the original document.

1

A general view of the English economy

This chapter illustrates some of the permanent features and recurrent problems facing the English economy. It is a complicated and dynamic economy that is revealed. Even before industrialisation had taken place or towns grown large (*A3*) the pattern of employment was varied and migration was a relatively common event. The modern pre-occupation with the loss of highly trained manpower should not blind us to the long continuing drain of unskilled and poor men from the British Isles to foreign parts (*A8*, *A13*). Within the country too, there was a drift to the large towns and especially to London, which always held a high place among the industrial as well as commercial centres of the country (*A10*). Already a rich and envied country in 1760 (*A1*), England continued to enjoy economic growth interrupted from time to time by periodic depression. For much of this period economic growth was taken for granted (*B1*, *B4*), based as it was on a flexible economic system that could quickly shift resources of men and capital into profitable ventures (*B5*).

In the twentieth century England has lost her economic supremacy and her belief in the inevitability of progress. But provided government and business exercise forethought, progress is still the general expectation (*B7*, *B11*). The economy has long suffered from recurrent bouts of economic depression (*B3*, *B6*, *B9*, *B10*) and in more recent years from a rate of economic growth substantial in itself but slow in comparison with that of Germany, Japan, Italy or Russia. In part slow economic growth reflects a slow growth of population (*A11*, *A12*); in part, too, a faulty economic structure which may be curable by means of new departures in policy (Chapter 10, *12*).

A. THE STRUCTURE OF SOCIETY

1 Voltaire looks at England, 1762

(Theodore Besterman, ed. *Voltaire's correspondence* (Geneva, 1959) L, p. 125.)

Voltaire to Louis René Caradeuc de la Chalotais　　*6 November* [*1762*]

The English have long been more stupid than us, it is true, but see how they improve themselves. They have neither monks nor convents, but victorious fleets; their clergy beget books and children; their peasants have brought wastelands under the plough; their commerce encircles the world; and their philosophers have taught us undoubted truths. I confess to a feeling of envy when I think of England.[†]

2 A village census, 1778

(Devon Record Office, 1165Z/Z1. From place-name evidence and the parish registers, the place can be identified as Wembworthy and the document dated to 1778. Wembworthy lies off the main Exeter-Barnstaple road, about twenty miles north of Exeter. Its population rose to a peak of 444 in 1851 and at the census of 1961 had fallen to 210.)

A list of all the inhabitants of this parish

Estates

The barton of heighwood [Haywood]	Mr John Saunders, Joan Saunders his wife 55, his sons John 40, Anthony 34; grandchildren William Saunders 10, John Saunders 8, Elizabeth Saunders 6; his servants Abraham Rowe 28, Abraham Heighwood 14, William Nickles 17, Margaret Stocky 35, Rebecca Mocksy 17. This famelly in all are 12.
Sales	Mr Andrew Saunders 52, his sons Andrew 29, William 18, Robert 17, John 15; daughters Ann 27, Elizabeth 25. These are 7.
Raishly [Rashleigh]	Mr John Passmore 48, Elizabeth Passmore his wife 50; his sons John 17, Richard 12; his daughters Elizabeth 15, Ann 14; his servants William Page 48, Francis Simmins 27, Willmot Triger 31, Elizabeth Short 16, John Bradford aprintis 17. These are 11.
Labdon	Mr John Osman 23, Susanna his wife 20; Elizabeth Pateridge 20; his servants George Dillin 17, John Saunders 15. These are 5.

---

† Translation

Stone	Mr Samuel Snell 24, Ann his wife 26; his daughter Ann 1; his servants George Lange 16, his printis Grace Born 9. These are 5.
In a cot house	Stafford Heighwood 40, Ann his wife 35; his daughters, Mary 9, Elizabeth 7, Ann 6. These are 5.
Upacott [Upcott]	Mr Robert Goss 49, Mary his wife 40; his sons William 8, Robert 7, John 4, Andrew 3; his daughter Mary 1; William Westcott aprintis 16. These are 8.
Heigher Upacott	Mr William Herd 40, Marget his wife 38; his sons William 6, John 5; his daughter Ann 2; Frances Luxton 9; his servant Joan Baker 27, printis William Lyle 16. These are 8.
Cumb [Combe]	Thomas Lange 50, Mary his wife 48; his daughter Mary 24, Elizabeth Kingsmill 23. These are 4.
Paddins	Mr William Saunders 50, Mary his wife 45; his son Richard 9; his servants Thomas Lange 18, Susanna Willcox 28. These are 5.
Bidbare	Mr Hugh Pyke 26, Ann his wife 25, his son William [no age given], John Pyke his brother 18; his prentisis John Webber 9, Grace Luxton 21. These are 6.
Yealands	Mr John Knight 63, Mary his wife 62, his grandson John Knight 7, servants Mary Westcott 26, William Westcott 20, and his printis John Mean 19. These are 6.
Wemworthy Down	Mr John Luxton 30, Ann his wife 28, Samuel Luxton 8, Elizabeth 7, Richard 5, Robert 3, his servants John Lee 40, Thomas Reed 29, Ales [Alice?] Webber 26, his printises Phillip Mean 17, Thomas Seage 12. These are 11.
Ally Mill	Mr Robert Knight 40, Ann his wife 30, his son Robert 3, his daughters Mary 5, Ann 2; his printises Thomas Standlake 15, Ann Trigger 14, These are 7.
Down	Mr Grigory Goss 64, Elizabeth his wife 60, his sons William 25, John 16; his daughters

Grace 21, Mary 19; his servant Thomas Mean 25; aprintis William Trigger 17. These are 8.

Kinny Down

Mr William Friend 45, his housekeeper Elizabeth Westcott 40; his servants Thomas Westcott 18, Thomas Rowe, 26, Mary Parkhouse 17; his printises William Friend 16, Phillip Westcott 10. These are 7.

Goses [Gosses?]

Mr George Nickles 50, Elizabeth Nickles, his wife 48; his sons George 27, John 26, Richard 17. These are 5.

Hole

Mr John Gridgeworthy 30, Ann his wife 28, his children Elizabeth 4, Mary 3, John 12 months old; his printises Thomas Chudly 16, Joseph Underhill 14. These are 7.

[No address given]

Elizabeth Underhill 50, her son Robert 8; her daughters Amey 23, Ann 17; her grandson William Underhill half one. These are 5.

Lavender

Mr John Mean 45, Joan Mean 40; his sons John 12, Richard 2; his daughters Mary 10, Ann 6; father in law, Thomas Galserry 84; living in the same house Ann Merrifield 50, Mary Call 34. These are 9.

in bridge beare

John Westcott 45, Judith Westcott his wife 40; his son John 4; his daughter Elizabeth 15. These are 4.

James Herding 40 [?], Elizabeth 36; his son James 1; his daughter Mary 3; his mother in law Mary Hellings 64; his Ephen Mary Hellings [no age given]. These are 6.

John Honney 34, Susanna his wife 32; his sons Richard 10, John 7; his daughter Mary 1; living in the same house Mary Gieve 84. These are 6.

Raishly Mill

James Gallyver 24, Isaac Gallyver 20. These are 2.

George Wrankmore 35, Joan his wife 30; his sons George 8, John 4, William 3 months old. These are 5.

George Mean 36, Elizabeth his wife [no age given], his son Richard 7; daughter

Mary 3; his father in law Richard Mean [sic] 80; in the same house, Mary Cann 56. These are 6.

Bridge beare

Michael Skinner 48, Elizabeth his wife 44; his sons Benjamin 16, Michael 9; his daughter Mary 11. These are 5.

John Trigger 64, Sarah his wife 66. These are 2.

Absalom Trigger 30, Grace his wife 28; his son Richard 4; his daughters Ann 6, Elizabeth 2. These are 5.

Catherine Arscott 66; her son Samuel 25; her daughter Mary 36. These are 3.

Also in a cot house

John Sture 40, Mary his wife 35; his sons John 7, Peter 3. These are 4.

Christian Corvett 50; her son John 19; her daughter Christian 16. These are 3.

Gallins Green

Bernard Short 45, Joan his wife 40; his daughter Susanna 16. These are 3.

Marget Petres 40, her son Simeon 18, her daughter Mary 20. These are 3.

Thomas Luxton 30, Elizabeth his wife 28, his children Mary 4, Elizabeth 3, John 6 months. These are 5.

Elizabeth Reed 50, her daughter Joan 20, her granddaughter Mary Peck 6. These are 3.

Speaks Cross

George Bird 40, Elizabeth his wife 36; his son Robert 6; his daughter Elizabeth 2. These are 4.

William Willky 30, Prue his wife 32; his son William 5. These are 3.

Daniel Cloke 46, Abigail his wife 40; his sons Daniel 5 and Jeremiah [no age given]. These are 4.

John Sloman 67, Mary his wife 50. These are 2.

Richard Hill 24, Elizabeth his wife 30; his daughter Elizabeth 1. These are 3.

3 England's provincial towns in 1781

(Adam Anderson, *An historical and chronological deduction of the origin of commerce from the earliest accounts to the present time, containing an history of the great commercial interest of the British Empire* (rev. ed. London, 4 vols. (1787–9), IV, p. 388)

The following is a correct account of the number of houses in certain towns, laid before the House of Commons, in this year, by the tax office.

Towns	Houses	Towns	Houses
Exeter	1,474	Newcastle	2,219
Plymouth	1,510	Bristol	3,247
York	2,285	Bath	1,173
Hull	1,370	Ipswich	1,244
Sheffield	2,022	Birmingham	2,291
Liverpool	3,974	Cambridge	1,925
Manchester	2,519	Oxford	2,316
Norwich	2,302	Dover	1,193
Lynn	662	Nottingham	1,533
Yarmouth	682	Northampton	706
Shrewsbury	904		

4 The checks to population growth, 1803

(T. R. Malthus, *An essay on the principle of population* (2nd ed. 1803), pp. 16, 300–2, 308–9)

But without attempting to establish in all cases these progressive and retrograde movements in different countries, which would evidently require more minute histories than we possess, the following propositions are proposed to be proved:

1 Population is necessarily limited by the means of subsistence.

2 Population invariably increases, where the means of subsistence increase, unless prevented by some very powerful and obvious checks.

3 These checks, and the checks which repress the superior power of population, and keep its effects on a level with the means of subsistence, are all resolvable into moral restraint, vice, and misery.

＊   ＊   ＊

The most cursory view of society in this country, must convince us, that throughout all ranks, the preventive check to population prevails in a considerable degree. Those among the higher classes, who live principally in towns, often want the inclination to marry, from the facility with which they can indulge themselves in an illicit intercourse with the sex. And others are deterred from marrying, by the idea of the expences that they must retrench, and the pleasures of which they must deprive themselves, on the supposition of having a family. When the fortune is large, these considerations are certainly trivial; but a preventive foresight of this kind, has objects of much greater weight for its contemplation as we go lower.

A man of liberal education, with an income only just sufficient to enable him to associate in the rank of gentlemen, must feel absolutely certain, that if he marry, and have a family, he shall be obliged, if he mix in society, to rank himself with farmers and tradesmen. The woman, that a man of education would naturally make the object of his choice, is one brought up in the same habits and sentiments with himself, and used to the familiar intercourse of a society totally different from that to which she must be reduced by marriage. Can a man easily consent to place the object of his affection in a situation so discordant, probably, to her habits and inclinations. Two or three steps of descent in society, particularly at this round of the ladder, where education ends and ignorance begins, will not be considered by the generality of people as a chimerical, but a real evil. If society be desirable, it surely must be free, equal, and reciprocal society, where benefits are conferred as well as received, and not such as the dependent finds with his patron, or the poor with the rich.

These considerations certainly prevent a great number in this rank of life, from following the bent of their inclinations in an early attachment. Others, influenced either by a stronger passion, or a weaker judgment, disregard these considerations; and it would be hard indeed, if the gratification of so delightful a passion as virtuous love, did not sometimes more than counterbalance all its attendant evils. But I fear that it must be acknowledged, that the more general consequences of such marriages are rather calculated to justify, than to disappoint, the forebodings of the prudent.

The sons of tradesmen and farmers, are exhorted not to marry, and generally find it necessary to comply with this advice, till they are settled in some business or farm, which may enable them to support a family. These events may not, perhaps, occur till they are far advanced in life. The scarcity of farms is a very general complaint; and the competition in every kind of business is so great, that it is not possible that all should be successful. Among the clerks in counting houses, and the competitors for all kinds of mercantile and professional employment, it is probable, that the preventive check to population prevails more than in any other department of society.

The labourer who earns eighteen-pence or two shillings a day, and lives at his ease as a singleman, will hesitate a little, before he divides that pittance among four or five, which seems to be not more than sufficient for one. Harder fare, and harder labour, he would perhaps be willing to submit to, for the sake of living with the woman that he loves; but he must feel conscious, that, should he have a large family, and any ill fortune whatever, no degree of frugality, no possible exertion of his manual strength, would preserve him from the heart-rending sensation of seeing his children starve, or of being obliged to the parish for their support. The love of independence is a sentiment that surely none would wish to see eradicated; though the parish law of England, it must be confessed, is a system of all others the most calculated gradually to weaken this sentiment, and in the end will probably destroy it completely.

The servants who live in the families of the rich, have restraints yet stronger to break through in venturing upon marriage. They possess the necessaries, and even the comforts, of life, almost in as great plenty as their masters. Their work is easy, and their food luxurious, compared with the work and food of the class of labourers; and their sense of dependence is weakened by the conscious power of changing their masters, if they feel themselves offended. Thus comfortably situated at present, what are their prospects if they marry. Without knowledge, or capital, either for business, or farming, and unused, and therefore unable, to earn a subsistence by daily labour, their only refuge seems to be a miserable alehouse, which certainly offers no very enchanting prospect of a happy evening to their lives. The greater number of them, therefore, deterred by this uninviting view of their future situation, content themselves with remaining single where they are.

If this sketch of the state of society in England be near the truth, it will be allowed, that the preventive check to population operates with considerable force throughout all the classes of the community. And this observation is further confirmed by the abstracts from the registers returned in consequence of the late Population Act. The results of these abstracts shew, that the annual marriages in England and Wales, are to the whole population as 1 to $123\frac{1}{5}$, a smaller proportion of marriages than obtains in any of the countries which have been examined, except Norway and Switzerland.

✳   ✳   ✳

There certainly seems to be something in great towns, and even in moderate towns, peculiarly unfavourable to the very early stages of life; and the part of the community on which the mortality principally falls, seems to indicate, that it arises more from the closeness and foulness of the air, which may be supposed to be unfavourable to the tender lungs of children, and the greater confinement, which they almost necessarily experience, than from the superior degree of luxury and debauchery, usually, and justly, attributed to towns. A married pair, with the best constitutions, who lead the most regular and quiet life, seldom find that their children enjoy the same health in towns as in the country.

In London, according to former calculations, one half of the born died under three years of age; in Vienna and Stockholm under two; in Manchester, under five; in Norwich, under five; in Northampton, under ten. In country villages, on the contrary, half of the born live till thirty, thirty-five, forty, forty-six, and above. In the parish of Ackworth, in Yorkshire, it appears, from a very exact account kept by Dr Lee of the ages at which all died there for 20 years, that half of the inhabitants live to the age of 46, and there is little doubt, that, if the same kind of account had been kept in some of those parishes before mentioned, in which the mortality is so small as 1 in 60, 1 in 66, and even 1 in 75, half of the born would be found to have lived till 50 or 55.

5 The division of labour, machinery, and economic growth, 1822

(J. Lowe, *The present state of England*, 1822, Appendix, pp. 63–4, 67–8)

Subdivision of Employment in great Cities

To mark this subdivision in all its extent, the observer must repair to the French, or rather to the English capital, where the mercantile, the manufacturing, the mechanical professions, all assume the most simple form. A London banker, different from his provincial brethren, issues no notes, and keeps no interest account with his customers: a merchant confines his connexions to a few foreign sea-ports, perhaps to a particular colony or town; and the name of general merchant, though not yet disused, is hardly applicable even to our greatest houses. But it is in the mechanical arts that the subdivision of employment takes a form the most familiar and most intelligible to ordinary observation. In London the class of shoemakers is divided, says Mr Gray, into makers of shoes for men, shoes for women, shoes for children: also into boot-cutters, boot-closers, boot-makers. Even tailors, though to the public each appears to do the whole of his business, are divided among themselves into makers of coats, waistcoats, breeches, gaiters. In other lines an equally minute repartition takes place: and as to the ornamental or elegant arts, such as those of jeweller, painter, engraver, nothing would be more easy than to exhibit a long list of professions limited to large towns, and wholly unknown in a thinly-peopled district.

Effect of this Subdivision

What, it may be asked, is the practical result of this minute sub-division, this nice distinction of employment? By fixing the attention of the workman on a single part of his business, it renders him sur-prizingly correct and expeditious: his performance gains equally in quality and in dispatch. This is the result of a mechanical dexterity, acquired without any particular effort of the mind; for we must by no means infer that the quickness characteristic of the inhabitants of a large town, that promptitude which distinguishes the Londoner and the Parisian from the hesitation and circumlocution of the country-man, is the consequence of any innate superiority: those who walk in a crowd must adopt the step of others, and advance with the rapidity of the moving mass. The attainments of these persons, meaning such attainments as they possess accurately and thoroughly, are often confined to a few branches; but these are the objects of their profession or business; and the result is, that their work proceeds straight forward, very little time being lost by them in planning, altering, or correct-ing. ...

Effect of Machinery on the Condition of the working Classes

The effect of mechanical improvement in adding to the income of a community admits of no doubt, its result being to afford a commodity frequently of better quality, and always at a cheaper rate. To be satisfied of the latter, we have merely to compare the prices of either our cottons or hardware of the present day with those of similar articles made by us thirty years ago, or with those made at present on the Con-tinent, where machinery is as yet but partially adopted. But what, it will be asked, is the effect of machinery on the income and comfort of the workman? At first injurious, bringing with it the evils of transition, which are very serious in a time marked like the present, by a great reduction in the demand for hands for the public service.

To take an instance familiar to those of our countrymen who have resided in France: in that country coal is very little used, and the general fuel in town, as in country, is wood: the trees after being felled, are cut into short but thick blocks, carted into the towns, sold in the public markets, and broken up by men who make a business of it, but whose labour, aided only by the wedge and saw, is tedious and fatiguing, adding nearly ten per cent to the cost of the article. To break these solid blocks by machinery would cause a considerable saving of both time and expence, but in the present stagnation of the demand for labour, it would be harsh, and indeed unsafe to resort to such an alternative, without providing for the thousands who would thus be deprived of employment.

Such, in a greater or less degree, is the case in almost every transition of importance. Eventually, however, the hardship is overcome, and the use of machinery becomes productive of great additional comfort to the lower orders. To prove that its beneficial effects are general, it is not enough to cite the prosperity of a few manufacturing districts, as the success of these may be accompanied by distress in other parts; the prosperity of Lancashire may cause embarrassment in Saxony, Flanders, or the Banks of the Rhine. The advantage, then, arising from the use of machinery, rests on a broader basis; on that law in productive industry which makes every *real* reduction of cost an addition to individual income, or, what is the same thing, to the comforts procured by that income. The benefit of such reduction is enjoyed by the public at large: the evil, on the other hand, is partial, being confined to the manufacturer. He, however, is benefited in his capacity of consumer, and experiences relief from his distress as soon as it is found practicable to transfer to a new branch a portion of the capital and industry hitherto employed on his own. Such transfers are, it is true, tasks of great time and difficulty: we have felt them to be so in our own country, while in others less advanced, they can hardly be accomplished in the life-time of a generation.

6 Birmingham and Manchester compared, 1835

(Alexis de Tocqueville, *Journeys to England and Ireland* ed. J. P. Mayer (Faber, 1958), pp. 94, 104–5)

Birmingham (*25th–30th June*)

We found as much goodwill here as in London; but there is hardly any likeness between these two societies. These folk never have a minute to themselves. They work as if they must get rich by the evening and die the next day. They are generally very intelligent people, but intelligent in the American way. The town itself has no analogy with other English provincial towns; the whole place is made up out of streets like the rue du Faubourg St-Antoine. It is an immense workshop, a huge forge, a vast shop. One only sees busy people and faces brown with smoke. One hears nothing but the sound of hammers and the whistle of steam escaping from boilers. One might be down a mine in the New World. Everything is black, dirty and obscure, although every instant it is winning silver and gold.

*   *   *

Manchester (2nd July 1835)
Peculiar character of Manchester

The great manufacturing city for cloth, thread, cotton... as is Birmingham for iron, copper, steel.

Favourable circumstances: ten leagues from the largest port in England, which is the best-placed port in Europe for receiving raw materials from America safely and quickly. Close by the largest coal-mines to keep the machines going cheaply. Twenty-five leagues away, the place where the best machines in the world are made. Three canals and a railway quickly carry the products all over England, and over the whole world.

The employers are helped by science, industry, the love of gain and English capital. Among the workers are men coming from a country where the needs of men are reduced almost to those of savages, and who can work for a very low wage, and so keep down the level of wages for the English workmen who wish to compete, to almost the same level. So there is the combination of the advantages of a rich and of a poor country; of an ignorant and an enlightened people; of civilisation and barbarism.

So it is not surprising that Manchester already has 300,000 inhabitants and is growing at a prodigious rate.

Other differences between Manchester and Birmingham

The police are less efficient at Manchester than at Birmingham. More complete absence of government; 60,000 Irish at Manchester (at most 5,000 at Birmingham); a crowd of small tenants huddled in the same house. At Birmingham almost all the houses are inhabited by one family only; at Manchester a part of the population lives in damp cellars, hot, stinking and unhealthy; thirteen to fifteen individuals in one. At Birmingham that is rare. At Manchester, stagnant puddles, roads paved badly or not at all. Insufficient public lavatories. All that almost unknown at Birmingham. At Manchester a few great capitalists, thousands of poor workmen and little middle class. At Birmingham, few large industries, many small industrialists. At Manchester workmen are counted by the thousand, two or three thousand in the factories. At Birmingham the workers work in their own houses or in little workshops in company with the master himself. At Manchester there is above all need for women and children. At Birmingham, particularly men, few women. From the look of the inhabitants of Manchester, the working people of Birmingham seem more healthy, better off, more orderly and more moral than those of Manchester.

Exterior appearance of Manchester (2nd July)

An undulating plain or rather a collection of little hills. Below the hills a narrow river (the Irwell), which flows slowly to the Irish sea. Two streams (the Medlock and the Irk) wind through the uneven ground and after a thousand bends, flow into the river. Three canals made by man unite their tranquil, lazy waters at the same point. On this watery land, which nature and art have contributed to keep damp, are scattered palaces and hovels. Everything in the exterior appearance of the city attests the individual powers of man; nothing the directing power of society. At every turn human liberty shows its capricious creative force. There is no trace of the slow continuous action of government.

7 The railway age, 1842

(Inscription on headstone, St Peter's churchyard, Newton-le-Willows. I owe this document to the kindness of Mr David H. Pill.)

SACRED

TO THE MEMORY OF

PEERS NAYLOR

ENGINEER

WHO DEPARTED THIS LIFE

10th DECEMBER, 1842,

AGED 29 YEARS

My *engine* now is cold and still
No water does my *boiler* fill;

My *coke* affords its flame no more,
My days of usefulness are o'er.
My *wheels* deny their noted speed,
No more my guiding hand they heed.
My *whistle*, too, has lost its tone,
Its shrill and thrilling sounds are gone.
My *valves* are now thrown open wide,
My *flanges* all refuse to guide.
My *clacks*, also, though once so strong,
Refuse to aid the busy throng.
No more I feel each urging breath,
My *steam* is now condensed in death.
Life's railway's o'er, each *station* past,
In death I'm stopped, and rest at last.
Farewell, dear Friends, and cease to weep,
In Christ I'm safe; in Him I sleep.

8 The emigrant's passage to America

(*i*) *The hardships of the voyage, c. 1842*

(Charles Dickens, *Martin Chuzzlewit*, Chapter XXIV)

... 'How do *you* find yourself this morning, sir?'.

'Very miserable,' said Martin, with a peevish groan. 'Ugh! This is wretched, indeed!'

'Creditable,' muttered Mark, pressing one hand upon his aching head and looking round him with a rueful grin. 'That's the great comfort. It *is* creditable to keep up one's spirits here. Virtue's its own reward. So's jollity'.

Mark was so far right, that unquestionably any man who retained his cheerfulness among the steerage accommodations of that noble and fast-sailing line-of-packet ship, 'The Screw', was solely indebted to his own resources, and shipped his good humour, like his provisions, without any contribution or assistance from the owners. A dark, low, stifling cabin, surrounded by berths all filled to overflowing with men, women, and children, in various stages of sickness and misery, is not the liveliest place of assembly at any time; but when it is so crowded (as the steerage cabin of 'The Screw' was, every passage out), that mattresses and beds are heaped upon the floor, to the extinction of everything like comfort, cleanliness, and decency, it is liable to operate not only as a pretty strong barrier against amiability of temper, but as a positive encourager of selfish and rough humours. Mark felt this, as he sat looking about him; and his spirits rose proportionately.

There were English people, Irish people, Welsh people, and Scotch people there; all with their little store of coarse food and shabby clothes; and nearly all with their families of children. There were children of all ages; from the baby at the breast, to the slattern-girl who was as much a grown woman as her mother. Every kind of domestic suffering that is bred in poverty, illness, banishment, sorrow, and long travel in bad weather, was crammed into the little space.

(ii) Cheap fares, 1851

(Select Committee on the Passengers' Act, *B.P.P.*, 1851, XIX, qq. 734–48, 6094–114)

What is the average freight now paid in the case of a liner for the lowest class of passengers?—Their price is 5*l*; but they do not always adhere to that price.

Does that include provisions?—It includes the provisions according to the Act.

What portion of the 5*l* do you believe now is paid for the provisions?—The bread-stuffs cost about 14*s* or 15*s*.

In case an emigrant vessel does not sail at the time appointed, do they allow the emigrants money or rations?—Money, in general.

At what rate per day?—At the rate of 1*s* a day, which is provided for in the Act.

Then if the emigrant were to be compelled to be there one day before the sailing of the vessel for the facility of your examination, the effect of that would be to increase the charge by 1*s*? Yes, if it was claimed by the passengers.

If an emigrant arrived 24 hours before the vessel sailed, he must be sustained either by himself, or by somebody else?—If he claimed it; but if the people were examined on the Tuesday, and the ship got away before 12 o'clock on the Wednesday, I do not apprehend that he would have any claim for subsistence.

He must support himself then?—He must support himself.

(*Chairman*) If they were on board they would not receive the shilling, but would be fed with the provisions of the ship?—They would claim their provisions.

And the provisions do not cost a shilling a day?—No. I should say that the bread-stuffs cost 14*s* or 15*s* for the whole voyage.

And therefore the cost per day would be what?—I should suppose about 4*d* or 5*d*; not more, if so much.

What is the average length of the voyage?—The average length of the voyage to New York is, I think, about five weeks.

That would be more than 4*d* then; it would be 5*d*; but whether it was 4*d* or 5*d*, or whatever the sum was, it would be an additional expense to the ship?—It would; I have given the average for the liners.

Your examination all refers to the liners: what is the charge by other vessels?—£4.

And do they give the same rations?—The same; they are obliged to do so by law.

*   *   *

Since you have been in the trade has there been a great variation in the price of passage?—A very great variation.

Has its tendency been gradually to diminish?—It has not gradually either increased or diminished.

But it has fluctuated?—It has fluctuated; at the present moment it is less than it ever has been.

If you take the average for the last six or seven years, leave out the year before the famine, the price of passage has surely diminished, has it not?—No.

Could the people go then for 3*l*, 5*l* and 4*l*?—Yes; it ought to have increased, for the law has imposed increased restrictions upon the shipowners.

Not of a nature to increase the cost of passage much, has it?—To the extent of the provision which has been added to the dictary scale.

Was not the dietary greater before?—There was none.

But they ate?—They found themselves; the expense of carrying them has been increased 1*l* at least.

(*Mr Divett*) And you do not think the expense of the passage has increased in proportion to those regulations?—No.

(*Chairman*) Taking the year 1848, when the Passenger Act was passed, and taking the first year of the working of the Act, has it diminished?—I think it has stood about the same.

Provisions are cheaper, are they not, now than they were then?—They were cheaper, but they are now advancing.

The price of oatmeal is high at present, is it not?—It is extremely high in proportion to what it has been; for what you paid 20*s* a short time ago, you would now have to pay 30*s*.

Meat is cheaper, is it not?—We do not give them meat; we entirely adopt the English scale.

I thought the custom was to survey upon the English scale, and to adopt the American?—We adopt entirely the English scale.

Which is the best scale?—I think the English is.

Which is the cheapest?—The American.

Just a turn?—Perhaps two or three shillings cheaper.

A little less bread-stuff and a couple of pounds of meat instead would probably bring it to about the same thing, would it not?—Yes.

Would it not be an advantage if the two scales could be assimilated? —We have always entertained that opinion. If the two Governments would assimilate their passenger law it would be very much better.

If they were to assimilate the law with reference to space also, would not that, in your opinion be an advantage?—Yes.

But in England the shipowners would object, would they not, to the larger space being allowed, because it would interfere so much with the price of the Canadian passage?—No, I do not think it would. The price of the Canadian passage now is much higher than the price of the passage to America, because there is not the same amount of trade. The price of the passage to America is just now, and has been for the last few weeks, extremely low; it has been as low as 2*l* 5*s*, or even lower than that.

9 The pattern of jobs in 1851—an incomplete industrial revolution

(The Census of 1851, Report, *B.P.P.*, 1852-3, LXXXVIII, Part I)

Table 41—Occupations in Great Britain, and Number of Persons engaged in them (arranged in the order of the Numbers) in 1851

Occupations	Persons
Agricultural Labourer	1,460,896
Farm Servant, Shepherd	1,038,791
Cotton Calico, manufacture, printing, and dyeing	561,465
Labourer (branch undefined)	376,551
Farmer, Grazier	306,767
Boot and Shoe maker	274,451
Milliner, Dressmaker	267,791
Coal-miner	219,015
Carpenter, Joiner	182,636

n Table 41, on page 17, for
Farm Servant, Shepherd',
ead 'Domestic Servant'.

Occupations	Persons
Army and Navy	*178,773
Tailor	152,672
Washerwomen, Mangler, Laundry-keeper	146,091
Woollen Cloth manufacture	137,814
Silk manufacture	114,570
Blacksmith	112,776
Worsted manufacture	104,061
Mason, Pavior	101,442
Messenger, Porter, and Errand Boy	101,425
Linen, Flax manufacture	98,860
Seaman (Merchant Service), on shore or in British Ports	89,206
Grocer	85,913
Gardener	80,916
Iron manufacture, moulder, founder	80,032
Innkeeper, Licensed Victualler, Beershop keeper	75,721
Seamstress, Shirtmaker	73,068
Bricklayer	67,989
Butcher, Meat Salesman	67,691
Hose (Stocking) manufacture	65,499
School,—master, mistress	65,376
Lace manufacture	63,660
Plumber, Painter, Glazier	62,808
Baker	62,472
Carman, Carrier, Carter, Drayman	56,981
Charwoman	55,423
Draper (Linen and Woollen)	49,184
Engine and Machine Maker	48,082
Commercial Clerk	43,760
Cabinet maker, Upholsterer	40,897
Teacher (various), Governess	40,575
Fisherman, woman	38,294
Boat, Barge, Man, Woman	37,683
Miller	37,268
Earthenware manufacture	36,512
Sawyer	35,443
Railway Labourer	34,306
Straw-plait manufacture	32,062

* This is the Army and Navy of the *United Kingdom*, exclusive of the Indian Army and Navy.

Occupations	Persons
Brick maker, dealer	31,168
Government Civil Service	30,963
Hawker, Pedlar	30,553
Wheelwright	30,244
Glover	29,882
Shopkeeper (branch undefined)	29,800
Horsekeeper, Groom (not Domestic), Jockey	29,408
Nail manufacture	28,533
Iron-miner	28,088
Printer	26,024
Nurse (not Domestic Servant)	25,518
Shipwright,Shipbuilder	25,201
Stone Quarrier	23,489
Lodging-house Keeper	23,089
Lead-miner	22,530
Copper-miner	22,386
Straw Hat and Bonnet maker	21,902
Cooper	20,215
Watch and Clock maker	19,159
Brewer	18,620
Dock Labourer, Dock and Harbour Service	18,462
Clergyman of Estab. Church	18,587
Protestant Dissenting Minister	9,644
Police	18,348
Plasterer	17,980
Warehouse,—man, woman	17,861
Saddler, Harness maker	17,583
Hatter, Hat manufacture	16,975
Coachman (not Domestic Servant), Guard, Postboy	16,836
Law Clerk	16,626
Coachmaker	16,590
Cowkeeper, Milkseller	15,526
Ropemaker	15,966
Druggist	15,643
Surgeon, Apothecary	15,163
Tin-miner	15,050
Paper manufacture	14,501
Coalheaver, Coal Labourer	14,426
Greengrocer, Fruiterer	14,320
Muslin manufacture	14,098

Occupations	Persons
Confectioner	13,865
Tinman, Tinker, Tin-plate worker	13,770
Staymaker	13,690
Solicitor, Attorney, Writer to the Signet	13,256
Dyer, Scourer, Calenderer	12,964
Currier	12,920
Builder	12,818
Farm Bailiff	12,805
Hair-dresser, Wig-maker	12,173
Coal merchant, dealer	12,092
Glass manufacture	12,005
Carpet and Rug manufacture	11,457
Goldsmith, Silversmith	11,242
Brass founder, moulder, manufacture	11,230
Maltster	11,150
Railway Officer, Clerk, Station Master	10,948
Bookbinder	10,953
Road Labourer	10,923
Wine and Spirit Merchant	10,467
Fishmonger	10,439
Merchant	10,256
Ribbon manufacture	10,074

10 Internal migration

(*i*) *The industrial regions attract country people*

(The Census of 1851, Report, *B.P.P.*, 1852–3, LXXXVIII, pp. cvii–cviii)

Upon comparing the actual increase in the population of each county with the numbers by which the births exceeded the deaths, in the ten years 1841-51, the proportion of the increase that is due to natural causes and immigration is apparent. Thus the births that were registered in London exceeded the deaths by 144,688; while the increase of numbers in the same time, as shown by the Censuses, was 413,819; so that, had all the births been registered, 269,131 of the latter numbers must have been referable to immigration. In Lancashire and Cheshire the increase by births was 218,443; by immigration, 205,375. In Sussex, Hants, and Bedford a small portion of the increase was due in the ten years to immigration, and a much larger proportion in the counties of Stafford, Worcester, Warwick, the West and the East Ridings of York, Durham, Northumberland, Monmouth, and South Wales. The other counties, if we may borrow a phrase from Natural History, send out swarms of their population every year. Thus the births in the Eastern Counties were 118,574 in the ten years; the increase, as determined by the Census, was 73,366; so that 4,521 of the youth of Norfolk, Suffolk, and Essex leave their native counties every year to reap elsewhere the fruits of the education, skill, and vigour which they have derived, at great expense* from their parents at home. The district in which they labour is the district in which they contribute directly or indirectly to the poor rate; and in it they should receive relief.

A free circulation of the people is now necessary in Great Britain, to meet the varying requirements of the Public Industry.

Such is a brief digest of the answers which have been received from the inhabitants of Great Britain to the question: 'Where were you born?'

The separation of families which is inevitable in a population like that of Great Britain is in some respects painful; but the facilities of travelling, of meeting, and of intercourse by letters, have happily increased faster than the population, so as to mitigate the evil; and the whole of the inhabitants will gradually grow acquainted with the different parts of their native land, to which, as well as to the town or village of their birth, it is desirable that the people of the United Kingdom should be attached.

Hitherto the population has migrated from the high or the comparatively healthy ground of the country to the cities and seaport towns, in which few families have lived for two generations. But it is evident that henceforward the great cities will not be like camps—or the fields on which the people of other places exercise their energies and industry—but the birth-places of a large part of the British race.

About *seventy-seven thousand children* are born in London annually. Such arrangements of the houses, and of the squares and open spaces,

---

* The present value of the future earnings of an agricultural labourer in Norfolk is about 482 *l* at the age of 20; the present value of his subsistence from that age is 248 *l*; leaving 234 *l* as the net value of his services. Consequently the 4,526 emigrants of this class carry away a large amount of capital which they have acquired in their native counties.

should therefore be progressively made, as it is known, by experience, are conducive to the health, vigour, and efficient training of children. Facilities for the distribution over wider areas, and for the periodical concentration of the town population, can be made by the agency of the railways; and as the working people go and return to the shops at regular hours, they may evidently be conveyed at as little cost as any kind of merchandise; and thus we may hope that the worst of all Birth-places—the crowded room, or the house of many families—will never be the *Birth-place* of any considerable portion of the British population.

(ii) London still a magnet, 1871–81

(Charles Booth, *Labour and life of the people in London, Vol. I East London* (3rd ed. London, 1891), pp. 504–5)

...What do we mean by the influx into London?

In 1881, out of every 1000 persons living within the metropolitan district, 629 were born in the district, 343 in other parts of the United Kingdom, and 28 abroad. These facts would seem at first to be conclusive evidence of a considerable inflow of population from other parts. But a very large part of this admixture of population merely results from the ordinary ebb and flow of labour, set up by numberless industrial causes in all parts of the kingdom alike. Taking the whole of England and Wales, we find that in 1881 only 720 out of 1000 persons were living in the county of their birth. If we take the seven largest Scotch towns (the only towns for which statistics are published) the result is still more striking; for only 524 out of every 1000 inhabitants were natives by birth of the towns in which they were living.

There are districts in London where as many as a quarter of the inhabitants change their addresses in the course of a year. Every part of England shows a similar shifting backwards and forwards of population to a greater or less degree. All this internal movement, though usually confined to short distances, indicates the existence of migratory habits among the people, which must in the long run produce a considerable admixture of population, though only by a straining of language could we class it as 'influx'.

It is when we turn from the consideration of the mere numbers of outsiders living in London to a comparison between these numbers and the number of Londoners living elsewhere—in other words, when we compare the volume of inflow and outflow—that we see the real significance of the influx.

There were in 1881 nearly double as many natives of other parts of England and Wales resident in London as natives of London living in other parts of England and Wales. In other words (leaving out for the moment the question of foreign immigration and emigration) London was, at that time, recruited from England and Wales to the extent of 579,371 persons, the excess of inflow over outflow. We may look at the question statistically from another point of view. The population of London in 1871 was 3,254,260. The excess of births over deaths in London in the next ten years was 454,475. Thus the population should in 1881 have been 3,708,735. It actually was 3,816,488, showing an unaccounted-for excess of 107,753, which is the nett direct result of the process of recruiting from the country and abroad during ten years. Thus London gains *directly* at the rate of rather more than 10,000 a year from its contact with other places, a number which would be largely increased if we included in London such rapidly growing districts as West Ham in Essex. There is also probably a considerable *indirect* gain, which will be spoken of later.

Just as changes of temperature represent the balance of gain or loss due to a far larger and constant exchange of heat by radiation and absorption, so this comparatively small annual gain to London is the index of a much more extensive interchange of population between London and the country.

11 The declining birthrate, 1907

(Sidney Webb, *The decline in the birth rate* (Fabian Society, tract no. 131, 1907), pp. 16–18)

...Apart from some mystic idea of marriage as a 'sacrament', or, at any rate, as a divinely instituted relation with peculiar religious obligations for which utilitarian reasons cannot be given, it does not seem easy to argue that prudent regulation differs essentially from deliberate celibacy from prudential motives. If, as we have for generations been taught by the economists, it is one of the primary obligations of the individual to maintain himself and his family in accordance with his social position and, if possible, to improve that position, the deliberate restriction of his responsibilities within the means which he has of fulfilling them can hardly be counted otherwise than as for righteousness. And when we pass from obligations of the 'self-regarding' class to the wider conception of duty to the community, the ground for blame is, to the ordinary citizen, no more clear. A generation ago, the economists, and, still more, the 'enlightened public opinion' that caught up their words, would have seen in this progressive limitation of population, whether or not it had their approval, the compensating advantage of an uplifting of the economic conditions of the lowest grade of laborers. At any rate, it would have been said, the poorest will thereby be saved from starvation and famine. To those who still believe in the political economy of Ricardo, Nassau Senior, Cairnes and Fawcett—to those, in fact, who still adhere to an industrial system based exclusively on the pecuniary self-interest of the individual and on unshackled freedom of competition—this reasoning must appear as valid to-day as it did a generation ago.

To the present writer the situation appears in a graver light. More accurate knowledge of economic processes denies to this generation the consolation which the 'Early Victorian' economists found in the limitation of population. No such limitation of numbers prevents the lowest grade of workers, if exposed to unfettered individual competition, from the horrors of 'sweating' or the terrors of prolonged lack of employment. On the other hand, with Factory Acts and trade union 'collective bargaining' maintaining a deliberately fixed national *minimum*, the limitation of numbers, however prudent it may be in individual instances, is, from the national standpoint, seen to be economically as unnecessary as it is proved to be futile even for the purposes for which McCulloch and Mill, Cairnes and Fawcett so ardently desired it.

Nor can we look forward, even if we wished to do so, to the vacuum remaining unfilled. It is, as all experience proves, impossible to exclude the alien immigrant. Moreover, there are in Great Britain, as in all other countries, a sufficient number of persons to whom the prudential considerations affecting the others do not appeal, or appeal less strongly. In Great Britain at this moment, when half, or perhaps two-thirds, of all the married people are regulating their families, children are being freely born to the Irish Roman Catholics and the Polish, Russian and German Jews, on the one hand, and to the thriftless and irresponsible—largely the casual laborers and the other denizens of the one-roomed tenements of our great cities—on the other. Twenty-five per cent of our parents, as Professor Karl Pearson keeps warning us, is producing 50 per cent of the next generation. This can hardly result in anything but national deterioration; or, as an alternative, in this country gradually falling to the Irish and the Jews. Finally, there are signs that even these races are becoming influenced. The ultimate future of these islands may be to the Chinese!

Thus, modern civilization is faced by two awkward facts; the production of children is rapidly declining, and this decline is not uniform, but characteristic of the more prudent, foreseeing and self-restrained sections of the community. It is only in mitigation of the first of these facts that it can be urged that the death-rate is also declining, so that in most countries the net annual increase of population exhibits little sign of slackening. This, indeed, affords but slight ground of satisfaction. The probable diminution in the death-rate has very narrow limits; whilst that in the birth-rate is cumulative and limitless. What is of far greater social importance is that a diminished death-rate among those who are born in no way mitigates the evil influence of an adverse selection—it even intensifies its effects.

The conclusion which the present writer draws from the investigation is, however, one of hope, not of despair. It is something to discover the cause of the phenomenon. Moreover, the cause is one that we can counteract. If the decline in the birth-rate, had been due to physical degeneracy, whether brought about by 'urbanization' or otherwise, we should not have known how to cope with it. But a deliberately volitional interference, due chiefly to economic motives, can at any moment be influenced, and its adverse selection stopped, partly by a mere alteration of the economic conditions, partly by the opportunity for the play of the other motives which will be thereby afforded.

What seems indispensable and urgent is to alter the economic incidence of child-bearing. Under the present social conditions the birth of children in households maintained on less than three pounds a week (and these form four-fifths of the nation) is attended by almost penal consequences. The wife is incapacitated for some months from earning money. For a few weeks she is subject to a painful illness, with some risk. The husband has to provide a lump sum for the necessary medical attendance and domestic service. But this is not all. The parents know that for the next fourteen years they will have to dock themselves and their other children of luxuries and even of some of the necessaries of life, just because there will be another mouth to feed. To four-fifths of all the households in the land each succeeding baby means the probability of there being less food, less clothing, less house room, less recreation and less opportunity for advancement for every member of the family. Similar considerations appeal even more strongly to a majority of the remaining 20 per cent of the population, who make up the 'middle' and professional classes. Their higher standard of life, with its requirements in the way of culture and refinement, and with the long and expensive education which it demands for their children, makes the advent even of a third or fourth child—to say nothing of the possibility of a family of eight or twelve—a burden far more psychologically depressing than that of the wage-earner. In order that the population may be recruited from the self-controlled and foreseeing members of each class rather than of those who are reckless and improvident, we must alter the balance of considerations in favour of the child-producing family.

12 Large families unfashionable, 1945

(Mass Observation, *Britain and her birthrate* (Murray, 1945), pp. 74–5. The authors are quoting a middle-aged Lancashire housewife.)

My own opinion is that people wish to have a small family on account of public opinion which has now hardened into custom. It is customary—and has become so during the last twenty-five years or so—to have two children and no more if you can avoid it. A family of five or six children loses in prestige and, some think, in respectability. It is on behalf of their children that parents feel this most keenly. If anyone doubts this let him walk along a seaside promenade with five or six children of varying ages and see the attention they attract. Some years ago I sat in a train and heard a woman exclaim to her daughter: 'Look at that lady on the platform with six children! If I had six children I would never take them out.' The daughter remarked charitably: 'Perhaps they aren't all hers'.

Recently in a bus queue a young man said to his girl: 'See that lady with the child in her arms, she has another who can only just walk, and a third of about four or five.' (Actually they were a little older) 'Isn't she well supplied!' Inside the bus, the conductor said in an amiable way 'You ought to be allowed free on the bus with this lot'. Fortunately the woman had a sense of humour for she remarked to the elder child: 'It's a good thing we didn't bring the others'.

Two is certainly the correct number of children in a British Family. People do not approve of 'only ones', besides one may be deprived of an only child by death. With two, things are more certain. Nobody seems to desire three children, and I know of only five families with three children. One of these mothers said to me: 'I was so ashamed when the third was expected, I wouldn't go out if I could help it'. This is an extreme case perhaps, but I have heard of others similar.

13 The brawn drain, 1968

(*Express and Echo*, Exeter, 29 July 1968)

AUSTRALIA

QUEENSLAND

Mount Isa Mines Limited

required

LABOURERS

(Single men—aged 21-35)

for work underground at Mount Isa

Must be used to manual labour in heavy industry or work with heavy earthmoving equipment. Long-term jobs for good men.

Passages arranged for successful applicants.

Starting wages are $A69.19 (£32 5s 10d) per 40-hour week, including bonuses.
Full board and lodging at Company hostel $A14.75 (£6 17s 0d) per week.

Please Note: Officers of the Company will be available to interview interested men at the ROYAL CLARENCE HOTEL CATHEDRAL CLOSE, EXETER, on WEDNESDAY, JULY 31st, anytime between 9 a.m. to 1 p.m., and 2 p.m. to 6 p.m.
Please ask for Company Representative at reception.

B. ECONOMIC PROBLEMS AND PROSPECTS

1 A resilient economy, 1776

(Adam Smith, *An enquiry into the nature and causes of the wealth of nations* (ed. E. Cannan, Methuen, 1904), I, 326–8)

The annual produce of the land and labour of England, for example, is certainly much greater than it was, a little more than a century ago, at the restoration of Charles II. Though, at present, few people, I believe, doubt of this, yet during this period, five years have seldom passed away in which some book or pamphlet has not been published, written too with such abilities as to gain some authority with the public, and pretending to demonstrate that the wealth of the nation was fast declining, that the country was depopulated, agriculture neglected, manufactures decaying, and trade undone. Nor have these publications, been all party pamphlets, the wretched offspring of falsehood and venality. Many of them have been written by very candid and very intelligent people; who wrote nothing but what they believed, and for no other reason but because they believed it.

The annual produce of the land and labour of England again, was certainly much greater at the restoration, than we can suppose it to have been about an hundred years before, at the accession of Elizabeth. At this period too, we have all reason to believe, the country was much more advanced in improvement, than it had been about a century before, towards the close of the dissensions between the houses of York and Lancaster. Even then it was, probably, in a better condition than it had been at the Norman conquest, and at the Norman conquest, than during the confusion of the Saxon Heptarchy. Even at this early period, it was certainly a more improved country than at the invasion of Julius Caesar, when its inhabitants were nearly in the same state with the savages in North America.

In each of those periods, however, there was, not only much private and public profusion, many expensive and unnecessary wars, great perversion of the annual produce from maintaining productive to maintain unproductive hands; but sometimes, in the confusion of civil discord, such absolute waste and destruction of stock, as might be supposed, not only to retard, as it certainly did, the natural accumulation of riches, but to have left the country, at the end of the period, poorer than at the beginning. Thus, in the happiest and most fortunate period of them all, that which has passed since the restoration, how many disorders and misfortunes have occurred, which, could they have been foreseen, not only the impoverishment, but the total ruin of the country would have been expected from them? The fire and the plague of London, the two Dutch wars, the disorders of the revolution, the war in Ireland, the four expensive French wars of 1688, 1702, 1742, and 1756, together with the two rebellions of 1715 and 1745. In the course of the four French wars, the nation has contracted more than a hundred and forty-five millions of debt, over and above all the other extraordinary annual expence which they occasioned, so that the whole cannot be computed at less than two hundred millions. So great a share of the annual produce of the land and labour of the country, has, since the revolution, been employed upon different occasions, in maintaining an extraordinary number of unproductive hands. But had not those wars given this particular direction to so large a capital, the greater part of it would naturally have been employed in maintaining productive hands, whose labour would have replaced, with a profit, the whole value of their consumption. The value of the annual produce of the land and labour of the country, would have been considerably increased by it every year, and every year's increase would have augmented still more that of the following year. More houses would have been built, more lands would have been improved, and those which had been improved before would have been better cultivated, more manufactures would have been established, and those which had been established before would have been more extended; and to what height the real wealth and revenue of the country might, by this time, have been raised, it is not perhaps very easy even to imagine.

But though the profusion of government must, undoubtedly, have retarded the natural progress of England towards wealth and improvement, it has not been able to stop it. The annual produce of its land and labour is, undoubtedly, much greater at present than it was either at the restoration or at the revolution. The capital, therefore, annually employed in cultivating this land, and in maintaining this labour, must likewise be much greater. In the midst of all the exactions of government, this capital has been silently and gradually accumulated by the private frugality and good conduct of individuals, by their universal, continual, and uninterrupted effort to better their own condition. It is this effort, protected by law and allowed by liberty to exert itself in the manner that is most advantageous, which has maintained the progress of England towards opulence and improvement in almost all former times, and which, it is to be hoped, will do so in all future times. England, however, as it has never been blessed with a very parsimonious government, so parsimony has at no time been the characteristical virtue of its inhabitants. It is the highest impertinence and presumption, therefore, in kings and ministers, to pretend to watch over the œconomy of private people, and to restrain their expence, either by sumptuary laws, or by prohibiting the importation of foreign luxuries. They are themselves always, and without any exception, the greatest spendthrifts in the society. Let them look well after their own expence, and they may safely trust private people with theirs. If their own extravagance does not ruin the state, that of their subjects never will.

2 The sources of prosperity, 1792

(Pitt's budget speech, 17 February 1792, in *The speeches of the Rt. Hon. William Pitt in the House of Commons*, 1806, III, pp. 40–7)

If we compare the revenue of last year with that of the year 1786, we shall find an excess in the last year of 2,300,000*l*. If we go back to the year 1783, which is the first year of peace, we shall find the increase since that period, including the produce of the additional permanent taxes which have been imposed in the interval, to be little less than four millions. We shall, I believe, also find, that, with the exception of the year 1786, in which the suspense of trade, occasioned by the negociation for the commercial treaty with France, naturally affected the revenue, there is hardly any one year in which the increase has not been continual.

In examining the branches of revenue, we shall find that rather more than one million has arisen from the imposition of new taxes; about one million more in those articles in which particular and separate regulations have been made for the prevention of fraud; and that the remaining sum of two millions appears to be diffused over the articles of general consumption, and must therefore be attributed to the best of all causes—a general increase in the wealth and prosperity of the country.

If we look more minutely into the particular articles on which the revenue arises, we shall still find no ground to imagine, that any considerable part of it is temporary or accidental, but shall have additional reason to ascribe it to the cause which I have just now stated. In the revenue of the customs there is no material article where an increase might be supposed to proceed from the accident of seasons, but that of sugar, and it appears that, upon the average of the four years on which I have formed my calculation, that article has not produced beyond its usual amount. Many of the articles under the head of customs, in which the augmentation is most apparent, consist of raw materials, the increasing importation of which is, at once, a symptom and a cause of the increasing wealth of the country. This observation will apply, in some degree, even to the raw material of a manufacture which has generally been supposed to be on the decline—I mean that of silk. In the article of wool, the increase has been gradual and considerable. The quantity of bar-iron imported from abroad is also increased, though we all know how considerably our own iron works have been extended during the period to which I have referred. There is hardly any considerable article in which there is any decrease, except that of hemp in the last year, which is probably accidental, and that of linen, the importation of which from abroad may be diminished by accidental causes, or perhaps in consequence of the rapid increase of the manufacture of that article at home.

On looking at the articles composing the revenues of excise, the same observations will arise in a manner still more striking. There is, indeed, one branch of that revenue, the increase of which may in part be attributed to the accident of seasons—I mean that which arises from the different articles of which malt is an ingredient; but I am inclined to believe that this increase cannot be wholly ascribed to that cause, because, during all the four years, the amount of the duty upon beer and ale has uniformly been progressive. In the great articles of consumption which I will shortly enumerate, without dwelling on particulars —in home-made and foreign spirits, wine, soap, tobacco, the increase has been considerable and uniform. In the articles of bricks and tiles, starch, paper, and printed goods, there has also on the whole been a considerable increase, although there has been some fluctuation in different years. Almost every branch of revenue would furnish instances of a similar nature. The revenue raised by stamps has increased in the produce of the old duties, while at the same time new duties have been added to a large amount, and the augmentation is on this head, on the whole, near 400,000*l*, a sum which is raised in such a manner as to be attended with little inconvenience to those who pay it. The amount of the duty upon salt during the same period has been progressive. The revenue of the post-office is another article, comparatively small, but which furnishes a strong indication of the internal state of the country. No additional duty has been imposed since the year 1784. In 1785, it yielded 238,000*l*, and in the last year 338,000*l*. I mention all these circumstances as tending to throw additional light on the subject, and serving to illustrate and confirm the general conclusion to which they all uniformly tend.

4*

If from this examination of the different branches of the revenue, we proceed to a more direct enquiry into the sources of our prosperity, we shall trace them in a corresponding increase of manufacture and commerce.

The accounts formed from the documents of the custom-house are not indeed to be relied upon as shewing accurately the value of our imports and exports in any one year, but they furnish some standard of comparison between different periods, and in that view I will state them to the committee.

In the year 1782, the last year of the war, the imports, according to the valuation at the custom-house, amounted to 9,714,000*l*; they have gradually increased in each successive year, and amounted, in the year 1790, to 19,130,000*l*.

The export of British manufactures forms a still more important and decisive criterion of commercial prosperity. The amount in 1782 was stated at 9,919,000*l*; in the following year, it was 10,409,000*l*; in the year 1790, it had risen to 14,921,000*l*; and in the last year (for which the account is just completed as far as relates to British manufactures), it was 16,420,000*l*. If we include in the account the foreign articles re-exported, the total of the export in 1782 was 12,239,000*l*; after the peace it rose, in 1783, to 14,741,000*l*; and in the year 1790, it was 20,120,000*l*. These documents, as far as they go, (and they are necessarily imperfect) serve only to give a view of the foreign trade of the country. It is more than probable, that our internal trade, which contributes still more to our wealth, has been increasing in at least an equal proportion. I have not the means of stating with accuracy a comparative view of our manufactures during the same period; but their rapid progress has been the subject of general observation, and the local knowledge of gentlemen from different parts of the country, before whom I am speaking, must render any detail on this point unnecessary.

Having gone thus far, having stated the increase of revenue, and shown that it has been accompanied by a proportionate increase of the national wealth, commerce, and manufactures, I feel that it is natural to ask, what have been the peculiar circumstances to which these effects are to be ascribed?

The first and most obvious answer which every man's mind will suggest to this question, is, that it arises from the natural industry and energy of the country: but what is it which has enabled that industry and energy to act with such peculiar vigour, and so far beyond the example of former periods?—The improvement which has been made in the mode of carrying on almost every branch of manufacture, and the degree to which labour has been abridged, by the invention and application of machinery, have undoubtedly had a considerable share in producing such important effects. We have besides seen, during these periods, more than at any former time, the effect of one circumstance which has principally tended to raise this country to its mercantile pre-eminence—I mean that peculiar degree of credit which, by a twofold operation, at once gives additional facility and extent to the transactions of our merchants at home, and enables them to obtain a proportional superiority in markets abroad. This advantage has been most conspicuous during the latter part of the period to which I have referred; and it is constantly increasing, in proportion to the prosperity which it contributes to create.

In addition to all this, the exploring and enterprising spirit of our merchants has been seen in the extension of our navigation and our fisheries, and the acquisition of new markets in different parts of the world; and undoubtedly those efforts have been not a little assisted by the additional intercourse with France, in consequence of the commercial treaty; an intercourse which, though probably checked and abated by the distractions now prevailing in that kingdom, has furnished a great additional incitement to industry and exertion.

But there is still another cause, even more satisfactory than these, because it is of a still more extensive and permanent nature; that constant accumulation of capital, that continual tendency to increase, the operation of which is universally seen in a greater or less proportion, whenever it is not obstructed by some public calamity, or by some mistaken and mischievous policy, but which must be conspicuous and rapid indeed in any country which has once arrived at an advanced state of commercial prosperity. Simple and obvious as this principle is, and felt and observed as it must have been in a greater or less degree, even from the earliest periods, I doubt whether it has ever been fully developed and sufficiently explained, but in the writings of an author of our own time, now unfortunately no more, (I mean the author of a celebrated treatise on the Wealth of Nations) whose extensive knowledge of detail, and depth of philosophical research, will, I believe, furnish the best solution to every question connected with the history of commerce, or with the systems of political economy. This accumulation of capital arises from the continual application, of a part at least, of the profit obtained in each year, to increase the total amount of capital to be employed in a similar manner, and with continued profit in the year following. The great mass of the property of the nation is thus constantly increasing at compound interest; the progress of which in any considerable period, is what at first view would appear incredible. Great as have been the effects of this cause already, they must be greater in future; for its powers are augmented in proportion as they are exerted. It acts with a velocity continually accelerated, with a force continually increased.

Mobilitate viget, viresque acquirit eundo.

It may indeed, as we have ourselves experienced, be checked or retarded by particular circumstances—it may for a time be interrupted, or even overpowered; but, where there is a fund of productive labour and active industry, it can never be totally extinguished. In the season of the severest calamity and distress, its operations will still counteract and diminish their effects;—in the first returning interval of prosperity, it will be active to repair them. If we look to a period like the present, of continued tranquillity, the difficulty will be to imagine limits to its operation. None can be found, while there exists at home any one object of skill or industry short of its utmost possible perfection; —one spot of ground in the country capable of higher cultivation and improvement; or while there remains abroad any new market that can be explored, or any existing market that can be extended. From the intercourse of commerce, it will in some measure participate in the growth of other nations, in all the possible varieties of their situations. The rude wants of countries emerging from barbarism, and the artificial and increasing demands of luxury and refinement, will equally open new sources of treasure, and new fields of exertion, in every state of society, and in the remotest quarters of the globe. It is this principle which, I believe, according to the uniform result of history and experience, maintains on the whole, in spite of the vicissitudes of fortune, and the disasters of empires, a continued course of successive improvement in the general order of the world.

Such are the circumstances which appear to me to have contributed most immediately to our present prosperity. But these again are connected with others yet more important.

They are obviously and necessarily connected with the duration of peace, the continuance of which, on a secure and permanent footing, must ever be the first object of the foreign policy of this country. They are connected still more with its internal tranquillity and with the natural effects of a free but well regulated government.

What is it which has produced, in the last hundred years, so rapid an advance, beyond what can be traced in any other period of our history? What but that, during that time, under the mild and just government of the illustrious Princes of the family now on the throne, a general calm has prevailed through the country, beyond what was ever before experienced; and we have also enjoyed, in greater purity and perfection, the benefit of those original principles of our constitution, which were ascertained and established by the memorable events that closed the century preceding? This is the great and governing cause, the operation of which has given scope to all the other circumstances which I have enumerated.

It is this union of liberty with law, which, by raising a barrier equally firm against the encroachments of power, and the violence of popular commotion, affords to property its just security, produces the exertion of genius and labour, the extent and solidity of credit, the circulation and increase of capital; which forms and upholds the national character, and sets in motion all the springs which actuate the great mass of the community through all its various descriptions.

The laborious industry of those useful and extensive classes (who will, I trust, be in a peculiar degree this day the object of the consideration of the house) the peasantry and yeomanry of the country; the skill and ingenuity of the artificer; the experiments and improvements of the wealthy proprietor of land; the bold speculations and successful adventures of the opulent merchant and enterprising manufacturer; these are all to be traced to the same source, and all derive from hence both their encouragement and their reward. On this point therefore let us principally fix our attention, let us preserve this first and most essential object, and every other is in our power! Let us remember, that the love of the constitution, though it acts as a sort of natural instinct in the hearts of Englishmen, is strengthened by reason and reflection, and every day confirmed by experience; that it is a constitution which we do not merely admire from traditional reverence, which we do not flatter from prejudice or habit, but which we cherish and value, because we know that it practically secures the tranquillity and welfare both of individuals and of the public, and provides, beyond any other frame of government which has ever existed, for the real and useful ends which form at once the only true foundation and only rational object of all political societies.

3 A cure for post-war distress, 1817

(Canning Papers, Liverpool Public Reference Library, quoted in Charles R. Fay, *Huskisson and his age* (Longmans Green, 1951), pp. 252-4)

Geo. Canning

Liverpool 4th Mo. 10 1817.

Respected Friend,

Since my return from London it has afforded me great pleasure to learn that Govt. had decided on lending money on Public Works, and as this is a subject which has occupied much of my attention for several months, I shall request their attention to a few additional remarks on this subject.

What I have before stated went in substance to prove that we had plenty of everything in the country (in good seasons) and that we only wanted something to give circulation and this would be better done by domestic improvements than it had been by war.

What I have now to state is that I consider the great improvements in machinery for saving manual labour as one of the chief causes for the unparalleled exertions of the last war as by affording the necessaries and comforts of life with less manual labour a greater portion of our Population were set at liberty for military employment etc. and that which was so great a benefit in war may, if it is not our own fault, be equally beneficial in peace, for surely it is a consolation to consider that our present distress arises not from a wasted or exhausted country nor anything which if rightly considered should cause despondency, but quite the contrary. It is the mighty surplus power of the country now lying dormant and waiting to be directed to some useful object. It is a power which, if turned to domestic improvement will astonish the world, as much as it did in Military exertions.

If it is true, as I believe it to be, that we may enjoy as much foreign trade as ever (tho' not precisely of the same kind), if we enable our population to become large consumers of foreign articles, then all our Public Works are a clear gain to the country from a state of peace, and if, as I have before stated, we cannot suffer a diminution of our population either from emigration or from any other cause without great general suffering, it forms an additional inducement for giving employment.

There is another point of view in which I wish to place this apparently surplus population. Nothing can be more natural than to suppose that if we have too many people we shall cure the evil if we allow them to be reduced, but I trust I can show that nothing can be more fallacious. All our present wants can be supplied with a less portion of manual labour than they were before the war. We must therefore invent some new *wants* or we cannot employ them. If our present population are one-tenth part more than we want and that one-tenth part were to emigrate or be in another way reduced those who were left would still be in the same relative situation. There would still seem to be one-tenth too many and so on until a ruinous depopulation followed.

What then are these new wants to be? If the surplus wealth of the country was generally in the hands of the landowners, it is probable an increased consumption of domestic or foreign luxuries would give additional employment to the Poor, but this surplus wealth being more in the hands of men accustomed to accumulate, our *new wants* must be something in their way, something likely to bring in an income—as Roads, Bridges, etc.

Thus if we are willing to enjoy the luxury of fine roads comparatively short and level, to see the country improved and ornamented, we shall with it see *immediately* the sufferings of the poor relieved and the weight removed from the parishes which have supported them; we shall see a *gradual* improvement in agriculture and manufactures from an increasing demand, a *gradual* increase in our foreign trade from an increasing consumption of foreign articles, and a *gradual* increase in our revenues from all these causes united, and we may trust to see at no very distant period the present depression and distress turned into a state of prosperity and happiness of which we have known no parallel.

One of the principal objects of my writing now is to state that views directly the opposite of these are very prevalent amongst men of all political parties, but I have also found that these feelings in a great proportion at least may be removed if proper means are used, and I have been glad that in this respect thy views accord with mine, because if in introducing the proposed measure, its prospective advantages were stated fully to the House of Commons, it would be incomparably the best means of recommending the measure and of removing the present groundless despondency.

A very grand plan for a chain bridge at Runcorn was produced, and if we had been fully satisfied of its practicability, it is probable the Committee would have endeavoured to obtain leave to bring in a bill this session. The expectation that Govt. might probably grant a loan towards the completion of the undertaking was received with general satisfaction—the Bridge alone is estimated at about £100,000.

I am very respectfully,

thy friend James Cropper
[Quaker corn merchant]

4 The inevitability of economic progress, 1847

(J. S. Mill, *Principles of political economy*, 1847, ed. William J. Ashley (Longmans, 1923), pp. 696–7)

...Of the features which characterize this progressive economical movement of civilized nations, that which first excites attention, through its intimate connexion with the phenomena of Production, is the perpetual, and so far as human foresight can extend, the un-limited, growth of man's power over nature. Our knowledge of the properties and laws of physical objects shows no sign of approaching its ultimate boundaries: it is advancing more rapidly, and in a greater number of directions at once, than in any previous age or generation, and affording such frequent glimpses of unexplored fields beyond, as to justify the belief that our acquaintance with nature is still almost in its infancy. This increasing physical knowledge is now, too, more rapidly than at any former period, converted, by practical ingenuity, into physical power. The most marvellous of modern inventions, one which realizes the imaginary feats of the magician, not metaphorically but literally—the electro-magnetic telegraph—sprang into existence but a few years after the establishment of the scientific theory which it realizes and exemplifies. Lastly, the manual part of these great scientific operations is now never wanting to the intellectual: there is no difficulty in finding or forming, in a sufficient number of the working hands of the community, the skill requisite for executing the most delicate processes of the application of science to practical uses. From this union of conditions, it is impossible not to look forward to a vast multiplication and long succession of contrivances for economizing labour and increasing its produce; and to an ever wider diffusion of the use and benefit of those contrivances.

Another change, which has always hitherto characterized, and will assuredly continue to characterize, the progress of civilized society, is a continual increase of the security of person and property. The people of every country in Europe, the most backward as well as the most advanced, are, in each generation, better protected against the violence and rapacity of one another, both by a more efficient judicature and police for the suppression of private crime, and by the decay and destruction of those mischievous privileges which enabled certain classes of the community to prey with impunity upon the rest. They are also, in every generation, better protected, either by institutions or by manners and opinion, against arbitrary exercise of the power of government. Even in semi-barbarous Russia, acts of spoliation directed against individuals, who have not made themselves politically obnoxious, are not supposed to be now so frequent as much to affect any person's feelings of security. Taxation, in all European countries, grows less arbitrary and oppressive, both in itself and in the manner of levying it. Wars, and the destruction they cause, are now usually confined, in almost every country, to those distant and outlying possessions at which it comes into contact with savages. Even the vicissitudes of fortune which arise from inevitable natural calamities, are more and more softened to those on whom they fall, by the continual extension of the salutary practice of insurance.

Of this increased security, one of the most unfailing effects is a great increase both of production and of accumulation. Industry and frugality cannot exist where there is not a preponderant probability that those who labour and spare will be permitted to enjoy. And the nearer this probability approaches to certainty, the more do industry and frugality become pervading qualities in a people.

5 Mobility of capital and labour, 1870s

Walter Bagehot, (i) *Lombard Street: a description of the money market* (London, 1873), p. 10; (ii)'Postulates of political economy' in *Economic studies* (Longmans, Green, 1880), pp. 22–3)

(i) The enterprising spirit

...No country of great hereditary trade, no European country at least, was ever so little 'sleepy', to use the only fit word, as England; no other was ever so prompt at once to seize new advantages. A country dependent mainly on great 'merchant princes' will never be so prompt; their commerce perpetually slips more and more into a commerce of routine. A man of large wealth, however intelligent, always thinks, more or less—'I have a great income, and I want to keep it. If things go on as they are I shall certainly keep it; but if they change I *may* not keep it'. Consequently he considers every change of circumstance a 'bore', and thinks of such changes as little as he can. But a new man, who has his way to make in the world, knows that such changes are his opportunities; he is always on the look-out for them, and always heeds them when he finds them. The rough and vulgar structure of English commerce is the secret of its life; for it contains 'the propensity to variation', which, in the social as in the animal kingdom, is the principle of progress.

(ii) A flexible economy

... A rise in the profits of capital, in any trade, brings more capital to it with us nowadays—I do not say quickly, for that would be too feeble a word, but almost instantaneously. If, owing to a high price of corn, the corn trade on a sudden becomes more profitable than usual, the bill-cases of bill-brokers and bankers are in a few days stuffed with corn bills—that is to say, the free capital of the country is by the lending capitalists, the bankers and bill-brokers, transmitted where it is most wanted. When the price of coal and iron rose rapidly a few years since, so much capital was found to open new mines and to erect new furnaces that the profits of the coal and iron trades have not yet recovered it. In this case the influence of capital attracted by high profits was not only adequate, but much more than adequate: instead of reducing these profits only to an average level, it reduced them below that level; and this happens commonly, for the speculative enterprise which brings in the new capital is a strong, eager, and rushing force, and rarely stops exactly where it should. Here and now a craving for capital in a trade is almost as sure to be followed by a plethora of it as winter to be followed by summer. Labour does not flow so quickly from pursuit to pursuit, for man is not so easily moved as money—but still it moves very quickly. Patent statistical facts show what we may call 'the tides' of our people. Between the years shown by the last census, the years 1861 and 1871, the population of

The Northern counties increased	23 per cent
Yorkshire increased	19 per cent
North-Western counties increased	14 per cent
London increased	16 per cent

While that of

The South-Western counties only increased	2 per cent
Eastern counties only increased	7 per cent
North Midland counties only increased	9 per cent

—though the fertility of marriages is equal. The set of labour is steadily and rapidly from the counties where there is only agriculture and little to be made of new labour, towards those where there are many employments and where much is to be made of it.

No doubt there are, even at present in England, many limitations to this tendency, both of capital and of labour, which are of various degrees of importance, and which need to be considered for various purposes. There is a 'friction', but still it is only a 'friction': its resisting power is mostly defeated, and at a first view need not be regarded.

6 'The great depression', 1886

(Royal Commission on the depression of trade and industry, Final Report, *B.P.P.*, 1886, XXIII, pp. x, xv, xx)

(27) Summarising very briefly the answers which we received to our questions, and the oral evidence given before us, there would appear to be a general agreement among those whom we consulted—

(*a*) that the trade and industry of the country are in a condition which may be fairly described as depressed;

(*b*) that by this depression is meant a diminution, and in some cases, an absence of profit, with a corresponding diminution of employment for the labouring classes;

(*c*) that neither the volume of trade nor the amount of capital invested therein has materially fallen off, though the latter has in many cases depreciated in value;

(*d*) that the depression above referred to dates from about the year 1875, and that, with the exception of a short period of prosperity enjoyed by certain branches of trade in the years 1880 to 1883, it has proceeded with tolerable uniformity and has affected the trade and industry of the country generally, but more especially those branches which are connected with agriculture.

(28) As regards the causes which have contributed to bring about this state of things there was, as might be expected, less unanimity of opinion; but the following enumeration will, we think, include all those to which any importance was attached: (1) over-production; (2) a continuous fall of prices caused by an appreciation of the standard of value; (3) the effect of foreign tariffs and bounties, and the restrictive commercial policy of foreign countries in limiting our markets; (4) foreign competition, which we are beginning to feel both in our own and in neutral markets; (5) an increase in local taxation and the burdens on industry generally; (6) cheaper rates of carriage

enjoyed by our foreign competitors; (7) legislation affecting the employment of labour in industrial undertakings; (8) superior technical education of the workmen in foreign countries.

*　　*　　*

The diminution in the return on capital would have had a much more
serious effect if it had not been accompanied by a heavy fall in the
prices of nearly all articles of ordinary consumption, which has enabled those with fixed incomes payable in gold to maintain a position
not less prosperous than that which they enjoyed in the years of inflated
trade and high prices.

(51) We may therefore sum up the chief features of the commercial
situation as being—

(a) a very serious falling off in the exchangeable value of the produce of the soil;

(b) an increased production of nearly all other classes of commodities;

(c) a tendency in the supply of commodities to outrun the demand;

(d) a consequent diminution in the profit obtainable by production;
and

(c) a similar diminution in the rate of interest on invested capital.

(52) The diminution in the rate of profit obtainable from production,
whether agricultural or manufacturing, has given rise to a widespread
feeling of depression among all the producing classes.

Those, on the other hand, who are in receipt of fixed salaries or who
draw their incomes from fixed investments have apparently little to
complain of; and we think that, so far as regards the purchasing power
of wages, a similar remark will apply to the labouring classes. ...

It remains to indicate the causes which have assisted to produce the
state of things above described.

We have shown that the production of the more important classes of
commodities has on the whole continued to increase; and there can
be no doubt that the cost of production tends to diminish. It is difficult,
therefore, to understand how the net product of industry, which
constitutes the wealth of the country, can have failed to increase also.
There is, moreover, sufficient evidence that capital has on the whole
continued to accumulate, throughout the period which is described as
depressed, though there has been a sensible depreciation in the value
of some kinds of capital.

How then are we to account for the general sense of depression which undoubtedly exists and is becoming perhaps more intense every year?

(55) We have observed above that the complaint proceeds chiefly from the classes who are more immediately and directly concerned in production; and there can be no doubt that of the wealth annually created in the country a smaller proportion falls to the share of the employers of labour than formerly.

The view, therefore, which we are disposed to adopt is that the aggregate wealth of the country is being distributed differently, and that a large part of the prevailing complaints and the general sense of depression may be accounted for by the changes which have taken place in recent years in the apportionment and distribution of profits.

*   *   *

(75) The increasing severity of this competition both in our home and in neutral markets is especially noticeable in the case of Germany. A reference to the reports from abroad will show that in every quarter of the world the perseverance and enterprise of the Germans are making themselves felt. In the actual production of commodities we have now few, if any, advantages over them; and in a knowledge of the markets of the world, a desire to accommodate themselves to local tastes or idiosyncracies, a determination to obtain a footing wherever they can, and a tenacity in maintaining it, they appear to be gaining ground upon us.

(76) We cannot avoid stating here the impression which has been made upon us during the course of our inquiry that in these respects there is some falling off among the trading classes of this country from the more energetic practice of former periods.

Less trouble appears to be taken to discover new markets for our produce, and to maintain a hold upon those which we already possess; and we feel confident that, if our commercial position is to be maintained in the face of the severe competition to which it is now exposed, much more attention to these points must be given by our mercantile classes.

(77) There is also evidence that in respect of certain classes of products the reputation of our workmanship does not stand so high as it formerly did. The intensity of the competition for markets, while in many respects is has legitimately diminished the cost of production, has also tended to encourage the manufacture of low-priced goods of inferior quality, which have not only failed to give satisfaction

themselves, but have also affected the reputation of other classes of goods to which no such exception could be taken.

(78) The reputation of British workmanship has also suffered in another way by the fraudulent stamping of inferior goods of foreign manufacture with marks indicating British origin.

This appears to be particularly the case with the hardware goods of Birmingham and Sheffield which have secured so wide a reputation in the markets of the world.

We regret, however, to be obliged to add that the practice of fraudulent marking appears from the evidence before us to be not unknown in this country.

7 Britain's prospects in the new century

(i) A complacent view, 1904

(Robert Giffen, *Economic inquiries and studies* (Bell, 1904), II, pp. 420–3)

The final point for our inquiry according to the programme above laid down is the suitability of the United Kingdom as a place of residence and industry, assuming that what is required from abroad can be obtained casily. As already stated, a favourable answer on this head may be taken for granted in the case of an old country like the United Kingdom; but a more formal treatment is proposed, as the wonderful combination of circumstances in our favour is inadequately realised.

Climate is a condition on which much might be written, but the historical opinion that England is a country where you can be out of doors more days than in any other sums up generally the climatic conditions in our favour. Temperateness is the characteristic which our climate possesses in greater degree than that of any of our western European neighbours that come nearest to us in the matter. That other communities, like the United States and Canada, find a drawback in climatic conditions to many industrial advantages they possess appears undoubted. They are countries of extremes, where it costs more for food, shelter and clothing to permit of the same work to be done than it costs in the United Kingdom; and this difference of cost is a considerable advantage to us. Our place of residence has been improved, moreover, by generations of workers, who have executed drainage and sanitary improvements, built roads, streets, walls and fences; created parks, gardens and lawns; and generally increased the amenities of life for a huge town population, such as a population must be that brings its food and raw materials mainly from a distance.

The next advantage we possess in addition to climate and the artificial amenity of the land for residence is compactness of situation. All the different parts of the country are close together, well connected by railways, road and sea, while the sea, of course, affords perfect access to the rest of the world. The United Kingdom is more like a single huge city than a country of districts and towns separated by wide intervals. This is no small advantage for local industry, as it places consumers and producers side by side, just as if they were in a small village with its own neighbourhood, complete in itself.

The equipment of machinery and buildings for the local industries to be carried on has likewise every facility. There is, perhaps, a temptation to carelessness in obtaining the newest and best equipment, as so much can be done with less, owing to our favourable conditions; but it is entirely our own fault if the best is not always done. Hereditary skill and training are likewise consequences of the past which we must long retain, and along with this a perfection of subsidiary industries which facilitates the great industries themselves, as those who attempt to set up manufactures in new countries will understand. There are many daily wants of great manufactures which are supplied by subsidiary industries in all our large manufacturing centres, and these are not brought into existence in a day.

A fourth advantage we possess in following our own home industries is the large importation of raw materials to be used up in our export trade. This helps to make a better market here for all similar materials used in the home trade itself. The one market helps the other, with the result that if there is abundance and variety anywhere it is here. We always get the first offer. This applies, it should be understood, not merely to the raw material we obtain for the manufactures for export so-called; but to the raw materials we import in order to carry on such industries as coal mining, and those other industries, such as shipowning and shipbuilding, which enable us to make invisible exports. Because we have this enormous importation we are in a better position to practise to the fullest advantage all those industries where we work and exchange amongst ourselves.

The fifth advantage is of the same kind, though it requires special mention. This is that the United Kingdom is a country of free imports, not merely as regards raw material, but as regards articles of food and manufactured articles of all kinds. 'Free imports' may not be the same thing as 'free trade', as we are frequently told; but it is of no small advantage in some ways to be a 'free port'. A free port means abundance and variety of everything at the lowest market price. However some individuals may suffer at times, the compensating advantage to the community as a whole, in which the sufferers participate, is immense, and ought not to be lightly regarded. In passing, however, I must say that I for one do not believe that many individuals have suffered by free imports who did not deserve to do so by reason of their own indolence and lack of foresight and intelligence.

Finally, we have an immense advantage in the United Kingdom in the development of our free banking system, and the bankers' tradition of assisting trade by advances to manufacturers and tradesmen. Our banks are full of money which comes to them from all parts of the earth, and every man who has anything to attempt in trade, if he has any property of his own, or can persuade people with property to trust him, may have ready money lent him by the bank. The tales I have heard of a different condition of things elsewhere are simply astonishing to business people in the United Kingdom, who do not know what it is to go without money to carry on their business because the banks themselves have got none to lend. Yet what is unknown here is frequent in the Far West of the United States and in all new countries, and is not unknown on the Continent, ready as Continental banks are to do financial business with which the prudent trader is not concerned.

These are all assets to the good which facilitate every form of home industry, so that the United Kingdom, if it is not so well placed as other nations for the primary industries of agriculture and mining, has yet advantages of its own as a place of residence and for carrying on all other industries, in which no other community excels it.

In all respects, then, the conditions for the industrial future of the country into which we have been inquiring are satisfactory. The amount and proportion of the things we require to import from abroad are not exceptionally large, and we are not in fact put to any exceptional strain in obtaining them. By means of our foreign investments; our exports of certain raw materials of which we have a monopoly; our exports of manufactures *eo nomine*; and the earnings of our ships and other earnings—our invisible exports, as they are called—we do obtain easily enough all that we require, and the conditions up to the latest date remain unchanged.

(ii) Academic caution, 1908

(Alfred Marshall, 'The fiscal policy of international trade', 1903 and 1908, in *Official papers by Alfred Marshall* (Macmillan, 1926), pp. 404–7)

... That combination of liberty with order, and of individual responsibility with organised discipline, in which England excelled, was needed for pioneer work in manufactures; while little more than mere order and organised discipline will go a long way towards success, where the same tasks are performed by modern machinery 'which does most of the thinking itself'. Thus England is at a steadily increasing relative disadvantage in trading not merely with people like the Japanese, who can assimilate every part of the work of an advanced factory; but also with places where there are abundant supplies of low-grade labour, organised by a relatively small number of able and skilled men of a higher race. This is already largely done in America, and it certainly will be done on an ever-increasing scale in other continents.

Consequently, England will not be able to hold her own against other nations by the mere sedulous practice of familiar processes. These are being reduced to such mechanical routine by her own, and still more by American, ingenuity that an Englishman's labour in them will not continue long to count for very much more than that of an equally energetic man of a more backward race. Of course, the Englishman has access to relatively larger and cheaper stores of capital than anyone else. But his advantage in this respect has diminished, is diminishing, and must continue to diminish; and it is not to be reckoned on as a very important element in the future. England's place among the nations in the future must depend on the extent to which she retains industrial leadership. She cannot be *the* leader, but she may be *a* leader.

The economic significance of industrial leadership generally is most clearly illustrated just now by the leadership which France, or rather Paris, has in many commodities which are on the border line between art and luxury. New Parisian goods are sold at high prices in London and Berlin for a short time, and then good imitations of them are made in large quantities and sold at relatively low prices. But by that time Paris, which had earned high wages and profits by making them to sell at scarcity prices, is already at work on other things which will soon be imitated in a like way. Sixty years ago England had this leadership in most branches of industry. The finished commodities and, still more, the implements of production, to which her manufacturers were giving their chief attention in any one year, were those which would be occupying the attention of the more progressive of Western nations two or three years later, and of the rest from five to twenty years later. It was inevitable that she should cede much of that leadership to the great land which attracts alert minds of all nations to sharpen their inventive and resourceful faculties by impact on one another. It was inevitable that she should yield a little of it to that land of great industrial traditions which yoked science in the service of man with unrivalled energy. It was not inevitable that she should lose so much of it as she has done.

The greatness and rapidity of her loss is partly due to that very prosperity which followed the adoption of Free Trade. She had the full benefit of railways, and no other country at that time had. Her coal and iron, better placed relatively to one another than elsewhere, had not begun to run short, and she could afford to use largely Bessemer's exacting but efficient process. Other Western nations partially followed her movement towards Free Trade, and in distant lands there was a rapidly increasing demand for manufactures, which she alone was able to supply in large quantities. This combination of advantages was sufficient to encourage the belief that an Englishman could expect to obtain a much larger real income and to live much more luxuriously than anybody else, at all events in an old country; and that if he chose to shorten his hours of work and take things easily, he could afford to do it.

But two additional causes of self-complacency were added. The American Civil War and the successive wars in which Germany was engaged, partially diverted the attention of these countries from industry: it checked the growth of their productive resources; and it made them eager to buy material of war, including railway plant and the more serviceable textile materials, at almost any cost. And lastly, the influx of gold enriched every English manufacturer who could borrow money with which to buy materials, could apply moderate intelligence in handling them, and could then sell them at a raised level of prices and discharge his debt with money of less purchasing power than that which he had borrowed.

This combination of causes made many of the sons of manufacturers content to follow mechanically the lead given by their fathers. They worked shorter hours, and they exerted themselves less to obtain new practical ideas than their fathers had done; and thus a part of England's leadership was destroyed rapidly. In the 'nineties it became clear that in the future Englishmen must take business as seriously as their grandfathers had done, and as their American and German rivals were doing: that their training for business must be methodical, like that of their new rivals; and not merely practical, on lines that had sufficed for the simpler world of two generations ago; and lastly that the time had passed at which they could afford merely to teach foreigners and not learn from them in return.

This estimate of leadership is different from, almost antagonistic to, measurement of a country's leadership by the *volume* of her foreign trade without reference to its *quality*. Measurement by mere quantity is misleading.

Of course, the statistics of foreign trade are specially definite and accessible; and since the fluctuations of business confidence and activity are reflected in foreign trade among other things, the habit has grown up of using export statistics as a *prima facie* indication of the time and extent of such fluctuations. For instance, vital statisticians have frequently pointed to the parallel movements of exports and the marriage-rate, when the real parallelism to be indicated was between fluctuations of credit and of those of the marriage-rate. Even for this purpose export statistics are not very trustworthy; while for broader purposes they are quite untrustworthy.

Other things being equal, an increase in the efficiency of those industries in which a country is already leading will increase her foreign trade more than in proportion. But an increase in the efficiency of those in which she is behind will diminish her foreign trade.

England has recently [1903] been behind France in motorcar building, and behind Germany and America in some branches of electrical engineering. A great relative advance on her part in those industries would enable her to make for herself things which she had previously imported, and would thus diminish her foreign trade. On the other hand, even a small advance in her power of spinning very high counts of cotton yarn would increase her foreign trade considerably; because that is a thing for which other nations have an elastic demand, and are at present almost wholly dependent on England.

England's export trade, though still very much larger in proportion to population than that of Germany and America, is not [in 1903] increasing as fast as theirs. But this fact is not wholly due to causes which indicate relative weakness.

The chief cause of it is that the improvements in manufacture and in transport, aided by Free Trade, enable England to supply her own requirements as regards food, clothing, etc. at the cost of a continually diminishing percentage of her whole exports. Her people spend a constantly diminishing percentage of their income on material commodities; they spend ever more and more on house-room and its attendant expenses, on education, on amusement, holiday travel, etc. Present censuses show a progressive increase in the percentage of Englishmen who earn their living by providing for these growing requirements. That is to say, the number of Englishmen who devote themselves to producing things which might be exported in return for foreign products increases very slowly. Of course, if her foreign trade be measured by the quantity of things exported and imported, it is increasing fast; for a man's daily labour now deals with a much larger volume of goods in almost every industry than formerly. But still it is not increasing [in 1903] as fast as that of Germany and America.

8 Business looks up, 1896–1914

(Hartley Withers, *Poverty and waste* (Smith, Elder, 1914), pp. 65–7)

Scarcity and dearness of capital are a commonplace complaint whenever men of business are gathered together. At a meeting of the Union of London and Smiths Bank last July, Sir Felix Schuster said that 'owing to the continuous growth of trade and of new countries, the demands for capital had been on an enormous scale. Until the end of the half-year these fresh issues of capital were comparatively well taken up by the public, but then it became manifest that the supply exceeded the demand, and the Stock Exchanges were no longer able to absorb the multitude of new issues that were being offered'. This inability of investors to meet the demands on their power to absorb new issues has been a frequently recurring symptom in recent years. The price of capital, or the rate of interest that investors have been able to obtain, has risen gradually and steadily from 1896 until the end of last year. In 1896 London and North Western 4 per cent Preference Stock touched $162\frac{1}{2}$, at which price it returned less than $2\frac{1}{2}$ per cent to the buyer. In 1913 the same security was dealt in at $97\frac{5}{8}$, yielding the buyer more than 4 per cent.

It is true that 1896 was an exceptional year, in which there was, or seemed to be, a glut of capital such as had never been seen before. This, I believe, only happened because, owing to the shock to confidence following the crises of 1890 and 1893, there was a long pause in the development of new countries, and for a time accumulation went ahead of development. It may be that, owing to the shock to confidence due to Mexico's default and mistrust of the financial position of certain South American states, a similar pause in the development of new countries may take place now and that there may for a time be an apparent glut of capital. But it seems hardly likely, when we consider the enormous demands of the civilized and uncivilized world for capital, that the seeming glut can last long, and, if it does, it will only do so because capital cannot be had at any price by borrowers whose credit is impaired, not because it is not wanted. The normal condition of the financial world is now one in which capital is scarce and dear.

Four chief causes of this scarcity and dearness of capital stand out. One is its wholesale destruction by wars. The second is the opening up of the uttermost parts of the earth to cultivation and development by improved means of communication, which increases the world-wide demand for capital to be put into production, which takes some years to bear fruit on a great scale. The third is the huge expenditure of the nations, especially on armaments and preparations for war. The high taxation that is now exacted by our rulers has little or no effect on the personal comfort of the wealthier classes, but it very seriously curtails their saving power. The fourth is the high level of personal expenditure and extravagance that modern fashion prescribes.

9 Mixed fortunes of the nineteen-twenties

(*Britain's industrial future* ([Liberal Party Yellow Book], 1928), pp. 5, 12–13)

The simple explanation that attributes our troubles to the War is obviously some part of the truth. But it is far from being the whole truth. Certainly it is unsound to suppose either that we shall drift comfortably back into the situation of 1913 if only we have patience or that all would have been well if there had been no war. Indeed, many of the acutest difficulties which the War has thrust upon us were not created by it. They are rather the result of developments which had begun much earlier, but were very greatly speeded up by the economic upheaval of 1914–18. Moreover, the pre-war situation is only idealised in retrospect by those who have forgotten that the decade before the War was one of greater industrial strife than ever before, and that poverty, overcrowding, sweating, and many other social evils, were the occasion of serious and justified discontent. We cannot hope to prescribe remedies until we have decided how far the problems we have set out to solve are new, how far they are due to temporary and passing causes, and how far they have deep-seated roots in the past. It will also help us if, in passing, we can note how far our difficulties are peculiar to Great Britain and how far they are shared with other countries. We propose, therefore, first to mention very briefly some of the main tendencies in the economic history of Britain during the last hundred years; secondly, to give a picture of the economic conditions of the country at the present time; thirdly, to direct attention to the black spots and note how they affect the whole economic life of the community; and finally, to see what answer we can give to the questions we have posed.

*   *   *

(*vi*) The rise of prices, which continued until the wholesale index number rose to over 300 in 1920 and was accompanied by an allround reduction in the hours of labour, was followed by a fall of unprecedented rapidity and extent. This involved a corresponding readjustment of wage-rates. The wide use of the cost of living index number, improved machinery of wage negotiation, and the prevention of distress by a wide extension of Unemployment Insurance enabled us to weather the severest economic storm of the last hundred years without a social upheaval. But the fall of money wages, which inevitably involved a certain amount of friction, did not take place as rapidly in trades 'sheltered' from foreign competition as in those which had to face the direct effects of adverse international conditions. A disparity of wages in these two groups of trades continues to the present day and gives rise to many economic difficulties, especially for the export industries.

(*vii*) These troubles of readjustment, the frustration of the hopes entertained at the end of the War, and the survival from pre-war days of the dissatisfaction with the existing order of things, which was dormant during but not removed by the War, have combined to produce—particularly in the coal-fields and the railways—a series of bitter industrial disputes which have repeatedly set back the process of industrial reconstruction.

In view of these facts it is not surprising that ever since the brief boom which followed the Armistice, the great basic exporting industries of Great Britain—coal, metallurgy, and textiles—have been in a bad way. Instead of expanding rapidly as they used to do, they have been now for some seven years in a more or less stagnant condition. Instead of leading the way in wages and hours and standards of livelihood, they now, for the most part, contrast unfavourably with other occupations. Instead of being the chief providers of employment for a rapidly growing population, they are now the chief contributors to obstinate post-war unemployment.

These great groups of industries stand out so prominently in our economic life and played so preponderant a part in our development during the last century that they are apt to monopolise attention. It is common, accordingly, to speak of the post-war 'trade depression' as though British trade and industry as a whole were experiencing the same misfortunes as the basic trades. This, however, is far from being true. There is not today, and there has not been for several years, a general trade depression in the sense in which economists are accustomed to use the term. In many directions, on the contrary, there has been remarkable expansion. New industries have sprung up, or have grown from small beginnings, which have provided compensation, in no small degree as regards both employment and the national income, for the decline in the basic trades. Meanwhile there has been a steady expansion in a great variety of miscellaneous occupations, catering mainly for the home market. For these and other reasons our economic position as a people is not so unfavourable as would be concluded from considering only the state of the basic export trades.

10 Long-term unemployment, 1938

(Pilgrim Trust, *Men without work* (Cambridge University Press 1938), pp. 5–9)

(ii) *Long Unemployment before the Slump*

In November 1936, when the sample enquiry was undertaken, a cycle of three years' growing depression, followed by four of recovery, had just been completed. In November 1929 the economic life of the country had been for some time in a state of equilibrium. These were not years of real prosperity, though they appeared so in the light of the 'great depression' that was going to follow. Of ten English workers willing to contribute their share to national production, only nine were doing so at any time during that period. The tenth man was registered at an Employment Exchange as looking for work. Thus, of Great Britain's working population insured against the risk of unemployment,[1] gradually increasing from 11,000,000 men and women

---

[1] About three-quarters of the total working population in a social sense but representative for nearly the whole *industrial* working class.

to 12,000,000, rather more than 1,000,000 were involuntarily out of work. This state of affairs, though puzzling to the mind of those used to the 'good old times' before the War (and incidentally, before disturbing figures of unemployment began to be officially published), had gradually come to be accepted as a normal state.

There was no considerable section of the population which was permanently 'tenth man'. In February 1930 when the depression had already begun to make itself felt and the number of unemployed had risen to 1,500,000, only 131,000 of this number had failed, between February 1928 and February 1930, to make at least thirty contributions to the Unemployment Insurance Scheme, while the number of people who had had no work at all during this time was again very much smaller. In September 1929, of a total unemployed army of 1,150,000 people, only 53,000 (less than 5%) had been continuously out of work for a year or more. Apart from this small stagnant pool, we must think of pre-depression unemployment as a fairly rapidly moving stream. A figure of 53,000 means that in a typical middle-sized community of 40,000 people there were not more than some fifty long-unemployed men and women. The percentage of the total industrial population which persistently failed to get absorbed into employment (less than $\frac{1}{2}$%) was the same as that which we found seven years later in the most prosperous of our six towns, the Borough of Deptford. Although there is no information whatsoever available about the composition of this small number of long-unemployed men in 1929, because no one bothered about a problem of this extent, some light may be thrown on who the long unemployed were before the depression by the description which will be given of long unemployment in Deptford, and the reader who tries to imagine pre-depression conditions in the light of our Deptford study will not be very far wrong.

There are reasons for thinking, however, that even the situation as we found it in Deptford was worse than that of the normal community in 1929. Of the 53,000 unemployed for more than a year at that time, not less than 38,000 were coal-miners, most of them thrown out of work in the 1926 lock-out and never reabsorbed. The problem in 1929 was thus mainly a localized abnormality of coal-mining districts dependent on mines abandoned or permanently closed. In all other industries taken together, the number of long unemployed shrinks to the small figure of 15,000 and the number of long-unemployed families in our normal community of 40,000 people, if it does not happen to be a mining community, shrinks from 50 to 15, unnoticeable, and an easy object for individual help and supervision.

(iii) Depression and Recovery

Within three years of the beginning of the depression the number of unemployed had risen from 1,281,000 in the last quarter of 1929 to 2,757,000 in the last quarter of 1932. There were four men for every three available jobs. 60% of all workers in shipbuilding and allied industries were out of work, 46% of all workers in the iron and steel industries. As the depression went on, the number of men and women out of work for a year or more increased even more rapidly than the general tide, as more and more men who had been thrown out or temporarily stopped work by firms closing or contracting failed to get work and began to be reckoned as long unemployed.

At the beginning of 1932 the number of long-unemployed[1] men alone passed the 300,000 mark, and during that year, when the depression gradually spent its force and not many more men were turned out of work afresh, the number of long-unemployed men continued to increase rapidly, passing the 350,000 mark in the summer of 1932, the 400,000 mark in autumn, and the 450,000 mark early in 1933. Their number continued to increase right into the summer of 1933, when the general tide of unemployment had already definitely begun to recede.

In July 1933 the total number of long-unemployed men and women was over 480,000. Of 100 men queuing up at the Exchange in September 1929, only 5 or 6 had been persistently unemployed. Of 100 men queuing up three years later, when the depression was at its worst, 20 had had no work for a year or more. In the middle of 1933, when the last of the men thrown out by the depression and not reabsorbed had passed on into the long-term class, 25 of every 100 unemployed had had no work during the last year. Instead of 50 families in our 'average community' there was now a crowd of 480 families. A social problem of the first order had arisen. We have seen that, with the exception of some mining districts, it was wholly a creation of the economic depression. The question was whether recovery would undo what the depression had done, and whether the new problem was thus going to be automatically solved.

The three years of depression had already been followed by four years of recovery when the sample of long unemployment which forms the basis of this report was taken. Thus it illustrates a state of recovery which had progressed considerably both in duration and

---

[1] By 'long unemployed' we shall from now on mean men continuously unemployed for more than a year.

extent, though it was by no means yet completed; for the year which has passed since it was made brought continued recovery, and we found a considerable number of long unemployed in work again when they were visited in the course of the six months following the sample date. The problem that we saw and now describe is, therefore, what was left by the great depression. Some of these long unemployed may still be absorbed into normal industrial society, but they have failed to be reabsorbed for a long time, during which expanding industrial activity has brought back to work many unemployed and has absorbed many newcomers as well. Economic forecasts are uncertain, though perhaps less so than they used to be. Conditions during the slump showed that the industrial system can fare worse than when we analysed its least used resources of labour in 1936, and subsequent improvements have shown that it can fare better. We may perhaps say that by November 1936 a stage had been reached which represented fair prosperity for the country as a whole under post-War economic conditions.

Of the rise in unemployment from 1,281,000 to 2,757,000 persons between the last quarters of 1929 and 1932, a great portion has been made good in the four following years. In the last quarter of 1936, 1,621,000 persons were out of work. Of 100 men and women thrown out of work between 1929 and 1932, the great majority, 77, have returned to employment, if we neglect for the moment those who died, those who retired with an Old Age Pension, or the women who withdrew from the labour market into family life. By the third quarter of 1937 the majority of the remaining 23, namely 17, had returned to employment as well. But, in this recovery, what had become of the long-unemployed man, this new phenomenon of the depression?

We saw that before the depression (September 1929), among 100 men queuing up at the Exchange, there would have been no more than five or six who had not had work for a year past, and we noticed the rapid rise of this figure to 20 at the depth of the depression, and to 25 in the summer of 1933 when firms had ceased to close down, but the last batch of dismissals during the depression had increased the number of long-unemployed men to the new high level of 460,000. When the recovery took place, the proportion of long unemployed did not go back to the old level; it did not even show a tendency to decline again, but it went on increasing.

During the first year after the peak in the summer of 1933, the actual number of men out of work for a year or more fell considerably. It had dropped by 90,000 to 370,000 in October 1934, and at that date we would have found 24 long-unemployed men among 100 registering at the Exchange, and some optimistic observers might have thought that unemployment might both return to its old level, and, in the process, revert to its old character as mainly an industrial turnover of human labour, a fairly rapidly moving stream with only small stagnant pools here and there. But the following year, from October 1934 to October 1935, showed that nothing of the sort was going to happen. While the recovery went on rapidly and the total number of unemployed men fell by a considerably larger number than the year before, the number of the long unemployed remained nearly stationary, declining by a bare 10,000 to 360,000, so that in the autumn of 1935, 26 out of 100 men at the Exchange had been out of work for a full year or more.

11 The case for planning, 1947

Sir Oliver Franks, *Central planning and control in war and peace* (Longmans, 1947), pp. 20–3)

... I am not advocating a doctrine of nationalization. Public ownership may indeed be a form of control and it may be argued whether as such it is good or bad. But this argument is irrelevant to the general thesis of the inevitability of planning and control by the State as generally directing the policies of industry and trade. It would be a very good thing if the main issue could be separated once and for all from the controversies which surround the topic of nationalization and its application to particular industries or trades.

What then are the grounds for my conclusion? First there is the argument from military security. Politically the world is still composed of sovereign nation states. So long as they are many, there remains the permanent possibility of war, and each and every nation goes in fear of it. Each nation maintains armed forces and prepares for the possibility of war in the name of military security. The extent and the thoroughness of the preparations at any time depend very largely on the general prospects of international relations. But each nation as soon as it examines the issues raised for it by military security at once discovers that they have an economic side of the greatest importance. The general size and efficiency of industry is the basic war potential. Any nation examining its economic war potential discovers deficiencies of varying importance and takes measures to deal with all or some of them. As a result each nation feels compelled to grow things, acquire things, store things, manufacture things which otherwise it would not or which are not worth while on purely economic grounds. The fear of war directly affects the economy of every nation state including the United Kingdom. But for the present argument the important effects of this element of national policy are not mainly to be found in the direct interventions made by each nation state in its own economy but rather in more general and more indirect results. The fact that every state does things for fear of war in its own economy which otherwise it would not do and which are uneconomic, forces it also to erect barriers to keep out the products which would otherwise flow in from elsewhere, cheaper or better in quality, but taking work away from enterprises and industries to be protected for reasons of military security. Cumulatively the effect of these barriers against the flow and expansion of international trade is considerable. Their effect is very large in the world to-day. They represent because of the permanence of their causes a persistent and steady opposition to general prosperity. And in particular they militate against the economic prospects of the United Kingdom which depends so vitally on the volume of its international trade and can only maintain it if the general flow is free and large.

Then there is the argument from fear of unemployment. The citizens and Governments of most nation states fear the occurrence of large-scale unemployment more than any other disaster that can happen in peace. No Government of the future can hope to stay for long in power if it lets large-scale unemployment destroy the physical and spiritual conditions of life among the people. The great depression of 1929–35 made in many ways a deeper and more permanent mark on this country than the war of 1939–45. It is noticeable in the present generation of undergraduates at my University. They are nearly all men demobilized from the Forces and four to six years older than usual. As one would expect, they are mature even for their age. They work very hard and play very little. Almost all of them are deeply worried about securing employment. They take no encouragement from the almost universal shortage of workers in all walks of life. They do not expect their own wide experience to carry them far. They view their university course as giving them additional qualifications which may help in the struggle to secure a job. The world to which they look forward seems insecure, impoverished and uncertain. Their frame of mind looks back beyond the forties to the thirties. What they fear is large-scale unemployment.

The Governments of most nations, including that of the United Kingdom, intervene in their domestic economies, attempting by various measures to lessen the risks of scarcity of work to their people. The measures may be indirect or direct. They may take the form of financial subsidies to keep undertakings or whole industries going which otherwise could not withstand ordinary economic competition. They may be designed to discourage imports and secure home markets for home-made products. They may even include nationalization if it is thought public ownership would enable industries to plan ahead, take risks and spend money on capital investment in a way in which private enterprise could not or would not. Once again it is the total effect of these policies, as applied by almost all nations, that is important rather than what is done for these reasons in the United Kingdom. The effect of each nation trying to diminish scarcity of work for its own people leads to the erection of further barriers to the expansion of international trade. Each nation by trying to be more self-sufficient than is economic *pro tanto* impoverishes the world and lessens the profitable exchange of goods. For the reasons already stated this tendency is especially damaging to the United Kingdom. Nor is it likely soon to diminish in strength; the fear of unemployment is too deep rooted.

Lastly there is the argument from the general condition of the world. It took five or six years to re-establish economic activity in Europe after the 1914–18 war. This time the collective disorganization and dislocation is indefinitely greater. Far more of the world is involved. Military operations were incomparably more destructive. Great industrial and trading nations have been eliminated, at least for many years. The economies of many of the combatants have been shattered so that it must be the work of years to achieve even a foundation for a prosperous life. There have been political uprisings and revolutions and there may be more. And there is the conflict of political ideas and systems which threatens to divide the world as it is dividing Europe into separate blocks between which the passage of ideas or goods is a matter of special contrivance and real difficulty. All these events have impoverished and continue to impoverish the world. They do more: they conspire to prevent the growth of that confidence, regularity and order without which business and trade cannot flourish. International trade depends upon and presupposes a general acceptance of political, social and economic habits and conventions. The nations are still growing apart rather than together. When the seller's market is over, this is the world in which the United Kingdom will find itself, a hard world for developing and maintaining a large export trade. A world, too, in which the adverse factors will too often be obdurate and slow to disappear.

If the United Kingdom is to enjoy in the future anything like pre-war standards of living, it has to increase the volume of its export trade, we have been told, by about 75 per cent over prewar. If this is not done, the foreign currencies required to pay for food and raw materials will not be there. So far the export drive has forced the volume of exports up to a little more than the prewar figure. This progress has been possible through the direct intervention of the Government making countless arrangements with industries and particular undertakings as a result of which they export a large proportion of their production and limit sales at home. It may be expected that this year and next year will see a further expansion of exports. It may even be that the target will be touched. If it is, it will be the crown of a very great and highly concerted effort. But what matters is not so much attaining the target as maintaining it, and doing so year after year. The present seller's market will be of all too short duration. It will then be much harder to sell and foreign competition for export markets will be fierce. The United Kingdom is by no means alone in striving to increase exports of manufactured goods.

The aim of the United Kingdom is to secure and keep a large slice of the cake of international trade. Let it be assumed that the slice is secured. It will not be kept unless the cake itself is so large that the slice looks reasonable and does not provoke reprisals. For this reason an essential long-term interest of the United Kingdom must be to promote the general freedom and increase of trade between nations. But when the considerations adduced in previous paragraphs are taken into account, it becomes obvious that no sudden, large and permanent expansion will take place. International conferences on trade will find as they move from aspirations and principles to applications and practice that progress slows down and agreements are much harder to get. It will take an effort to win every foot of ground. It will be a struggle to lower any barrier to international trade. This gives no reason for despair and does not lessen the importance of the end pursued. It means that the job is a long-term job and will be accomplished by the accumulation of small gains rather than by a quick resounding triumph.

2

Agriculture

Agriculture has been marvellously transformed in the past two hundred years. In the reign of George III large tracts of land remained unenclosed, farmed on a system of immemorial antiquity. High prices and food shortages speeded the process of consolidation (*A2*) which has continued to the present day. Paradoxically, in the twentieth century as farms get bigger the great estates have tended to disappear (*C7*, *C10*). Nineteenth-century radicals had insisted on compensation to farmers for unexhausted improvements to the land but failed in their wider attacks on landowners as a class (*C6*, *C8*). (Not that all farmers resented the obligation to pay rent to the landlord (*C4*, *C9*).) The lure of high prices for land and the burden of death duties have done the work that single taxers and land nationalisers failed to perform. Despite the consolidation of holdings the business of agriculture is still carried on by a very large number of firms operating on a scale small by the standards of manufacturing industry. And this remains true after the necessary qualification has been made that farming is not one but several industries (*B3*, *B4ii*, *B7*). The technical transformation of farming has largely been the work of the last thirty years. Enclosure, the displacement of oxen by horses and the beginnings of mechanisation and agricultural chemistry are all much older, but high output per acre and per man is essentially a recent development and one that has yet to yield its full fruits (*A5*, *8*). In the middle of the eighteenth century English agriculture fed about six million Englishmen; today it feeds more than twenty million and feeds them much better. But agriculture has greatly declined in popular esteem: from being the greatest interest in the country, identified by Adam Smith with the consumer himself, it sank in the nineteenth century to being one among several competing sources of food supply. At the end of the nineteenth century it was often justified on social and political as much as on economic grounds. Today the farmer operates under a system of subsidy rather than of tariff protection, at fairly high cost to the public, the expense being justified on the prosaic ground that it saves foreign exchange! As a minority sector often at odds with the predominant industrial interests agriculture has suffered periodic bouts of low profits and considerable distress for landowners and tenants (*B4, 5*). On the other hand, especially in wartime and its immediate aftermath, farming has enjoyed periods of high prosperity (*B1, 8*).

A. THE GROWTH OF OUTPUT

1 The Norfolk rotation of crops, 1787

(William Marshall, *The rural economy of Norfolk comprising the management of landed estates and the present practice of agriculture in that county* (London, 2 vols. 1787), I, 132–6.)

In Norfolk, as in other arable countries, husbandmen vary more or less in the succession of crops and fallows to each other.—But if we confine ourselves to *this* District; namely, the north-east quarter of the county; we may venture to assert, without hazard, that no other District of equal extent in the kingdom is so invariable in this respect; commonfield Districts excepted.

It is highly probable, that a principal part of the lands of this District have been kept invariably, for at least a century past, under the following course of cultivation:

> Wheat,
> Barley,
> Turneps,
> Barley,
> Clover,
> Raygrass, broken up about Midsummer,

and fallowed for wheat, in rotation.

Thus, supposing a farm to be laid-out, with nineteen or twenty arable divisions, of nearly equal size, and these to be brought into six regular shifts, each shift would consist of three pieces; with a piece or two in reserve, at liberty to be cropped with oats, peas, tares, buck; or to receive a thorough cleansing by a whole-year's fallow.

This course of culture is well adapted to the soil of this District, which is much more productive of barley than of wheat; and is in every other respect, as will hereafter appear, admirably adapted to that excellent system of management of which it is the basis.

The soil of the southern parts of the District being stronger and deeper than that upon which the foregoing course of crops is prevalent, it is better suited to wheat; and there the round of

> Wheat,
> Turneps,
> Barley,
> Clover,

is common; though not in universal practice.

This difference in soil and management renders it necessary to consider the southern Hundreds of Fleg, South-Walsham, and Blow-field, as appendages, rather than as parts, of the District most immediately under description: which is furnished with a less genial soil; namely, that shallow, and somewhat lightish, sandy loam, which may be called the common covering of the county; broken, however, in some places, by a richer, stronger, deeper soil; and in others, by barren heaths and unproductive sands; from which even the Hundreds of Erpingham, Turnstead, and Happing, are not entirely free; though, perhaps, they enjoy a greater uniformity of soil than any other District of equal extent in the county.

This therefore, is the site best adapted to the study of the system of management which has raised the name of Norfolk husbandmen, and which is still preserved, inviolate, in this secluded District. For a shallow sandy loam, no matter whether it lie in Norfolk or in any other part of the kingdom, there cannot, perhaps, be devised a better course of culture; or, taken all in all, a better system of management, than that which is here in universal practice*.

But excellent as this succession of crops undoubtedly is, it cannot be invariably kept up; for even a Norfolk husbandman cannot command a crop of turneps or a crop of clover; and when either of these fail, the regularity of the succession is of course broken into.

If his turneps disappoint him, he either lets his land lie fallow through the winter, and sows it with barley, in course, in the spring; or more frequently, though less judiciously, sows it with wheat in autumn;

---

* If any improvement of the present system can be made, it would perhaps be by adopting the practice of a judicious husbandman in the northern part of the District (Mr Edmund Bird, of Plumstead); who divides his farm into seven, instead of six, shifts; his course of *crops* are the same as those of his neighbours; his seventh shift being a whole-year's *fallow* for wheat.

sometimes, though not always, sowing it with clover and ray grass in the spring; by this means regaining his regular course.

If the clover miss, the remedy is more difficult; and no general rule is in this case observed. Sometimes a crop of peas is taken the first year; and the next, buck plowed under: or perhaps a crop of oats are taken the first year, and over these clover sown for the second: in either of these cases, the soil comes round for wheat the third year, in due succession.

It has already appeared in the *Heads of a Lease*, p. 75, that the Norfolk farmers are restricted from taking more than two crops of corn successively. At the close of a lease this restriction may sometimes have a good effect; for ill-blood between landlord and tenant too frequently leads a farmer to do what he knows will, in the end, be injurious both to himself and his farm. The crime of taking more than two crops of corn successively is, however, held, by farmers in general, in an odious light, and is never practised by a good farmer, unless 'to bring into course' a small patch, with some adjoining piece;—or to regulate his shifts.

2 The case for enclosure

(i) Consolidation of holdings, 1797
(Enclosure Act for Croydon, Surrey—37 Geo. III c. 144, pp. 1, 3. Excerpts transcribed from a copy in the Goldsmiths Library, University of London.)

Whereas there are within the Parish of *Croydon*, in the County of *Surrey*, several Open and Common Fields and Common Meadows, containing together Seven hundred and Fifty Acres, or thereabouts; also several Commons, Marshes, Heaths, Wastes, and Commonable Woods, Lands, and Grounds, containing together Two thousand and Two hundred Acres, or thereabouts; and inclosed Lands and Grounds, containing together Six thousand and Two hundred Acres, or thereabouts:

*   *   *

And whereas the Lands in the said Open and Common Fields and Common Meadows lie so dispersed and intermixed that the same cannot be commodiously or advantageously enjoyed; and the same, as also the said Commons, Marshes, Heaths, Wastes, and Commonable Woods, Lands, and Grounds, are capable of great improvement by an Inclosure, and it would be for the Advantage of all Persons interested in the said Open and Common Fields and Common Meadows, and in the said Commons, Marshes, Heaths, Wastes, and Commonable Woods, Lands, and Grounds, if the same were divided and inclosed and allotted to and amongst the several Persons interested therein, according to their respective Properties, Rights of Common, and other Interests: But the same cannot be effected without the Aid and Authority of Parliament: . . .

(*ii*) *Reclamation of wastes*, 1795
('Report of select committee [on] means of promoting the cultivation and improvement of the waste, uninclosed and unproductive lands of the kingdom, Appendix B', *House of Commons Journal*, LI (23 December 1795), pp. 257–8, 264–5)

Address to the Members of the Board of Agriculture, on the Cultivation and Improvement of the Waste Lands of *Great Britain*: By the President [Sir John Sinclair].

INTRODUCTION

At the Conclusion of the last Session, I had the Honour of stating to the Board, my Intention of laying before it some Observations on the Cultivation and Improvement of the Waste Lands of the Kingdom, a Subject at all Times of great Importance, but peculiarly so at the present Moment, when the Nation is under the Necessity of looking to Foreign Countries for a Part of its Subsistence. Fortunately, however, we have Resources in our Power, if properly called forth, more than sufficient to prevent the Necessity of depending in future, upon other Countries, for any of the Necessaries of Life. To point out the Means of bringing such Resources into Action, and to explain the Advantages to be derived from them, is the Object of this Address. . . .

This leads me briefly to state the Objections which have been made to the Improvement of Waste Lands, and the Obstacles which have hitherto prevented their Cultivation.

In the first Place it has been urged, that the Improvement of Wastes, has a Tendency to depopulate the Country, by diminishing the Number of Cottagers, who reside in their Neighbourhood; and who, in a great Measure, exist, as it is supposed, by the miserable Profits derived from them. Such an Idea, however, is as little justified by Experience, as it is evidently contrary to Reason and common Sense.

It is impossible to suppose that the Poor should be injured by that Circumstance, which secures to them a good Market for their Labour (in which the real Riches of a Cottager consists) which will furnish them with the Means of constant Employment, and by which the Farmer will be enabled to pay them better Wages than before. If a general Bill for the Improvement of Waste Lands were to be passed, every possible Attention to the Rights of the Commoners would necessarily be paid; and as Inclosures, it is to be hoped, will, in future, be conducted on less expensive Principles than heretofore, the Poor evidently stand a better Chance than ever of having their full Share undiminished. Some Regulations also must be inserted in the Bill, to secure the Accommodations they may have Occasion for, by enlarging, where Circumstances will admit it, the Gardens annexed to their respective Cottages, giving them a decided Preference with respect to Locality over the larger Rights, throwing the Burthen of Ring Fences upon the larger Commoners, and allotting, where it is necessary, a certain Portion of the Common for the special Purpose of providing them with Fuel; and thus the smallest Proprietor will in one Respect be obviously benefited, for any Portion of Ground, however inconsiderable, planted with Furze or quick growing Wood, and dedicated to that Purpose solely, would, under proper Regulations, be as productive of Fuel, as Ten Times the Space where no Order or Regularity is observed. If by such Means the Interests of the Cottagers are properly attended to, if their Rights are preserved, or an ample Compensation given for them, if their Situation is in every Respect to be ameliorated it is hoped that the Legislature will judge it proper and expedient, to take such Measures as may be the best calculated for bringing into Culture so large a Portion of its Territory, though it may not accord with the Prejudices of any particular Description of Persons, whose Objections evidently originate from the Apprehension, rather than the Certainty of Injury, and who will consider it as the greatest Favour that can be conferred upon them, when the Measure is thoroughly understood.

In the Second Place it is said, that Commons are an excellent Nursery for rearing young Cattle, and consequently ought to be preserved. No Idea however can possibly be more absurd. If any Person will take the Trouble of comparing the Stock on any Common, with those to be found in the neighbouring Inclosures, they will soon be satisfied of the contrary. That Commons are well calculated for stunting the Growth, or rather starving Animals of every Description, those who pasture their Stock upon them, have in general experienced.

—Where the Right of Common is unlimited, as Dean *Tucker* justly observes, the Ground is so overstocked with Numbers, that no large sized or generous Animal can be bred upon it, and even where the Right is limited, Frauds may be committed, and the Stint is in general so large, that in unfavourable Seasons, the Commons, though under Limitation, are of little Use. In regard to feeding Lambs, Colts, or Calves, it is apprehended, that Grounds free from Disease, and Inclosures properly watered and sheltered, can afford Grass better and more plentiful, and rear them on the Whole to more Advantage, than wild barren Commons over-run with Heath, Furze, Fern, or Brush-Wood.

* * *

...In general, those who make any Observations on the Improvement of Land, reckon alone on the Advantages which the Landlord reaps from an increased Income; whereas, in a National Point of View, it is not the Addition to the Rent, but to the Produce of the Country, that is to be taken into Consideration. It is for Want of attending to this important Distinction, that People are so insensible of the wonderful Prosperity that must be the certain Result of domestic Improvement.—They look at the Rental merely, which, like the Hide, is of little Value, compared to the Carcase that was inclosed in it. Besides, the Produce is not the only Circumstance to be considered—that Produce, by the Art of the Manufacturer, may be made infinitely more valuable than it originally was. For instance: If *Great Britain*, by improving its Wool, either in respect to Quantity or Quality, could add a Million to the Rent Rolls of the Proprietors of the Country, that, according to the common Ideas upon the Subject, is all the Advantage that would be derived from the Improvement: but that is far from being all—the additional Income to the Landlord could only arise from at least Twice the additional Produce to the Farmer; consequently, the Total Value of the Wool could not be estimated at less than Two Millions: and as the Manufacturer by his Art would treble the Value of the raw Material, the Nation would be ultimately benefited in the Amount of Six Millions *per Annum*. It is thus that internal Improvements are to infinitely superior, in Point of solid Profit, to that which Foreign Commerce produces. In the one Case, Lists of numerous Vessels loaded with Foreign Commodities, and the splendid Accounts transmitted from the Custom House, dazzle and perplex the Understanding; whereas, in the other

Case, the Operation goes on slowly but surely. The Nation finds itself rich and happy; and too often attributes that Wealth and Prosperity to Foreign Commerce and distant Possessions, which properly ought to be placed to the Account of internal Industry and Exertion. It is not meant by these Observations to go the Length that some might contend for; namely, to give any Check to Foreign Commerce, from which so much Public Benefit is derived, but it surely is desirable, that internal Improvement should at least be considered as an Object fully as much entitled to Attention, as distant Speculations, and when they come into Competition, evidently to be preferred.

There are some, however, who, although they are ready to acknowledge that the Improvement of the Soil is the best Source of National Wealth, yet have formed an Idea, that very little of the extensive Wastes in the Island are worth the cultivating, who are too apt to imagine that the Climate in which they are situated is hostile to Improvement; and were it better, that the Nature and Quality of the Soil are Obstacles not to be surmounted.

In Regard to the Climate of such Wastes, it is evidently worse in Consequence of the Want of Cultivation.—At the same Time, from the insular Situation of *Great Britain*, the Climate is infinitely milder and better than in any Part of the Continent of the same Latitude. It is stated in One of the Reports, on the most respectable Authority, that very fine Barley and Oats ripen in due Season, on the Summit of a Hill in *Forfarshire*, elevated 700 Feet above the Level of the Sea; and that in *Invernesshire*, at an Elevation of 900 Feet above the same Level, Wheat of a good Quality has been grown.—Hence it may be inferred, that Grain, and other Articles of a similar Nature, may be raised to such a Height upon the Sides and Summits of all the Hills in the Island; and, in regard to Grass, it is well known, that luxuriant Crops of Hay are obtained at the Lead Hills in *Lanarkshire*, elevated 1,500 Feet above the Sea. The Climate of this Country, therefore, can hardly be urged as an Objection to the Improvement of the greater Part of our Wastes, either for Grain or Grass; as to Trees, it is not to be questioned, that the Larch grows in *Italy* on higher Mountains that any we have in this Island.

In regard to the Soil, though the greater Part of the Wastes, having never received any Advantage from the Labour of Man, are at present of little Value, yet the Portion is not very considerable, that ought to be accounted totally barren and unprofitable, or incapable of yielding some useful and valuable Production: One-22d of the Whole, or One Million of Acres, is certainly an adequate Allowance.

This leads me shortly to state the various Purposes, to which the Remainder of such Wastes may be appropriated.

1. The higher situated and the most sterile Parts, ought undoubtedly to be devoted to Plantations.—There is scarcely any Spot, however rocky, or any Soil, however unproductive, that will not yield valuable Timber—an Article which at present we are under the Necessity of importing, at a great Expence, from Foreign Countries. At first Sight it may seem surprizing, that a Spot that would not produce a single Blade of Corn, will yet support the stately Pine, or the spreading Oak. But Trees draw their Nourishment from Sources beyond the Reach of smaller vegetable Productions, and by their Leaves are also supposed to derive additional Sustenance from the Air that surrounds them, or the Water they imbibe. By Plantations also, even barren Spots may in Process of Time be rendered fertile. The poorest Soils, if covered with Wood, from the Leaves which fall, and the Shelter they receive, improve every Year in Fertility, and when the Trees are ready for the Axe, become, in Process of Time, fit for Cultivation.

2. Many of the higher Wastes in the Island might easily be rendered perfectly dry, and soon converted into excellent Upland Pasture. There that valuable Article, fine cloathing Wool, might be grown in Perfection. The loftier the Situation, and the shorter the Herbage, the more valuable it would be; and the Price which the Article bears, joined to the Profit of the Carcase on which it grew, would amply compensate for all the Expence of the Improvement.

3. A much greater Proportion of the Wastes of this Country, than is commonly imagined, might be employed in Tillage.—The Surface may appear barren and unproductive, but Stratums may be found below, which, if incorporated with the Soil above, may render it sufficiently fertile. This is a Practice in Husbandry, which has not yet been carried, in any Degree, to the Extent of which it is capable. It is an Art pretty much in its Infancy, which when brought to Perfection, must be productive of the most important Consequences. As such, it will naturally call for the particular Attention of the Board of Agriculture, to ascertain the Principles on which it can best be conducted.

4. A considerable Proportion of the Wastes of *Great Britain*, consists of Land of a wet and boggy Nature, which it has been yet supposed was the most difficult to improve and cultivate. Fortunately, however, Discoveries have been made in the Art of draining such Bogs, by Mr *Joseph Elkington*, a Farmer of the County of *Warwick*, as renders the Improvement of swampy Land a Matter of much less

Difficulty or Expence than formerly. It is only necessary to add under this Head, that Mr *Elkington* has communicated his System of Draining to those Members of the Board, who were appointed to meet with him upon the Subject,—that he has undertaken to teach such Persons as may be appointed by the Board for that Purpose; and that there is Reason to imagine, that the Practice of his useful Art will be extended, in the Course of the ensuing Summer, from one End of the Island to the other—Bogs drained on Mr *Elkington*'s Principles soon become of very great Value as Meadows, and in many Cases may be converted into Arable Land.

Lastly, at least a Million of Acres of the Waste Lands in the Kingdom may certainly be brought to an astonishing Height of Produce by Watering or Irrigation.—This great Means of Improvement, though long established in some Parts of the Kingdom, yet in others has been unaccountably neglected. But when once that Art is extended as it deserves, the Advantages thence to be derived cannot easily be calculated,—for by it Land is not only rendered perpetually fertile without Manure, but the luxuriant Crops which it raises, produces Manure for enriching other Fields; and the Manure obtained from that Produce, is another Source of National Wealth, that could not otherwise be looked for.

Thus there is every Reason to believe, that the Wastes of this Kingdom, if planted—or appropriated for Pasture Lands—or cultivated for the Production of Grain—or converted into Meadow—or improved by Means of Irrigation; must necessarily be the Source of infinite Wealth and Benefit to this Country.

And if there is a Possibility of improving our Wastes, the Means for that Purpose are more abundantly in our Power, than perhaps in that of any other Country in the Universe. Without entering much at length into so wide a Field, it may be sufficient to remark, that there is none with such a Capital capable of being devoted to so useful and profitable an Object;—none where such a Spirit of Exertion exists, were all Obstacles to the Improvement of our Wastes removed;—none where there is such a Mass of Knowledge on agricultural Subjects;—none where such Abundance of Manures are to be found, particularly those of a fossil and mineral Nature, without the Aid of which it would be impossible to bring great Quantities of Waste Lands rapidly into Cultivation;—and lastly, none, where by Means of a Series of excellent Roads, and Canals every where rapidly extending, such Manures can be so easily and cheaply conveyed to the Lands they are calculated to fertilize. These are Advantages for improving Wastes, which no other Country enjoys in equal Perfection, and which would soon be the Means of Cultivation of a very large Proportion indeed of our at present useless Territory, if full Scope were given to the Industry and Exertions of the People.—Nor ought the Wealth to be derived from the Improvement of our Wastes to be alone taken into Consideration. The Increase of Population, and above all, of that Description of Persons who are justly acknowledged to be the most valuable Subjects that any Government can boast of, merits to be particularly mentioned. His Mind must indeed be callous, who feels uninterested in Measures, by which not only the barren Waste is made to smile, but of which the Object is, to fill the Desart with a hardy, laborious, and respectable Race of Inhabitants, the real Strength of a Country; being the fruitful Nursery, not only of our Husbandmen, but also of the Fleets, the Armies, and the Artists of the Nation.

3 Kentish farming, 1833

(Select committee on agriculture, *B.P.P.*, 1833, V, p. 297, evidence of William Taylor, farmer and valuer)

A Statement of the Expenditure and Receipts on 100 Acres of Arable Land in the Neighbourhood of Rochester, in Kent, under the Seven Tilth System, for One Year.

DISBURSEMENTS

		£	s	d
To the expense of four horses, including keep, tradesmen's bills, waggoner and mate, &c		200	—	—
To one extra horse		40	—	—
To three labourers at 13s 6d per week		105	6	—
Extra to ditto for haying, harvest and task-work		20	—	—
To a boy at 3s per week		7	16	—
Seed for 14 acres of turnips, one bushel		1	10	—
— 14 ditto barley, four bushels per acre, at 30s per quarter		10	10	—
— 14 ditto beans, four ditto – – ditto – 32s ...		11	18	—
— 28 ditto wheat, three ditto – ditto – 60s ...		31	10	—
— 14 ditto clover, two gallons – ditto – 50s per bushel		8	15	—
— 28 ditto wheat, three bushels ditto – 60s ...		31	10	—
— 14 ditto oats, five ditto – – ditto – 22s ...		9	12	6
Cutting and binding 72 acres of corn, 8s per acre		28	16	—
Ditto – – ditto 11 acres of clover, 4s		2	4	—
Beer for haying and harvest		10	—	—

	£	s	d
Poor-rate, at 5*s* per acre	25	—	—
Church and highway, ditto	5	—	—
Repairs	10	—	—
Thatching stacks, etc	5	—	—
Interest on capital, 1,000*l*	50	—	—
Wear of sheep-gates, tools, &c	5	—	—
Rent on 98 acres, having calculated only on that quantity, 28*s*	139	4	—
	£ 768	11	6

RECEIPTS

	£	s	d
By 14 acres of turnips, at 40*s* per acre	28	—	—
14 ditto – barley, five quarters per acre, 30*s*	105	—	—
14 ditto — beans, three – ditto – – 32*s*	64	4	—
28 ditto – wheat, three quarters two bushels, at 60*s*	273	—	—
14 ditto – clover, at 4*l* per acre	56	—	—
28 ditto – wheat, three quarters two bushels per acre, 60*s*	273	—	—
14 ditto – oats, six quarters per acre, 20*s*	84	—	—
Feed on stubbles, 1*s* 6*d* per acre	7	10	—
	890	14	—
Tithe taken in kind	89	1	—
	£ 801	13	—

This account is not framed from any actual data?—No, only framed from an idea of what the expenses of cultivation would be, and what it would be right to calculate that the produce would be.

4 Agricultural statistics, 1853

(Reports on the agricultural statistics of Norfolk and Hampshire, *B.P.P.*, 1854, LXV, pp. 16–17)

The following letter has been addressed by Lord Ashburton to the Statistical Committee of the Alresford Union:

My Dear Sir,   The Grange, October 25, 1853

Having been unfortunately prevented from attending the Board of Guardians on Friday last, when the subject of Agricultural Statistics was brought under discussion, I may perhaps be allowed to endeavour to remedy that omission by addressing a few words to you and to the other gentlemen appointed as a select committee to report upon the question. I am the more strongly induced to do so on account of the misapprehensions which seem to have prevailed, both as to the exact nature of the return desired of us, and as to the motives of the Government in desiring it.

Now, with regard to the first point, the Government does not seek to know the amount of each man's stock or the extent of each man's cultivation. Such a return would be too cumbrous for use, too expensive for publication. The Government wants the sum-totals, not the items of which these sum-totals are composed. It seeks no more to mark and distinguish the return of each occupier than we seek to mark and distinguish each brick of which our house is composed. The house must be put together brick by brick, and the return for the three kingdoms must be gathered item by item; but the items which compose the sum-total will be as much lost in the mass and aggregate of the whole as the bricks which compose the house are lost in the mass and magnitude of the building.

The next question is—why does the Government desire these statistics? What is its motive? It certainly is not with the view of turning corndealer itself, as some have supposed, for that would be not only absurd, but illegal. It assuredly has no notion of taxing our produce, for no Government, under a representative system, would dare to propose a tax upon the first necessaries of life. It evidently does not wish to pry into our secret concerns, for it provides that we may make our returns at our option, either jointly or severally.

It appears to me that the wonder is, not that the Government should now endeavour to collect Agricultural Statistics, but that it should never have sought to do so before. It has now for many consecutive years spent large sums in order to collect, digest, and publish, the statistics of trade, shipping, and manufactures for the good of the merchants, shipowners, and manufacturers. Why should not some little money have been spared to do as much for us? Is it consistent with common sense that every month the public should have paraded before their eyes, and canvassed in the newspapers, the tons of shipping and the pounds of cotton which have entered or quitted our ports and that no intimation should be given from year's end to year's end of the food prepared and preparing for a people's subsistence? Is our industry so unimportant, or capital so minute, that no note should be taken of its condition?

This is not the case in other countries. The United States of America make an annual return of the number of bushels of corn drawn, the quantity they require for their own consumption, and the quantity they can spare for export. The great corndealers have long felt the necessity of collecting some such information for their own guidance. Mr Saunders, of Liverpool, told the House of Commons' Committee in 1833 that he employed agents to travel over the corn districts and report to him both the cultivation and the yield.

Now, what is the consequence of this partial knowledge? Mr Saunders can operate on the market for many days before we, the bulk of the sellers, become aware of the true circumstances which regulate the price of what we have to sell.

Some ten years ago the same advantage was enjoyed by the great money jobbers on the London Stock Exchange. They kept their couriers travelling from city to city, and obtained information five or six days in advance of the ordinary post. They made rapid fortunes at the expense of the public; but now the electric telegraph has placed all upon a level. The publication of these statistics will produce the same good on the Corn Exchange.

There is a further consideration which should operate on our judgments, and I therefore mention it, though it may trench upon politics. Not only does the farmer suffer for want of statistics in his contest for price with the great dealer on the Corn Exchange, but he suffers also from the same want in his contest for consideration and political power with other classes on the great stage of life. I have no doubt in my own mind but that the capital we employ, and the produce we raise, exceed in value all the capitals and all the produce besides raised in this great manufacturing country; but I have no figures to appeal to—I can speak only from conjecture. When, therefore, next year, or when at any future time, it is proposed to make a new apportionment of power according to the importance and magnitude of the several industries, our claims will be most assuredly underrated.

These statistics would obtain for us justice in this respect. They would show that the contribution of the foreigner to the subsistence of this country is as nothing when compared to that furnished by us. They would prove that, instead of being a backward, unenterprising race, bigoted to ancient practices and incapable of improvement, we were bringing every year more and more acres into cultivation, and that we were every year investing more capital, however small might be the profit we derived from it. They would place the small farmer more upon an equality with the great dealer upon the Corn Exchange. They would further give to the trade such accurate information, as would diminish the danger of those fatal speculations which ransack the world for corn under mistaken apprehensions of scarcity, and bring ruin on all engaged.

Actuated by these impressions, I have long desired that we should be put on an equality with the other great industries of the country, and I have done my best at all times to induce the Government of the day to advance the money requisite for the experiment which is now before you.

I remain, my Dear Sir, yours faithfully,

Mr Edward Hunt   Ashburton

I should add that our Scotch neighbours, who are shrewd enough to detect what is and what is not for their advantage, have made their return without hesitation. A still more searching return has been made for two years from Ireland, without complaint on the part of a tenantry who are as quick in perceiving as they are skilful in producing a grievance.

5 Labour-saving machinery, 1871–2

(Joan Thirsk, *English peasant farming* (Routledge, 1957), pp. 325, 326, 329, quoting the manuscript diary of a Lincolnshire farmer)

Development is the order of the day amongst my implements of husbandry. Grant has considerably enlarged my corn drill this summer. It is now about $8\frac{1}{2}$ feet in width. He has also repaired the turnip cutters. Though this year may be one of considerable expense in the total, it is one of retrenchment in horses and labour....

The signs of the times in relation to the labour question are rather ominous. There is a strong outcry for an increase of wages. The working classes are combining in all parts of the country and demanding more money and less hours. I have paid mine 2/6 per day for two or three years past. Last winter when other farmers paid 2/3 mine had half a crown....

There is an ever-growing agitation going on for higher wages. What a kind Providence. As work increases and labour becomes scarcer, machinery develops and more widely adapts itself as a substitutionary power...these strikes will be overruled for good in proportion as they awaken mechanical ingenuity so as to make it more than ever fruitful in invention.

6 Mixed farming in Devon, 1892-3

(Royal Commission on agriculture [Eversley Commission], particulars of expenditure ... and farm accounts, *B.P.P.*, 1896, XVI, pp. 80–1)

Cash Accounts of a Farm of 245 Acres in the Barnstaple Union for the Years 1892 and 1893

1892

Expenditure	£	s	d
Rent	387	10	0
Rates	40	0	0
Tithe	32	5	0
Income tax, Schedule B	3	18	9
Land tax	12	19	11
Brewing licence	0	4	0
Duty on beer brewed, house being worth more than 10*l* per annum	1	15	4
Inhabited house duty	0	6	8
Interest on farm capital, 2,500*l* at 5 per cent	125	0	0
Cattle bought	124	16	6
Artificial manure	83	7	8
Cake	454	7	7
Corn bought	323	4	9
Labour	314	1	8
Seeds bought	19	0	7
Tradesmen's account:			
Saddler	7	8	4
Smith	25	16	9
Carpenter	19	17	11
Hire of steam threshing machine	9	18	0
Two female servants at 12*l* each and 8*s* per week each for food	65	12	0
Cooper, repairs to casks, cattle baskets, &c	1	16	0
Cattle drover	0	18	0
Coal for house and threshing	16	14	6
Dog licence	0	7	6
Sack hire	1	11	1
Cabbage plants	3	9	0
Plough reins, halters	0	9	2
Vet's account (as per contract)	5	5	0

Receipts	£	s	d
Cattle sold	769	17	4
Horses	265	0	0
Sheep	679	11	6
Pigs	16	16	0
Poultry, milk, butter, &c	149	18	0
Wheat	386	15	2
Wool	85	13	6
Potatoes and apples	7	12	0
Potato lands let to cottagers in lieu of allotments	8	0	0
Barm [yeast] sold from beer brewed	2	2	8
Prizes won	39	10	0

	£	s	d
Twine for binding straw and wheat	5	13	1
Sundries	1	3	0
	2,088	17	9
Total—profit	321	19	3
	2,410	17	0

2,410 17 0

1893

Expenditure	£	s	d	Receipts	£	s	d
Rent	387	10	0	Cattle	330	10	0
Rates	39	18	10	Horses	91	5	4
Tithes	31	5	0	Sheep	444	2	10
Income tax, Schedule B	3	18	9	Pigs	19	7	0
Land tax	12	19	11	Poultry, milk, butter,			
Brewing licence	0	4	0	&c	155	6	2
Duty paid on beer brewed	1	15	4	Wheat	222	7	0
Inhabited house duty ..	0	6	8	Wool	71	5	0
Interest on farm capital,				Potatoes, apples, &c	12	15	0
2,500*l* (at 5 per cent) .	125	0	0	Potato lands let in lieu			
Animal manure	94	7	5	of allotments	7	15	6
Cake	244	11	6	Barm sold from beer			
Corn bought	204	16	4	brewed	1	15	8
Labour	293	1	6	Prizes won	42	2	6
Tradesmen's account:					1,398	12	0
Saddler	5	4	2	Total—loss	216	14	7
Smith	31	5	3				
Carpenter	17	19	4				
Steam threshing	5	4	0				
Two servants and food at 8*s* per week each ...	65	12	0				
Coal for house and threshing	18	12	0				
Dog licence	0	7	6				
Vet's account, as per contract	5	5	0				
Twine for binding straw and wheat	2	7	6				
Seeds bought	21	6	3				
Sack hire	0	15	6				
Sundries	0	19	10				
Plough reins and halter, etc.	0	12	6				
	1,615	6	7		1,615	6	7

7 Agricultural credit and farmers' capital, 1927

(Ministry of Agriculture and Fisheries, *Report on agricultural credit* (Economic series, no. 8, H.M.S.O., 1927), pp. 31–2, 36–7)

Again, it has to be remembered that persons who compare the services to agriculture of the old private banks with those of the joint stock banks, are sometimes comparing a time when agriculture was more prosperous with the present time, when it is still suffering from the effects of depression. There seems to be no evidence that the present attitude of the banks does more than reflect the want of confidence in the industry to which farmers themselves give frequent expression.

To sum up, there is nothing in the evidence which has been available to suggest that the banks, within the limits of the present credit system, are unduly rigid or unsympathetic with the needs of agriculturists, nor is there anything to suggest that hasty marketing by the farmer is due to pressure being exerted upon him by the banks at harvest-time. On the contrary, there are many reasons to think that consistent with reasonable business prudence, the banks administer the present system as generously as the system permits. Whether the limitations it imposes both on them and their customers are too narrow, and whether the system itself is open to serious criticism is, however, a question which will be considered below.

Credit from Tradespeople

(*a*) *Auctioneers*—Auctioneers who have large connections in cattle, sheep and farm sales, sometimes allow farmers to buy live stock on credit, on condition that these cattle and sheep, or lambs born from the latter, are re-sold in their markets. By this means, however, the farmer is restricted in his market and consequently may be restricted in his profits. The auctioneer is stated to receive slightly higher rate of interest on the capital sum involved than the current bank rate, in addition to the commission he receives on the re-sales. In some districts auctioneers also give credit to butchers who buy at their sales; by this means not only is a constant entry at the cattle sales assured, but purchasers also. If the auctioneer is short of entries a farmer who is in his hands may be called upon to send cattle to the market, but it is obvious that by doing so he may have to accept a lower price than if he had been free to choose his own time, and what he ultimately pays for the credit may remain a matter of uncertainty. In consequence of this system the auctioneer's hold on some districts has been described as 'enormous'.

(*b*) *Corn Merchants, Seedsmen, etc.*—A similar system is in force in some districts in regard to corn, hay, etc. The farmer buys seeds on credit from a corn merchant subject to the condition that the latter takes the farmer's corn crops when ripe; in addition the farmer is sometimes charged interest on his account. This often results in a farmer having to buy in the dearest market and sell in the cheapest; he cannot choose his own time, and must accept the merchant's price for the corn. The actual financial basis of such arrangement varies according to the standing and character of the farmer, and there is seldom any settled basis of terms.

$$* \qquad * \qquad *$$

In the report of the Committee on Agricultural Credit, published in 1923, it is stated that the total loans to agriculturists from the five leading banks of England and Wales amounted to 46½ millions, of which 26 millions represented loans for the purchase of agricultural land and 20 millions normal loans for current trading. Through the courtesy of these banks, similar figures have been obtained for the beginning of the present year, which show that the total loans of the same banks to agriculturists amount to approximately 51 millions, of which about 25 millions is lent for current trading. In their report the Committee express surprise at the magnitude of such figures. If the sum of 25 millions represents loans mainly secured upon non-agricultural wealth or unsecured, there is perhaps ground for surprise, but how does it compare with the actual wealth on farms? Taking Great Britain as a whole, the Ministry of Agriculture make the following estimate:

The gross value of the output of farms (*i.e.*, produce sold off the farms) in Great Britain was estimated for the year 1922 at £261,250,000 made up as follows—

Farm Crops	£ 17,000,000
Live Stock	108,000,000
Dairy Produce	77,000,000
Wool	3,250,000
Poultry and Eggs	16,000,000
Miscellaneous Crops	10,000,000
	£ 261,250,000

7*

8 Mechanised farming: 1947 and 1961

(Michael A. Havinden and others, *Estate villages: a study of the Berkshire villages of Ardington and Lockinge* (Lund Humphries for the University of Reading, 1966), Appendix, p. 210)

Machinery Inventory, Lockinge Estate, 1947 and 1961*

		1947	*1961*
Vehicles	Tractors	19	24
	Lorries	1	3
	Vans and cars	4	7
Field equipment	Ploughs	16	11
	Cultivators	7	2
	Harrows	10	16
	Drags	8	3
	Horse Hoes	7	—
	Rollers	13	6
	Drills	14	9
	Broadcasters	2	6
	Binders	9	2
	Combines	4	6
	Mowers	10	5
	Rakes	20	15
	Sweeps	3	1
	Balers	5	5
	Elevators	4	2
	Green crop loaders	—	5
	Threshers	1	1
Transport equipment	Carts, wagons and floats	57	16
	Trailers	28	41
Barn equipment and plant	Root pulpers	11	2
	Chaff cutters	3	—
	Cake crushers	3	1
	Mixers	—	1
	Mills	2	3
	Cuber	—	1
	Engines	9	8
	Tanks	10	11

* Note that the acreage of farmed land on the estate has increased from 3,551 acres to 4,155 acres between these dates.

B. PROSPERITY AND DEPRESSION

1 Reclamation of waste in wartime, 1805

(J. W. F. Hill, *Georgian Lincoln* (Cambridge University Pres, 1966), pp. 174–5, quoting Banks MSS., National Library of Wales, MS. 12415, E34)

...We do not think much of the defeat of the Austrians [at Austerlitz] because the calamity is distant and the consequences to us remote; we think most on the progress of our drainage, which is very prosperous. Our land sold at first for somewhat more than £30 an acre. The second sale it brought £40. The sale of £75,000 worth lately brought £50, and a small sale since has brought £60 an acre, although there has been no particular convenience of situation, and in all cases the buyer pays auction tax, and all costs of enclosure, as we deliver only the surface of the commons marked out by surveyors' stakes.

The quantity of land which is to be won to the public from thistles in summer and wild ducks in winter as you have seen, and which did not, I am confident, on an average of years bring to the occupier 5s an acre or indeed 1s to the landlord, is 40,000 acres at £50 an acre, that is, £2,000,000 of money...

2 Post-war depression in perspective, 1821

(Select committee on ... agricultural distress, 1821, Report, *B.P.P.*, 1821, IX, pp. 6–7, 26–7)

... Your Committee cannot doubt, that, in so far as the alteration of our currency has contributed to lower the price of commodities, the productions of Agriculture have been, and must hereafter, in common with all other articles, be affected by the improved value of our money.

But Your Committee are also satisfied, by the result of their enquiries, that, in the present year, the price of corn has been further depressed by the general abundance and good quality of the last harvest, in all articles of grain and pulse; more particularly in Ireland, in which part of the United Kingdom the preceding harvest of 1819, was also uncommonly productive. Several of the witnesses examined have stated their belief that the prices of grain have further been depressed, in the present year, by the very large importations of foreign corn which took place before the ports were closed in the month of February

1819; but looking to the very high prices, and to the constant and brisk demand which prevailed in our markets so long as the ports continued open in 1817 and 1818, it may be inferred that the greatest part of those importations was necessary, and was disposed of during those years, to supply the daily wants of our consumption, and that it is therefore only in a remote degree that the present prices can be influenced by the occurrences of that period.

It can scarcely be necessary for Your Committee to remark, that the growth of wheat has been greatly extended and improved of late years, in all parts of the United Kingdom, but principally in Ireland, since the year 1807.

Your Committee feel it an important part of their duty to recall to the recollection of the House, and the Country, that, in the years 1804 and 1814, a depression of prices,—principally caused by abundant harvests, and a great extension of tillage, excited by the extraordinary high prices of antecedent years,—appears to have produced a temporary pressure and uneasiness among the owners, and occupiers of land, and a corresponding difficulty in the payment of rents and the letting of farms, in some degree similar to the apprehensions and embarrassments which now prevail; and, also, that in many earlier periods, similar complaints may be traced in the history of our Agriculture.

Among numerous instances of these complaints, which may be found in other publications, between the middle of the 17th century and the beginning of the late reign, two have been pointed out by one of the witnesses, in which the House will not fail to remark the great similarity between the arguments and alarms which were then current, with those which prevail in many quarters at this period.

That in these earlier and more remote stages of our Agriculture these alarms were only temporary, and that the fears of those who reasoned upon their continuance and increase, were ere long dissipated by the natural course of seasons and events, is now matter of history. And it is impossible to look back to the discussions of the years 1804 and 1814, and more especially to the evidence taken before the Committee appointed by the House on the latter occasion, without being forcibly struck with the conformity of the statements and opinions, then produced, respecting the ruinous operation and expected continuance of low prices, with those which will be found in the evidence now collected. Indeed these statements, in some instances, come from the mouths of the same witnesses. . . .

Your Committee trust that this reference to past experience will not be altogether useless and unavailing to allay the alarm, and to dispel some of the desponding predictions which, by unnecessarily increasing anxiety for the future, tend to aggravate the severe pressure of our present difficulties;—that the reflections which such a retrospect is calculated to excite, may lead the occupiers of the soil, as it has led Your Committee, to infer, that in Agriculture, as in all other pursuits, in which capital and industry can be embarked, there have been, and will be, periods of reaction; that such reaction is the more to be expected, in proportion to the long continued prosperity of the pursuit, and to the degree of previous excitement and exertion which that prosperity had called forth. They must add, as a further inference from the experience of former periods, to which the present crisis bears no distant resemblance, that there is a natural tendency in the distribution of capital and labour to remedy the disorders which may casually arise in society from such temporary derangements, and (without at all meaning to deny that it is the duty of the Legislature to do every thing in its power to shorten the duration, and to palliate the evils of the crisis) that it often happens that these disorders are prolonged, if not aggravated, by too much interference and regulation.

It is by no means with the expectation that the suffering of our own community can be alleviated by the contemplation of a corresponding pressure upon other nations, that Your Committee find themselves called upon to state, that many commodities of general and extensive demand, the staple productions of other countries, such as corn, cotton, rice and tobacco in the United States of America; sugar and rum in the West Indies; tallow, flax, hemp, timber, iron, wool, and corn, on the continent of Europe, appear to have fallen in price, in some instances more, and scarcely in any less, in proportion to the prices of those articles prior to 1816, than the fall in the price of grain in this country:—with regard to several of which articles, and the countries producing them, some of the causes which have principally affected the value of grain in this country cannot be considered as operating.

*　　　*　　　*

So far as the present depression in the markets of Agricultural produce is the effect of abundance from our own growth, the inconvenience arises from a cause which no legislative provision can alleviate; so far as it is the result of the increased value of our money, it is one not peculiar to the farmer, but which has been and still is experienced by many other classes of society. That result however is the more severely felt by the tenant, in consequence of its coincidence with an overstocked market, especially if he be farming with a borrowed capital and under the engagements of a lease; and it has hitherto been further aggravated by the comparative slowness with which prices generally, and particularly the price of labour, accommodate themselves to a change in the value of money.

From this circumstance, combined with other causes, the departure from our antient standard, in proportion as it was prejudicial to all creditors of money and persons dependent on fixed incomes, was a benefit to the active capitals of the country; and it cannot be denied that the restoration of that standard has, in its turn, been proportionally disadvantageous to many individuals belonging to the productive classes of the community, and especially to those who had engaged in speculative adventures, either of farming or trade.

That restoration must also be accompanied with embarrassment to the landowner, in proportion as his estate is encumbered with mortgages or other fixed payments, assigned upon it during the period when land and rents were raised to an artificial value, in reference to the impaired value of the money in which those encumbrances were contracted.

From the cessation of public loans, the probability of large accumulations of capital, and the constant operation of such a sinking fund, as in the present state of our finances, may, henceforward during the continuance of peace, be regularly appropriated to the reduction of the public debt, Your Committee trust that the rate of interest of money, may in a short time, be so far reduced below the legal *maximum*, as to make those encumbrances a lighter burthen upon the landed interests of the kingdom. It is an alleviation which former intervals of peace have produced, at periods in many respects less favourable to its attainment; and if, in the present instance, the want of that alleviation is become more urgent, Your Committee venture to hope that, from the greater accumulation of capital in the country, cooperating with the effects of a positive and steady reduction of the public debt, this salutary result will also be more speedily brought about. They look forward to this mode of easing the encumbrances of the landlord with the more anxiety, as, amidst all the injury and injustice which an unsettled currency,—an evil they trust never again to be incurred,—has in succession cast upon the different ranks of society, the share of that evil which has now fallen upon the landed interest, is the only one which, without inflicting greater injury and greater injustice, admits (now that we are so far advanced in the system of a restored currency) of no other relief. The difficulties, great as they unfortunately are, in which it has involved the farming, the manufacturing and trading interests of the country, must diminish in proportion as contracts, prices, and labour, adjust themselves to the present value of money. That this change is now in progress, and has already taken place to a considerable degree, is in evidence before Your Committee. They are satisfied that it will continue until that balance is restored, which will afford to labour its due remuneration, and to capital its fair return.

3 Corn and stock, 1770–1850

(Sir James Caird, *English agriculture in 1850 and 1851* (Longmans, 1851; Cass, 1968), pp. 475–6, 485)

	s	d	
In twenty-six counties the average rent of arable land, in 1770, appears from Young's returns to have been	13	4	an acre
For the same counties our returns in 1850–51 give an average of	26	10	an acre
Increase of rent in eighty years	13	6	or 100 per cent

Bushels

In 1770 the average produce of wheat was	23 an acre
In 1850–51 in the same counties it was	$26\frac{3}{4}$ an acre
Increased produce of wheat per acre	$3\frac{3}{4}$ or 14 per cent

	s	d	
In 1770 the labourers' wages averaged	7	3	a week
In 1850–51, in the same counties they averaged	9	7	a week
Increase in wages of agricultural labourers	2	4	or 34 per cent

Bread. Butter. Meat

In 1770 the price of provisions was	$1\frac{1}{2}d$	6d	$3\frac{1}{4}d$ per 1b
In 1850–51 it was	$1\frac{1}{4}$	1s	5 per 1b

	s	d	
In 1770 the price of wool was	0	$5\frac{1}{2}$	per 1b
In 1850–51 it was	1	0	per 1b

	s	d	
In 1770 the rent of labourers' cottages in sixteen counties averaged	36	0	a year
In 1850–51, in the same counties	74	6	a year

It thus appears that, in a period of 80 years, the average rent of arable land has risen 100 per cent, the average produce of wheat per acre has increased 14 per cent, the labourers' wages 34 per cent, and his cottage rent 100 per cent; while the price of bread, the great staple of the food of the English labourer, is about the same as it was in 1770. The price of butter has increased 100 per cent, meat about 70 per cent, and wool upwards of 100 per cent.

The increase of 14 per cent on the average yield of wheat per acre, does not indicate the total increased produce. The extent of land in cultivation in 1770 was, without doubt, much less than it is now; and the produce given then was the average of a higher quality of land, the best having of course been earliest taken into cultivation. The increase of acreable corn produce has therefore been obtained by better farming, notwithstanding the contrary influence arising from the employment of inferior soils. The increased breadth now under wheat, with the higher average produce, bear, however, no proportion to the increase of rent in the same period; and the price of wheat now is much the same as it was then. We must therefore look to the returns from stock to explain this discrepancy.

While wheat has not increased in price, butter, meat, and wool have nearly doubled in value. The quantity produced has also greatly increased, the same land now carrying larger cows, cattle which arrive at earlier maturity, and of greater size, and sheep of better weight and quality, and yielding more wool. On dairy farms, and on such as are adapted for the rearing and feeding of stock, especially of sheep stock, the value of the annual produce has kept pace with the increase of rent. With the corn farms the case is very different. In former times the strong clay lands were looked upon as the true wheat soils of the country. They paid the highest rent, the heaviest tithe, and employed the greatest number of labourers. But modern improvements have entirely changed their position. The extension of green crops, and the feeding of stock, have so raised the productive quality of the light lands, that they now produce corn at less cost than the clays, with the further important advantage, that the stock maintained on them yields a large profit besides. In all parts of the country, accordingly, we have found the farmers of strong clays suffering the most severely under the recent depression of prices.

*   *   *

Every intelligent farmer ought to keep this steadily in view. Let him produce as much as he can of the articles which have shown a gradual tendency to increase in value. The farms, which eighty years ago yielded 100*l* in meat and wool, or in butter, would now produce 200*l*, although neither the breed of stock nor the capabilities of the land had been improved. Those which yielded 100*l* in wheat then, would yield no more now, even if the productive power of the land had undergone no diminution by a long course of exhaustion.

4 The causes of depression, 1881 and 1897

(*i*) *A scientist-farmer*
(Royal Commission on agriculture, minutes of evidence, *B.P.P.*, 1881, XVII, qq. 57590–620)

Mr J. B. Lawes called in and examined

(*Chairman*) I believe you live at Rothamsted, near St Albans?—Yes.

Am I not right in saying that when you were at the university you were interested in chemical research, and on succeeding to your property you applied the knowledge there acquired to scientific agriculture, and since that time you have been connected with a business for the sale of artificial manures; and you are now engaged and have been engaged for many years in experimental research in connexion with agriculture?—Yes, ever since I left the university.

When did you commence experiments in connexion with agriculture?—About 1834 slightly; they were what you may call loose experiments, but still they were experiments.

And that led you to go into business, did it not?—Yes, soon after that, in 1841 or 1842.

How long were you connected with business?—About 30 years.

At the present moment have you anything to do with it?—For the last ten years I have ceased entirely to have anything to do with the business.

Will you describe in your own words the general features of your farm and its management at the present moment?—I hold about 500 acres of land, one half of which is under pasture and about half of that half has been laid down within the last few years. The rest is under arable cultivation, and about 50 acres of that is under very careful experiments. I keep the rest of the farm to attend to the experiments, that is to say, I keep 450 acres so as to have a staff of horses and men to bring to bear upon my experiments.

How many men have you employed in the farming operations, and how many in the experimental part of your farm?—I employ about

20 farming people, and I have about 10 probably entirely engaged in scientific investigations in connexion with the experiments.

Have you at all estimated what you have spent in experimental research?—Not in the gross; but I think that it costs between 2,000*l* and 3,000*l* a year at the present moment.

The laboratory built on your property was built by the contributions of agriculturalists who wished to bear testimony to the good work which you had done for agriculture?—Yes, it was.

Before we go into the question of the experimental part of your farm I would ask you whether the agricultural depression of the last few years has affected you to any extent?—In common with other agriculturalists it has certainly.

In what respect?—By the very bad crops that I have had, and the great injury done to my land by the wet, and by the growth of weeds.

What do you consider has been the primary cause of the depression? —The excessive wet and the low prices.

How many years do you consider that the agricultural depression has lasted?—Six years I consider it to be.

Speaking of the excessive wet, have you any information which you can put before the Commission which will show the excess of rainfall during a given period above the average?   I have, in one of my fields, a field of 30 acres, a large rain gauge containing the $\frac{1}{1000}$ part of an acre or about 43 cubic feet which we have for collecting a large quantity of rain water for analysis. In the 21 years from 1854 to 1874 the average rainfall was $26\frac{1}{2}$ inches; in 1875 it was 35 inches; in 1876 it was $34\frac{1}{2}$ inches; in 1877 it was $33\frac{1}{2}$ inches; in 1878 it was $32\frac{1}{3}$ inches; in 1879 it was 36 inches; and in 1880 it was 34 inches; giving an average for those six years of a little over 34 inches, being $7\frac{1}{2}$ inches more than the average of the previous 21 years. We have a similar rain gauge placed under the ordinary soil of my land, through the clay, at depths of 20 inches, 40 inches, and 60 inches; and I propose to give you the drainage, that is to say, the amount of water which does not evaporate into the air again, but which passes into the subsoil and into the drains during those six years compared with the previous years. During the four years previously $7\frac{3}{4}$ inches passed through the 60 inches rain gauge; in 1875 $15\frac{3}{4}$ inches passed through; in 1876 18 inches passed through; in 1877 $15\frac{1}{2}$ inches passed through; in 1878 15 inches passed through; in 1879 $19\frac{3}{4}$ inches passed through; and in 1880 $17\frac{1}{4}$ inches passed through; being an average of $16\frac{7}{8}$ inches of water which would pass into the subsoil, against $7\frac{3}{4}$ inches which was the average of the four previous years. In the field where I grow continous wheat we have had a number of drains on purpose to enable us to collect the water that passes through. The subsoil is chalk and does not require draining, but we gather water because we have found out that wherever water passes through land a large number of valuable ingredients pass with it, so that we pay great attention to the water passing out of the drains. Between 1867 and 1874 we caught water through the drains on the average nine times during the year; in 1875 we collected it 15 times; in 1876 we collected it 17 times; in 1877 we collected it 15 times; in 1878 we collected it 15 times; in 1879 we collected it 28 times; and in 1880 we collected it 22 times; the average being 19 times against nine times.

What do you mean by collecting the water?—My land is drained by the chalk? but if there is a glut of wet it runs through those drain pipes quicker than it can go through the chalk, and therefore we catch it. Whenever it rains we have a man to watch it, and we collect the water for analysis, therefore we watch it with a great deal of care.

Can you give any statistics with regard to the temperature?—I could do so, but I did not bring them with me.

Will you hand in any information that you can give on that subject?—Yes, certainly.

Do you consider that the effects of rainfall are more detrimental to agricultural production at one time of the year than at another?—If I was to separate the agricultural production I think I should say so, certainly; but speaking generally, I am not quite sure that I should say so, because what would be very beneficial to grass, and pasturage, and green crops would be very injurious to the corn crops.

Has the excess of rainfall during the last six years that you have spoken of come at a period when it was particularly damaging to farming?—Certainly as far as the corn crops are concerned.

When did it occur principally?—In two or three instances. Last year, for instance, we had a rainfall almost every day in July; in the previous year we had rain in May, June, and July almost every day; and in 1875 we had great floods in July. Those are the three years that I can recollect at this moment, in which rain came at a very inopportune time for the ripening of wheat or barley.

Then excessive rainfall at seed time would be very detrimental, would it not?—It would certainly. We would rather not have it, but we often know that wheat goes on in the very worst way, as it did last year, and gives a very good crop afterwards. In 1879 we had a glut of wet in May, June, and July, which was perhaps our typical worst year that we ever had, or ever shall have I hope.

The excess of rainfall would be very damaging at the time the crops were flowering?—It is more after they have flowered, when they are ripening.

To what other causes besides weather do you attribute the depression in agriculture?—To the low prices, and diseases of stock; those are the three great items, I believe.

The low prices being attributable to foreign competition?—Yes.

In which years have the low prices affected you most?—In the last four years.

Has the foreign competition affected the prices of the cereals and other farm produce, or the live stock more?—The wheat is the principal crop that has been affected by the imports, not the live stock.

Do you keep much stock?—I am sorry to say I do now. I am obliged to keep a great deal of stock. I keep a large dairy.

If the seasons should be more favourable than they have been, do you think there is any reason to fear the foreign competition?—That is a difficult question to answer. I think that if I was entering into farming as a business I might manage to get on somehow by adapting myself to the soil and locality and times altogether.

Assuming that you were farming as a farmer in the district where you live, what change should you make in the system of farming which usually prevails in order to meet the competition?—First of all I have laid down a great deal of land to permanent pasture during the last ten years, looking forward to what was coming, and I have established a dairy because foreign nations cannot so easily sell us milk as they can the other products; and I have given up a good deal of corn growing, which is not so profitable as it used to be.

I suppose you think that that is an example which ought to be followed?—In my district where there is heavy land and you depend upon corn growing very much.

(*ii*) *Retrospective view—the Eversley commission*, 1897

(Royal commission on agriculture, *Final report*, *B.P.P.*, 1897, XV, pp. 20, 85–86)

In the preceding paragraphs of this chapter, we have briefly reviewed the evidence put before us as to the distribution and effects of agricultural depression. It is clear that the depression has not equally affected all parts of Great Britain. In arable counties, where its presence is most manifest, it has entailed very heavy losses on occupiers and owners of land, in some districts considerable areas have ceased to be cultivated, and there has been a great withdrawal of land from the plough. These features of the crisis have been particularly marked on the strong clays and on some very light soils. Broadly speaking, it may be concluded that the heavier the soil, and the greater the proportion of arable land, the more severe has been the depression.

In England the situation is undoubtedly a grave one in the eastern, and in parts of some of the southern, counties. In the arable section of Scotland the position is in some respects not so serious; but there, also, great losses have been experienced during the past 12 years. In the pastoral counties of Great Britain the depression is of a milder character, but in most of them the depreciation of the value of live stock between 1886 and 1893, and the persistent fall in the price of wool, have largely diminished farming profits and rents. In districts suitable for dairying, market gardening, and poultry rearing, and in the neighbourhood of mines, quarries, large manufacturing centres, and towns, where there is a considerable demand for farm produce, there has been relatively less depression.

One prominent feature of the depression has been the great contraction of the area of land under the plough in all parts of the country. Its effects upon the distribution of the cultivated surface in Great Britain between 1875 and 1895 may be seen in the following statement extracted from the Agricultural Returns:

Year	Arable	Pasture	Total
	Acres.	Acres.	Acres.
1875	18,104,000	13,312,000	31,416,000
1880	17,675,000	14,427,000	32,102,000
1885	17,202,000	15,342,000	32,544,000
1890	16,751,000	16,017,000	32,768,000
1895	15,967,000	16,611,000	32,578,000

Commenting upon the important change in the agriculture of the country illustrated by these figures, Major Craigie remarks that the actual loss of arable area in the interval covered by the last two decades, which may be said to embrace the period of depression, is 2,137,000 acres, and that the diminution of the wheat acreage alone accounts for more than 1,900,000 acres of this loss.

*　　*　　*

From the foregoing analysis of the evidence put before us relating to foreign competition, it is clear that there has been a remarkable increase in the imports of all forms of agricultural produce during the past 20 years.

Of the various products of British agriculture, wheat has been the most affected by this development, the foreign supply of this grain having gradually displaced the home production until the latter now constitutes barely 25 per cent of the total quantity needed for consumption annually in this country. There has been no similar displacement of the other home-grown cereals, but in the case of barley it is worthy of notice that the low-priced varieties grown in Eastern Europe, which were imported in comparatively small quantities in 1876–80, now form the larger proportion of the foreign supply, and this change has been of some influence in the determination of the price of British barley. The price of feeding barley, as well as of oats, has also been affected to some extent by the large consumption of maize.

As regards meat we have been unable to trace any actual displacement of the home produce by the growth of the imports. The supply of foreign beef and mutton apparently meets a demand for cheap meat which has not hitherto been satisfied by the home production, and while it has undoubtedly seriously affected the price of the inferior grades of British produce, its influence on the superior qualities has been much less marked. Foreign competition has been, on the whole, perhaps more severe in pork than in other classes of meat, but it has been confined mainly to bacon and hams.

In the case of wool, the facts at our disposal show that there has been a progressive increase in the foreign supplies of this staple, and there has been some displacement of the home-grown product. The net imports form so large a proportion of the total supply that they must be an important factor in the determination of the price of British wool.

With respect to the extent of the foreign competition in dairy produce we have estimated that the importation of butter, margarine, and cheese represents more than 50 per cent of the total quantities of these articles available annually for consumption.

In dealing with cereals, meat, wool, and milk, we have attempted to gauge the intensity of the foreign competition by estimating approximately the dimensions of the volume of imports relatively to the aggregate supply of home and foreign production. The absence of satisfactory data has rendered it impossible for us to apply this method to the imports of vegetables, fruit, poultry, eggs, and other articles, but in view of the facts to which we have referred in paragraphs 280–287 above, it is evident that in most of these products the growth of the foreign supplies must be regarded as a serious element of competition with which the British producers have had to reckon.

It is, we think, important to note that in nearly every case, subject to the reference in paragraph 185 to the case of wheat, the expansion of the imports has been accompanied by a contraction in the price of the several products concerned, and that there has been a general correspondence between the fall of price and intensity of foreign competition.

An investigation as to the sources contributing to the increasing volume of imports of agricultural produce has shown that the United States has held the premier position throughout the last 20 years in the supply of wheat and meat, excluding mutton, while she has also contributed the major portion of the imports of maize, although her shipments in this article since 1890 have been exceeded by those of Roumania. Argentina has in recent years ranked next to the United States as an exporter of wheat and meat to this country. Other prominent contributors to the imports of cereals are Russia for wheat and barley and India for wheat alone, though the Indian supply has fallen off considerably. Australasia is responsible for the major portion of the imports of wool and mutton, and we have recently received large consignments of butter from this source. Denmark furnishes between 40 and 50 per cent of the butter imported annually; Canada and the United States practically monopolise the import trade in cheese; while Holland supplies nearly the whole of the margarine.

We have already discussed in some detail the conditions under which the exports from the United States have been maintained. The great feature of the production of wheat in that country has been the steady movement of the centre of cultivation towards the virgin soils of the comparatively newly settled States in the west and the reduction of the wheat acreage in the older eastern and central States. This change has proceeded concurrently with a progressive fall in the price of wheat, and may be explained to some extent by the fact that the cost of production in the more recently settled territories is considerably less than in the regions which have been settled and cultivated for a longer period. The rapid development of the exports of wheat from Argentina is also to be attributed partly to the circumstance that the expenses of cultivation on the 'extensive' system adopted on the virgin lands of that country are much below those incurred by producers in older centres of production where the exhaustion of the natural fertility of the soil has rendered necessary a greater application of labour and manures. Several estimates have been put before us of the outlay required to produce a bushel of wheat in North and South America, Australia, and Great Britain respectively, and although we should hesitate to attempt to appreciate exactly from such estimates the relative positions of the wheatgrowers in the several countries, we are convinced that the balance of advantage is decidedly in favour of the producers in Argentina, the Western States of America, and in the Colonies.

With reference to Argentina it has been held by several expert witnesses who have appeared before us that her exports of agricultural produce have been largely stimulated by the depreciation of the Argentine paper currency, and that a similar cause has operated in favour of the producer in Russia. It has been maintained also that the decline in the gold value of the rupee enabled the production of wheat for export to be continued in India in spite of the fall in the gold price of that cereal.

No detailed evidence has been placed before us as to how far the improvements in the means of transport have enabled producers abroad to maintain their consignments to British markets during a period of falling prices, but the witnesses who have referred to this subject have agreed that the development and improvement of the lines of communication by land and sea and the reduction of freight rates have facilitated the cultivation of the fertile areas in North and South America and in the Colonies, and have generally contributed in no small degree to the growth of foreign competition.

The circumstances which have attended the growth of foreign competition in dairy produce are entirely different from those to which we have referred in connection with the imports of other products. The ability of the foreign producer of grain and meat to compete with so much success in our markets is due, to a large extent, to the superiority of the natural and climatic conditions under which his business is carried on, but the successful competition of the foreigner in our butter and cheese markets is to be credited mainly to the fact that the dairy industry is better organised abroad than in Great Britain.

In connexion with the general subject of foreign competition, we have also to draw attention to a memorandum submitted by Sir Robert Giffen (see Appendix V), in which he suggests that the decline in the price of wheat itself may be partly attributed to the great increase in the supply and consumption of meat during the last 20 years, which has either diminished the demand for wheat per head among the people consuming wheat, or has checked the increase of that demand, which might have been expected to follow a great decline in price. This competition of article with article is, of course, a matter to be considered as well as the direct competition between foreign countries and Great Britain in the production of particular articles.

5 Rent in the depression

(Harriet S. Loyd Lindsay, Baroness Wantage, *Lord Wantage, V.C., K.C.B.* (Smith Elder, 1908), p. 378)

On Lord Overstone's Northamptonshire and other estates, where up to 1875 arrears of rent were practically unknown, the condition of things was no better. Reductions in rent, some temporary, but mostly permanent, had to be made to the extent of lowering the net rental on Lord Overstone's estates from about £44,000 to £12,000. To landlords of small resources this state of matters meant ruin. Lord Overstone and Lord Wantage, as capitalists, and not dependent entirely on land, were able to face the situation.

6 Return of prosperity, 1913

(A. D. Hall, *A pilgrimage of British farming, 1910–1912* (Murray, 1913), pp. 430–1)

After three years' wanderings from the Moray Firth to Cornwall, from Norfolk to Cork, one cannot help drawing some general conclusions about the state of British farming and some of the problems arising therefrom which have latterly been interesting a wider public. In this connexion it may be useful to state how little the questions that are most fiercely debated in other places seem to trouble the farming community. Amongst the farmers themselves there is no land question, no smouldering feeling nor general current of opinion that calls for a 'policy'; in the main they would ask to be let alone. But we should like to enter a preliminary apology for the use we shall have to make of the term farmers. Half the difficulties of controversy arise from the fact that each party has a different group in view when it speaks of farmers. One is thinking of the tenants of from 200 acres to 500 acres; another of men with 30 acres to 80 acres working for their daily bread; a third, and perhaps the most vocal, of the men who dominate the Farmers' Clubs and Chambers of Agriculture, men who may be owners or tenants, but are primarily business men connected with land, dealers

8*

in pedigree stock, valuers and agents, making the main part of their income by other means than sheer cultivation of the soil. In the first place we must recognize that the industry is at present sound and prosperous. The great depression touched its nadir about 1894; since that time prices have been moving upwards and methods improving. By degrees men learnt to cheapen production, in some cases by improved machinery and by savings in the actual husbandry, in others by a change of objective, as when grass and milk replaced wheat and beans. Rents were reduced, and before the century ended it began to be evident that a new race of farmers had grown up capable of making a living under the existing conditions. From that time all the advances in the price of corn and produce, of meat and milk and wool, have been so much clear gain, but it was not until about 1909 that there was any general recognition of returning prosperity. About that time it became difficult to obtain a farm which had not some patent disability attached to it; the advertisements of vacancies that prior to Michaelmas used to fill pages of the county newspapers in the nineties dwindled to a column or more, and agents found themselves able to pick and choose among would-be tenants. By 1912 the process has gone still farther, rents have definitely risen with the demand for land that cannot be satisfied, and in all parts of the country men are obtaining very large returns indeed on the capital they embarked in the business. Of course every farmer has not been making money; bad business habits and slipshod management are far too common, and nothing is more surprising than the way bad farming exists alongside good. We suppose that among grocers or gunmakers, solicitors or saddlers, the same inequality of performance exists, though the results are not set out so openly.

7 Intensive farming, 1941-3

(*National farm survey of England and Wales* (H.M.S.O., 1946). This survey was based on material collected in the years 1941-3.)

Type O Mainly Cash Crop Farming

The land in these type areas is generally the most fertile in the country; the holdings are mostly small in size (an average of 74 acres of crops and grass), but they are very intensively cropped, four-fifths of the area being arable and mainly devoted to potatoes (usually the key crop), wheat, sugar-beet and small fruit and vegetables. The main representaive of this Type is the Fens in Lincolnshire, Norfolk and the Isle of Ely. In the peaty portion of the Fen area, the main enterprises are the growing of wheat, sugar-beet and potatoes, but in the silt region further east, there is less sugar-beet and wheat, but more potatoes and some small fruit and vegetables. Other areas of this type are the Lancashire Plain around Liverpool; and South Yorkshire including the warp land. Livestock on farms in these type areas, where they are kept at all, are mainly to provide manure for the arable land.

Type P Market Gardening

This includes the growing of small fruit and vegetables, and also hops and orchard fruit, which between them in these areas constitute well over 25 per cent of the total of crops and grass. As with Type O, these holdings are small in size and very intensively worked. Most of the land is fertile but not as uniformly so as the Fens. The main representatives of this type are the Vale of Evesham; the Biggleswade-Potton area of Bedfordshire; the Wisbech area of the Isle of Ely and Norfolk; the Cottenham area of Cambridgeshire and West Huntingdonshire; the Botley area of Hampshire; and areas to the east and west of London. Two areas in North and Central Kent respectively, where hops and orchard fruit are particularly important are also included in this Type for the Farm Survey analysis, although they were classified as Type O in the Type of Farming Map.

8 High profits after the Second World War, 1948-53

(*i*) George Henderson, *Farmers' progress: a guide to farming* (Faber, 1950), pp. 131–3; (*ii*) J. Kirkwood, J. A. Gilchrist and J. M. Thomson, *The College Farm, Auchincruive* (Glasgow: West of Scotland Agricultural College, 1954), p. 32)

(*i*) Working on these principles, a capable and experienced young farmer took over a ninety-eight acre holding in September 1948, with a capital of only £1950. The tenant-right valuation was heavy at £1002, but included, by agreement, 27 acres of corn in the stack which the outgoer did not wish to thresh. This left the farmer with only just over £900 in cash by the time he had moved in and paid the incidental expenses involved.

He hire-purchased his tractor and implements under an agreement involving paying £200 and the balance spread over two years. He arranged six months' credit for his seed corn and artificial manure bills, but agreed to pay monthly for the 10cwt of feeding-stuffs to which he was entitled under his farm's allocation. This left him with £700 for the immediate stocking of the farm; wages and living expenses would be covered by the sale of potatoes and, he hoped, milk, although his buildings were unsuitable for milk production.

Weighing up the possibilities of the various kinds of stock, he bought first of all sixty store pigs at an average price of 85s, these to be fattened on the meal available, the corn to be threshed, and pig potatoes which were in plentiful supply; then fifty wether lambs at 90s to stock the grass and consume the roots; and with the balance of the money he bought two old, but good, in-calf, Jersey cows.

At the end of the second month he had to sell a few pigs, to pay for the meal consumed in the first month, and his threshing expenses which he had not allowed for. But from then on the pigs improved rapidly in value, until they were sold out in March, the total sales being £770, the net profit being £330, plus the ten best gilts kept back for breeding. The original capital was reinvested in pigs, the profit and the cashing-in of some of the corn taken over, and, through the pigs, used to pay the outstanding seed corn and artificial manure bill, which entitled him to a further six months' credit for what he would require for spring planting; and to buy 1000 day-old chicks at £80, and the essential appliances and materials necessary to rear them— assembled, of course, by his own labour.

Once again it was a tight squeeze to feed both the pigs and the chickens, but with the reserve of home-grown corn it was achieved. The cockerels realized a welcome £263 in late summer, the pigs coming in later with £890, owing to increased price per score, and finally wethers grading out to leave him a cheque for £457; and these, with their wool at £26, raised the gross receipts of the farm to £1949, or only a £1 short of his original capital, after deducting the original outlay on pigs and sheep and transport charges. But this year, 1950, with a very considerable increase in his valuation, he will be started in a much stronger position, with 250 laying pullets, more corn in hand, and a big allocation of feeding-stuffs earned on the scorage of bacon pigs sent in; and he is now all set to breed his pig replacements at half the cost he had to pay in the previous year. This is a typical example of making use of every penny all the time. Had he been starting in the spring instead of the autumn, poultry rearing would have been even better than the pigs, but he chose quite rightly in view of the enormous quantities of pig potatoes available in the year he started, and the opportunity to earn an extra allocation of feeding-stuffs, which will balance out to a certain extent the lack of potatoes in 1949. A lesser man would have moved heaven and earth to get his buildings passed for milk production, or have depended on arable crops, losing precious months on the one, or getting a poor return on the other; while here we have an example of making the best of the opportunity of the moment, but with an eye to the future, when the farm will no doubt be stocked up with dairy cattle; meanwhile fertility is being built up with pigs and poultry on the lines I have indicated in an earlier chapter. But even in the first year his business measures up to the required standards: gross output, less stock purchased, equal to capital invested, and twenty times calculated rent, while the labour charge of the farmer and the one youth he employed would not exceed one-quarter of the gross output.

✳ ✳ ✳

The Pig Account, in detail, for 1952–53 was:

(*ii*)　　　　*Pig Account* (*For Year Ended 31st March, 1953*)

Opening Valuation
Breeding Stock

5 Boars	£148	
31 Nursing and In-Pig Sows and Gilts at £20	620	
4 Nursing and In-Pig (Wessex) Sows at £35	140	
Other Pigs		
68 Sucklers at 10s	34	
205 Growing and Fattening Pigs	1,952	
		£2,894

Purchases and Expenses
Breeding Stock

2 Boars		107
Feeding		
Purchased: Meals	£3,274	
Swill	869	
Stockfeed		
Potatoes	47	
Home-Grown:		4,190
Grains	£1,129	
Potatoes	14	
Kale	29	
		1,172

Whey		62
Milk and Skim Milk		2
Charge for Pig Paddocks and Pens		19
Litter Straw		97
Labour and Power		
Man Labour	£881	
Horse Work	4	
Tractor Work	9	
		894
Rent and Maintenance Charge		150
Building Equipment Repairs, Depreciation, etc.		310
Coal, Oil, Light and Power		150
Sundry Expenses—Vet. Charges, etc.		157
Share of Farm 'Overheads'		169
Share of Salaries Charged Direct		305
		£10,678
Net Profit		2,642
		£13,320

Sales

439 Bacon Pigs	£10,436	
8 Gilts	162	
1 Young Boar	28	
7 Cast and Fat Sows	188	
2 Cast and Fat Boars	35	
		£10,849

Manurial Residues of Food
(Including litter straw) 267

Closing Valuation
Breeding Stock

4 Boars	£139	
24 Nursing and In-Pig Sows at £20	480	
3 Nursing and In-Pig (Wessex) Sows at £30	90	
9 In-Pig Gilts at £19	171	

Other Pigs

91 Sucklers at 10s	46	
150 Growing and Fattening Pigs	1,278	
		2,204
		£13,320

C. LANDOWNERSHIP AND FARM SIZE

1 The Church and the land, 1755

(Staffordshire Record Office. Transcript kindly provided by Mr E. R. Lloyd.)

The Terrier of the Glebe Lands and Tythes belonging to the Rectory of Norbury given in by Order of the Right Revd. Frederick Lord Bishop of Lichfield and Coventry at his visitation held at Stafford the 12th day of August 1755.

The parsonage House and Outbuildings thereto belonging and the Glebe Lands containing fifty five acres or thereabout

One house and Garden now in possession of William Glover

One house and Garden now in possession of Thomas Talbot

One house and Garden now in possession of Joseph Ashley

Corn is paid in kind of every sort

Wool Lambs Pigs Geese are paid one at seven, two at seventeen for Lamb and Wool one halfpenny for each under the tytheable number

The parish have a modus for their Hay some paying 4d some 6d some more or some less according as their Custom is

Herbage is paid according to the Statute

Mortuaries are paid according to the Statute

Hemp and flax are paid in kind

The Modus for cows and calves is each cow one penny each calf one halfpenny for a colt 2d

Bees for every Stall kill'd or removed 2d

Every Tradesman pays 2d for his hand

Burial fees and Churching fees are 10d each

The Church Yard fence and Church are repair'd by the parish

The Chancel repair'd by the Rector

The Clarks Wages are 20s Yearly and paid by the Churchwardens

for the Corn Mill at Weston Jones is paid 2s Yearly

Each householder pays 3d for Servants and Children each above sixteen yearly pays 2d

Edward Hughes   Rector

Charles Stokes ⎫
William Morris ⎭ Churchwardens

2 Enclosures and landholding, 1799–1809

(*i*) *Lincolnshire, 1799*

(Arthur Young, *General view of the agriculture of the county of Lincoln* (London, 1799), pp. 85–6)

Evidence of Elmhurst, a Commissioner under Enclosure Act

Another observation I at the first made, and ever after put in practice, was this, always to begin to line out and allot for the smallest proprietors first (whether rich or poor) in every parish, so as to make such allotment as proper and convenient for the occupation of such, or their tenant (as that might be) to occupy; and so on, from the smallest to the greatest: for it is for the advantage of the greatest and most opulent proprietors that a bill is presented and act passed; and at their requests, and not the small ones; and, as the little ones would have no weight by opposition, they must submit, was it ever so disadvantageous to them; as it very often happens; and, therefore, there can be no partiality in defending those who cannot help or defend themselves; and a little man may as well have nothing allotted to him, as to have it so far off, or so inconvenient for him, that it is not worth his having, as it would prevent his going to his daily labour; and, therefore, he must sell his property to his rich and opulent adjoining neighbours; and that, in some measure, decreases population.

(*ii*) *Norfolk, 1804*

(Arthur Young, *General view of the agriculture of the country of Norfolk* (London, 1804), pp. 82, 86, 94, 135, 156)

Bintrey and Twiford Enclosed 1795

Poor. There were 26 acres allotted for fuel, let by the parish. There were 46 commonable rights; the whole divided according to value; very few little proprietors; but small occupiers suffered.

Brancaster Enclosed 1755

Poor. Very well off; Barrow-hills, a common of 65 acres, allotted to them; and each dwelling-house has a right to keep the two cows or heifers; or a mare and foal; or two horses; and also to cut furze.

Cranworth, Remieston, Southborough Enclosed 1796

Poor. They kept geese on the common, of which they are deprived. But in fuel they are benefited; an allotment not to exceed $\frac{1}{20}$ let, and the rent applied in coals for all not occupying above 5*l* a year: this is to the advantage of those at Southborough, having enough allowed for their consumption; at Cranworth the poor are more numerous, and the coals of little use.

Ludham

The commons were enclosed in 1801 : all cottagers that claimed had allotments; and one for fuel to the whole; but the cottages did not belong to the poor; the allotments in general went to the larger proprietors, and the poor consequently were left, in this respect, destitute; many cows were kept before, few now. All the poor very much against the measure.

Sayham and Ovington Enclosed 1800

Poor.—An allotment of not less than 50*l* a year, for distributing to the poor in coals, was ordered by the act; it let for 98*l*. There were 100 commonable right houses. They used to sell a cottage of 3*l* a year, with a right, for 80*l*. For each, four acres were allotted: and the cottage with this allotment would now sell for 160*l*. And what is very remarkable, every man who proved to the Commissioners that they had been in the habit of keeping stock on the common, was considered as possessing a common-right and had an allotment in lieu of it. Nor was it an unpopular measure, for there were only two men against it from the first to the last.

(*iii*) *Gloucestershire, 1807*

(Thomas Rudge, *General view of the agriculture of the county o Gloucester* (London, 1807), pp. 92–3)

In all Acts of Inclosure, it might perhaps be proper, as it would certainly be equitable, to relieve the pressure which weighs on small proprietors, in a degree not proportioned to the advantages they derive from them: for it should be remembered, that the expence of fencing a small allotment is considerably greater than that of a larger one, according to the quantity; that is, a square piece of land containing ten acres will cost half as much as forty, though only of one-fourth value. This disproportion occasions much reluctance in the class of proprietors before-mentioned; and though it is frequently overcome by the superior influence of the great landholders, yet the injustice of it cannot but strike the considerate mind with conviction.

(iv) Northamptonshire, 1809

(William Pitt, *General view of the agriculture of the county of Northampton* (London, 1809), p. 70)

From the observations I have made in this county, I have no doubt but, if the average produce of common fields be three quarters per acre, the same land will, after a little rest as grass, and the improvements to be effected by enclosure, produce, on an average, four quarters per acre; and I believe that the produce of every common field may be increased in a like proportion by enclosure and an improved cultivation.

3 Form of a Cheshire lease, 1832

(C. Stella Davies, *The agricultural history of Cheshire, 1750–1850* (Manchester: Chetham Society, 3rd series, X, 1960, pp. 211–12)

General Covenant in leases granted by John Rd. Delap Tollemache Esq
August 1832

Memorandum of Agreement entered into the — day of — 183 — between John Richard Delap Tollemache of Picadilly in the Country of Middlesex Esq on the one part and — of — in the County of — of the other part as follows. The said John Richard Delap Tollemache doth hereby agree to demise and let and the said — doth hereby agree to take and to become tenant of all the messuage farm or tenement with the lands and appurtenances there to belonging situate in — in the county of Chester now in the possession of — and containing by estimation — of land of Statute measure or thereabouts for the term of — years to commence as to the mowing lands from the twenty fifth day of December — and as to the other lands (except an outlet for cattle to be set out by the said John Richard Delap Tollemache in the last year) from the 2nd day of February and as to the messuage and buildings and the said outlet from the Ist day of May — subject to the annual rent of — to be paid on the 24th day of June in each year and the first payment to begin to be made on the 24th day of June (except from the said farm to the said John Richard Delap Tollemache all timber and other trees woods underwood mines game rabbits woodcocks snipes and fish with all proper liberties to fell timber get mines and to hunt course fish and fowl and also to plant young trees in the hedgerows pitsteads or elsewhere making compensation in each case for waste of herbage). And the said — agrees to hold the said farm on the terms following.

To pay all leys and taxes and to serve all offices not to have in tillage in any one year more than one fourth part of the premises (summer work and potato ground included) and not to push plough pare or burn any part of the premises nor break up any of the ancient meadow land or sow any wheat or rye on the brush or without a summer fallow preceding or sow any hemp or flax. To break up the land (except the meadow land) in rotation and due course and lay down the same with good clover seed at the end of each tillage using after the rate of twenty pounds to a Cheshire acre. To eat and consume on the premises all the hay straw and fodder and not to sell or give any part thereof and to expend all the manure arising from or to be gathered on the premises on the meadow land or on the grass land last laid down for permanent pasture only and if any shall remain unspent at the end of the term to leave the same in the usual place for the benefit of the said John Richard Delap Tollemache. To keep all the buildings hedges ditches fences gutters drains and soughs in proper repair and condition during the term and so to leave the same at the expiration thereof (having timber in the rough bricks lime and slates allowed). To cut and plash all the hedges and fences in a husbandlike manner and to leave and preserve any young trees which may grow therein. To pay the said John Richard Delap Tollemache twenty pounds for every acre of ground and so in proportion for a less quantity that shall be converted into tillage or used contrary to the appointment before made and five pounds for every cwt of hay, thrave of straw or cart load of manure and so in proportion for a less quantity that shall be sold or taken from the premises during the term, such sums to be paid on demand after every breach and in default of payment to be considered as reserved rent and levied by distress and sale as rent in arrear may be levied and raised. To do two days boon work, to keep a dog and cock, to deliver a cheese yearly made in the month of June on the said premises, and to insure the house and buildings in one of the London Assurance offices in the sum of £ — against loss or damage by fire. Provided that if the said annual rent or the said advanced rent or penalties or any part thereof shall be unpaid for twenty days after the same shall become due or if the said — shall assign or underlet the said farm or any part thereof without the consent of the said John Richard Delap Tollemache first obtained or shall become bankrupt or insolvent or assign over his effects or any part thereof for the benefit of creditors or the same shall be taken in execution for debt or if he shall not truly perform the stipulations hereinbefore contained, then on the tender of one shilling by the said John Richard Delap Tollemache to the said — his executors or administrators these presents and the term before mentioned and every matter and thing herein contained shall be void to all intents and purposes. The said — further agrees that neither himself nor any part of his family shall hire any servant during the continuance of the tenancy for any term longer than 51 weeks and that in each contract of hiring with any servant the term of 51 weeks shall be expressed and that on the event of any servant continuing for two or more 51 weeks in his or her place the said—after the expiration of each 51 weeks agrees to hire such servant for the following 51 weeks. And in case of any breach in this agreement respecting the said hiring the said — hereby agrees to forfeit to the said John Richard Delap Tollemache the sum of £50 for each such breach to be recovered by distress as rent in arrear immediately or so soon after as the said John Richard Delap Tolle-mache may think proper on such breach being known to him or his agent. And the said — does hereby further agree when required to execute a lease in conformity and according to the terms of this agreement on such lease being presented to him for execution the costs and expenses of preparing and executing the same to be borne and defrayed by the said — witness our hands —

Signed in the presence of

4 A contented tenantry, 1859

(F. M. L. Thompson, *English landed society in the nineteenth century* (Routledge, 1963), p. 291, quoting Alnwick MSS. This song was sung at the annual dinner given for the Percy tenants.)

> Those relics of the feudal yoke
> Still in the north remain unbroke:
> That social yoke, with one accord,
> That binds the Peasant to his Lord...
> And Liberty, that idle vaunt,
> Is not the comfort that we want;
> It only serves to turn the head,

But gives to none their daily bread.
We want community of feeling,
And landlords kindly in their dealing.

5 The rewards of an improving landowner, 1869

(Robert E. Brown, *The book of the landed estate, containing direction for the management and development of the resources of landed property* (Edinburgh, 1869), pp. 3-4. The author was factor and estate agent at Wass, Yorkshire.)

To show to what extent both the present and prospective market value of land is affected by the judicious application of modern improvements, I shall give some particulars of one case out of many that have come under my own notice in dealing with estates.

The subject to which I refer is a farm of three hundred and forty acres—two hundred and eighty of which were arable before the improvements were commenced on it, and the remaining sixty old meadow pasture. The rent was £425 in all, or at the rate of £1 5s per acre. This farm is situated within about six miles of a thriving town, and the land consists of a light loam, a considerable portion of which, previous to the improvements, was wet, although it had been drained—but, as the result showed ,only imperfectly drained—seven years before. The lease of the farm having expired, the proprietor very properly resolved to take it into his own hand for improvement, and accordingly the farm-house and other buildings were all fitted up on the most improved principles; the fences thoroughly repaired and new ones substituted and added where these were found necessary. All the wet parts of the farm were thoroughly drained, and about one hundred acres of it were trenched, in order to deepen parts of the land, and to remove stones from it. In short, everything was done to the farm which modern experience could suggest and money accomplish for its improvement at the time, keeping strictly in view that the money expended should be so laid out as to return a good interest to the proprietor in the shape of yearly rent afterwards. The outlay in all was £2960, or at the rate of nearly £9 on each acre of the land embraced in the farm. After having the farm in his own hands for a period of three years, and having got it into the best possible condition for a tenant, the proprietor let it on a nineteen years' lease for the sum of £900 yearly rent, or at the rate of nearly £2 13s per acre per annum. From this it appears that the proprietor, by laying out £2960 on the improvement of the farm, secured an advance of yearly rent equal to £475, thus giving him fully fifteen per cent per annum as interest on the sum he expended on the improvements.

But besides the advantage stated in regard to the high interest the proprietor had obtained for the money he laid out in this manner, he had secured another, perhaps even more important—namely, that of having fully doubled the value of the farm in a commercial point of view; for, taking thirty-two years' rent as its value we have £13,590 as the probable sum the proprietor would have got for the subject had he put it into the market in its unimproved condition; whereas, taking the rent at which the farm was let after the improvements had been executed, and multiplying it by 32, on the same principle as in the other case, we have £28,800 as its improved market value; so that, by judiciously laying out about £3000 on the improvement of a farm, the proprietor added fully £15,000 to its commercial value.

This, then, is a matter of the first importance to all landed proprietors; for by judiciously laying out money on the improvement of their property, they not only reap an immediate advantage in an increased yearly rental, but also materially enhance the market value of their estates.

The following is another instance of the improved value of land from a judicious outlay of money upon it, but of a different description from the case already given:—A small farm on an estate under my management was, in 1862, let for the sum of £32; it extends to thirty-three acres of arable land, five acres of meadow, and twenty-two acres of rough pasture. The fences upon it were in a wretched state, and the buildings were very bad, and not sufficient for the requirements of the place. The land was also very poor and in a dirty condition. The tenant left the farm a bankrupt. I took it into our own management for two years, had the land thoroughly cleaned and put into good condition, the buildings remodelled, and a few additions made. As stone was plentiful on the farm, we replaced the old fences with strong and substantial stone walls or dykes. The total outlay on these improvements was £344; and the farm was then let to a good tenant at the yearly rent of £65, being an increased rental of £33 per annum, or double the old rent; which was at the rate of nearly $9\frac{1}{2}$ per cent upon the outlay, and at the same time doubling the commercial value of the farm.

A great deal more might be said illustrative of the importance of improvement as the means of enhancing the value of land, but the space which I have allotted myself does not admit of it. Suffice it to say, that whatever may be the present value of any piece of land, it is capable of being made very much more valuable by the judicious outlay of money in improving it, so as to increase its productive qualities, and hence its yearly rent as well as its prospective price in the market; unless, indeed, it is already cultivated and improved to the utmost limits which modern 'practice with science' will admit.

6 Compensation for improvements, 1846 and 1875

(*i*) *Advice from a land agent, 1846*

(John O. Parker, *The Oxley Parker papers* (Colchester: Benham, 1964), p. 114)

... It is highly necessary under present circumstances to advance as far as it is possible the onward progress of agricultural improvement and inasmuch as such improvement cannot be carried out except at a considerable outlay of capital it is highly desirable that some form of agreement should exist between the owner and occupier of the soil as should give to the latter in some measure a security for his expenditure. Whereas in your case there is good faith existing between the Landlord and Tenant the occupier may fairly expect that no one else will be allowed to reap the benefits of his outlay, and under the guarantee of such an impression he may be induced to farm *well*, but with a written agreement which shall ensure compensation for certain specific and permanent improvements, I think there can be no doubt that he would be induced to farm *better*, and at the same time that it has this good effect I am also inclined to believe that the power of referring on either part to certain definite terms of agreement may prevent in future those misunderstandings which sometimes arise in cases where the only mode of settlement (should any difference of opinion occur) is left to the undefined and often ill-understood rules which are applied by what is termed 'the general custom of the country'.

(*ii*) *The Agricultural Holdings Act, 1875*

(38 and 39 Vic., c. 92)

Compensation

5. Where, after the commencement of this Act, a tenant executes on his holding an improvement comprised in either of the three classes following:

First Class

Drainage of land
Erection or enlargement of buildings
Laying down of permanent pasture
Making and planting of osier beds
Making of water meadows or works of irrigation
Making of gardens
Making or improving of roads or bridges
Making or improving of water-courses, ponds, wells, or reservoirs, or of works for supply of water for agricultural of domestic purposes
Making of fences
Planting of hops
Planting of orchards
Reclaiming of waste land
Warping of land

Second Class

Boning of land with undissolved bones
Chalking of land
Clay-burning
Claying of land
Liming of land
Marling of land

Third Class

Application to land of purchased artificial or other purchased manure
Consumption on the holding by cattle, sheep, or pigs of cake or other feeding stuff not produced on the holding

he shall be entitled, subject to the provisions of this Act, to obtain, on the determination of the tenancy, compensation in respect of the improvement.

6. An improvement shall not in any case be deemed, for the purposes of this Act, to continue unexhausted beyond the respective times following after the year of tenancy in which the outlay thereon is made:

Where the improvement is of the first class, the end of twenty years:
Where it is of the second class, the end of seven years:
Where it is of the third class, the end of two years.

7 A modern domesday—landownership in Devon, 1873

('Return of owners of land, I Bedfordshire to Norfolk: Devon', *B.P.P.*, 1874, LXXII, part I pp. 5, 65. Devon had a 601,000 population in 1871. A [cre] 4840 sq. yds; R[ood] 1210 sq. yds; r[od] 30 $\frac{1}{4}$ sq. yds.).

Name of Owner	Address of Owner [3]	Extent of Lands			Gross Estimated Rental	
		A	R	r	£	s
Beavis, George	Withycombe Raleigh	1	1	1	21	5
Beavis, William	Sowton	38	3	9	84	3
Beayne, Charles	Morebath	112	1	22	98	5

		A	R	r	£	s
Beazley, Mrs F.	Newton Abbot	7	3	30	150	—
Beck, Nathan	Ashburton	1	3	5	80	3
Beck, Rev. William	Clannaborough	51	1	32	94	—
Bedford, Duke of	Woburn Abbey	22,607	2	31	45,907	4
Bedgood, C. H.	Cumberland Street, W	574	—	24	1,114	17
Beedel, William	Rackenford	7	1	14	13	7
Beedell, Thomas	Tiverton	892	—	32	934	11
Beedell, Thomas	Upton Pyne	4	1	13	3	—
Beer, John	Aylesbeare	35	3	6	37	7
Beer, John	Bridford	33	3	38	77	15
Beer, John	Devonport	58	2	12	47	10
Beer, Charles	Brixham	1	2	24	73	—
Beer, George	Bovey Tracey	3	3	29	15	7
Beer, Henry	Shaldon	27	2	14	38	10
Beer, Joseph	Exeter	1	1	14	20	—
Beer, Miss	Swansea	49	2	25	22	—
Beer, Philip	Membury	3	—	—	5	10
Beer, Robert	St. John's Wood, NW	128	1	14	57	5
Beer, R. W.	Swansea	70	1	16	30	15
Beer, Thomas	Brixham	1	—	8	20	—
Beer, William	Bideford	31	2	10	31	10
Beer, William	Munbury	2	2	—	12	10
Beer, William	Pimpley Northam	4	—	5	24	—
Beer, William R.	Kingsbridge	126	—	12	627	5
Beir, James	Bridford	15	—	16	23	—
Belfield, Hy.	Euston Road, N W	59	—	14	87	10
Belfield, John F.	Paington	585	2	8	2,894	5
Belfield, Thomas	Primley	209	3	39	452	—
Bell, Edward C.	Glen Elms, Northam	1	1	18	43	—
Bell, John	Brixham	2	3	33	19	—
Bellairs, Frederick	Alverdiscott	59	—	23	86	6
Bellamy, Rev. F. S. A.	Devonport	5	1	37	53	10
Bellamy, Robert	Cruwys Morchard	30	—	9	43	5
Bellamy, William	Northpetherwin	12	—	8	14	10
Bellew, H.	Ford Wiveliscombe	99	—	20	96	5
Bellew, Hy.	Oakhampton	63	2	29	61	10
Bellew, John F.	Stockleigh Crt., Credition	2,209	2	31	2,128	19
Bellew, Miss	Yarnscombe	419	2	29	304	—

*   *   *

9*

		Extent of lands	Gross estimated rental
Total owners of land of one acre and upwards	10,612	1,514,002 acres	£2,266,240
Total owners of less than one acre in extent	21,647	2,981 acres	£615,426
Estimated extent of any common or waste land in the county	—	77,868 acres	—

8 Radical programme for the land, c. 1890

(M. K. Ashby, *Joseph Ashby of Tysoe, 1859–1919* (Cambridge University Press, 1961), p. 152. The slogans here reproduced were painted on the sides of a red van that toured the Midland countryside bearing propagandists for the radical cause.)

FAIR RENTS

FAIR WAGES

THE LAND FOR ALL

JUSTICE TO LABOUR

ABOLITION OF LANDLORDISM

9 A tenant-smallholder, 1906

(Minutes of evidence taken before the departmental committee . . . [on] smallholdings in Great Britain, *B.P.P.* 1906, LV, qq. 8128–8275, 8170–83, 8186–275)

Mr Thomas Wright, of Duddon, Tarporley, Cheshire, was called; and Examined.

(*Chairman*) You are, I believe, a small holder on the estate of Mr Tomkinson, who has just been giving evidence?—Yes.

What is the size and description of your holding?—Twenty acres.

And what is your tenure?—Yearly.

What do you principally grow on your holding?—A mixture—corn, potatoes, cabbages and some market garden crops.

Where do you market?—Chester.

How far is that distant?—Eight miles.

Do you send your produce in by cart?—Yes.

Your own cart?—Yes; I take it myself.

What rent per acre do you pay?—£2 15s.

That, of course, covers the rent of the house and buildings?—Yes.

What is the nature of the house and buildings?—Very nice and convenient.

How many rooms are there in the house?—Four rooms.

What are the buildings?—There is room to tie about a dozen cows, and room also for young pigs.

How many cows do you keep?—Ten.

What do you do with the milk?—I convert it into cheese and butter.

You do not sell it as milk?—No. I make Cheshire Stiltons the last part of the season, from October to Christmas.

Do you grow potatoes?—Yes.

When do you get them on the market?—In July.

What is the nature of the soil?—A light loam.

How many horses do you keep?—One.

Do you do all your work with one horse?—Yes.

Does it take the produce to the market, too?—Yes.

Do you have any assistance or do you work the small holding entirely yourself?—I have one assistant.

How is the dairying done?—That is managed by my daughter.

How many acres of corn have you got?—It varies according to circumstances. I have sometimes from one to an acre and a half of oats.

Do you consider your holding a type of the small holdings in the district?—Yes.

Do most of the small holders crop their land in the same way that you do?—Not all.

Do any of them go in for market garden produce exclusively?—Yes.

Nothing else?—They may keep a cow or two for the manure.

Do any of them go in for dairying exclusively?—Not in the immediate neighbourhood.

Is your assistant a farm labourer?—Yes, he lives in the house with us.

Has he no cottage of his own?—No, he is a young man.

Has he got any land?—No, he is employed solely as my servant.

Do all the other small holders in your neighbourhood employ an outside hand?—No.

Do some of them work the entire holding themselves with their families?—Yes.

Do you not yourself do anything except farming?—No, I have no other occupation.

Is that the case with the other small holders?—They don't need any other occupation if they have twenty acres of land.

*   *   *

(*Mr Bidwell*) Out of your twenty acres how much represents grass? —There are twelve acres of permanent grass, but not all pasture, some of it being mown to provide hay for the winter.

I understand you have accommodation for ten cows on your holding. How many cows do you keep?—Ten.

Do not they go anywhere except upon your nine and a half acres of pasture;—I mean, is there no common land or anything of that sort for them?—No.

Have you any fruit trees or bushes planted on your twenty acres, or any orchard cultivation?—I have only a small garden.

Do you not grow fruit?—We have some fruit trees in the fences, chiefly damsons, apples, and pears.

During the hay and corn harvests do you ever work for other people?—Only when there is an arrangement that they will come to help me again.

Under such circumstances if a man has ten acres of hay you may help him with your horse and your own services if he returns the compliment?—Yes.

Do you arrange with your neighbours about jointly sending your produce to the market?—I take mine myself.

And sell it yourself?—Yes.

With no middleman?—No.

Do you use a separator for your milk?—No.

Do you raise any calves on this land with the surplus milk, the skim milk?—Yes.

About how many in a year, four or five?—Sometimes four, sometimes none.

Do you mean that when a cow calves you sometimes sell the calf rather than rear it?—Yes, if it is not a good type.

*   *   *

(*Mr Channing*) What quantity of hay have you?—Six acres of hay.

What quantity do you get per acre?—Sometimes four tons of clover including the second crop.

Do you get a good two tons of meadow hay off the grass?—Yes.

What quantity of land have you got in root crops?—About one

and a half acres in potatoes, half an acre in mangolds, and one and a half acres in cabbages, some for the market and some for the cattle.

Have you to buy a large quantity of feeding stuff for your cows?—Yes.

What do you make most of your profit out of?—Is it out of the cheese, the butter, or the fruit?—The prices vary; it all depends upon whether things are selling well.

What is your mainstay, out of which you chiefly expect to get your profit every year?—I expect a little out of everything.

But is not your dairy produce your mainstay?—Oh, yes.

What is the average quantity of cheese you produce and sell?—The cheese in 1904 yielded £137 7s 6d—that was from nine cows.

And what about the butter?—We do not go in for butter selling; we only make sufficient for our own use.

Your mainstay is the cheese?—Yes, in the dairy branch.

Do you keep many pigs?—Yes.

Have you any idea what you get one way or another as a gross return from the pigs in the course of a year?—At present I have eight fat pigs, thirteen young ones and a sow.

Do you have to buy food for your pigs? Or do you feed them with the produce of your twenty acres?—Their food is largely bought.

(*Mr Long*) Was your father a small holder?—Yes, but he did not depend on it altogether.

Did he work for anyone else?—He was a carrier.

But he occupied land?—Yes.

Did you obtain your skill and knowledge by working on his land?—He had not sufficient land to keep a family upon it and I went out as a farm servant for some time, and then when I got married I worked some years for Mr Tomkinson.

Then you obtained your knowledge of the management of a small holding by experience solely?—Yes.

Do you use artificial manures in addition to the dung produced on the farm?—Yes, a little.

What kind of manure?—I use bones and nitrate of soda.

Dissolved bones?—Chiefly.

Do you use super-phosphate or basic slag?—No.

You confine yourself mostly to the two manures you have mentioned?—Yes.

Would you mind telling us what you spend in artificial manure? And then how much on purchased foods for your stock?—I spent £132 on feeding stuffs in one year.

And how much in artificial manures?—I spent £1 15s on nitrate of soda, and £2 10s on bones, making a total of £4 5s.

Does that represent the whole of the artificial manure placed upon the land in a year?—Yes.

What do you have to pay in rates in the course of a year?—About £5.

Bearing in mind the fact that you keep a hired man can you tell us how much land you would require to live as comfortably as you do now if you had not a man to assist you?—I have neighbours who live on eight or nine acres of land, and make a good living. They chiefly grow strawberries or something of that sort.

What does your man cost you?—£10 a year in wages, in addition to which he has his board. I have sometimes been without a man.

Do you ever have your food stuffs or manures analysed as a guarantee of their purity?—No, I generally buy maize meal and bran.

But are you aware that Indian meal itself is sometimes adulterated? —I buy a good deal of it from a neighbouring mill where I can see it ground.

And you are satisfied as to its purity?—Yes.

In buying your seeds do you take any precautions to see that they are not only pure but germinate well?—I buy them on condition of their having been tested.

By whom?—The seedsman gives me a warranty, but not having large quantity I do not go to the expense of having it analysed.

Your principal produce, cheese, apparently represents about £15 per cow?—Yes.

Would the fact that your house is supplied bring it up to nearly £17 per cow?—I should think so.

Do you make Cheshire cheese?—In summer.

Early ripening or late?—Early at present.

I suppose you know it is not common for ten cows to be kept on so small an area as you devote to them?—I keep them because I think it pays me to keep them.

You have often heard the phrase 'three acres and a cow', it being assumed that three acres are required for a cow?—I could not live at that rate.

By feeding your cows well and manuring the land heavily you are able to keep more cows than you could otherwise?—Yes.

As a successful man in your own line, would you recommend other people like you to adopt the same plan?—I have done so.

Is the clover consumed by the cows or do you sell it?—I have it all consumed.

Is it mixed with rye grass?—Yes.

You would not grow clover alone?—It seems to be more profitable to grow both together.

Do you grow any forage crops such as crimson clover, early rye, or Indian corn for your cows?—No.

They simply get grass?—Yes.

Do you feed them on Indian meal and bran during the grazing season?—Yes.

How much per day?—I have never actually weighed the amount. Might it be about three or four pounds per day per cow?—Yes.

Does the half acre of mangolds provide all you require for your cows in the form of root crops for winter consumption?—We have the cabbages. They last till Christmas or after, and then the mangolds last till now.

How many tons to the half acre do you grow?—About twenty tons.

May I take it that your cows are exceptionally good?—I try to get the best I can.

What is the practice among the small farmers in your neighbourhood as to breeding. Is there any landlord or society that provides a bull? —Not particularly.

There is no combination for hiring a bull?—I take the cow to some adjoining farm and pick out the best bull I know.

Do you sell your produce at Chester market by retail or wholesale? —Wholesale.

To people who sell it again?—Yes.

How is your small holding cropped at present?—$9\frac{1}{2}$ acres of permanent grass; grass for mowing, 6 acres; oats, $1\frac{1}{2}$ acres; potatoes, $1\frac{1}{2}$ acres; mangolds, $\frac{1}{2}$ acre; vegetables, 1 acre.

Is any of the mowing grass permanent?—About three acres.

Do you take your cheese to Chester and sell?—A factor comes round and buys it.

At what price?—About 60s a hundredweight at present.

Do you take the produce of anybody else to market along with your own?—No.

How often do you go to market?—Twice a week at certain seasons of the year.

You mean in the cheese-making season?—Often twice, but sometimes once.

If you have a cart mare, do you breed a foal?—I have not done so of late years.

What value do you put approximately on your house and buildings as compared with the entire rent of your farm?—I think that is a question for the landlord. I have not studied it.

Is it not an exceptional house?—I think not. Such houses may vary in rental value from £8 to £20 per annum according to the accommodation and the land.

Taking your house, with buildings and garden, is it not superior to nineteen out of twenty occupied by small holders in Cheshire and elsewhere?—I don't think so. It was originally built for a smaller holding, and there have been some additions of land and buildings since then.

(*Mr Yerburgh*) Do you think you would have done better as a small freeholder than you have done as a tenant?—Not so long as one has a good landlord.

You think that, given a good landlord, tenancy is better than freehold?—I should fancy so.

I suppose that if a man has a tenancy he is more free to take advantage of any opportunity that occurs of taking a larger holding if he prospers?—Yes, a landlord generally gives the chance to one of his own tenants.

And so the tenant has a better chance of rising in the scale?—Yes.

Has it ever occurred to you that you might do better in buying what you want if you were to join with your neighbours and form a co-operative society?—That is a thing I have never considered. There has not been any society of the sort organised near me.

Could you not do better in the way of marketing if you were to join with your neighbours to send your stuff to market along with theirs?—I don't think so. In my position I can go to market when it suits me and have always something to take and to bring back.

(*Mr Brown*) Were you yourself brought up in a small town or in the country?—In the country.

When you grew up you became a farm servant?—Yes.

Were you ever a foreman or overseer on a farm?—No. I worked twelve years or so for Mr Tomkinson and then at my request he gave me the chance of a small holding.

Did he let you the small holding at a low rental on account of your having been a good servant?—No. I have improved the holding.

How old is your assistant?—About eighteen years of age.

Do you pay him no more than £10 a year?—That is his wage, over and above his board.

10 Break-up of the great estates, 1921

(*The Times*, 22 and 31 December 1921)

Messrs Knight, Frank and Rutley have sold 1,776,727 acres in the United Kingdom during the last four years.

∗ ∗ ∗

The Estate Market
Turnover in 1921
£15,000,000 Sales

The estate market has not escaped the general tendency this year towards a restriction of business and a lower level of values. The reasons for the lessened activity are common knowledge, and have been so exhaustively discussed that there is no necessity to reiterate them. The most encouraging thought, in reviewing the market for real property, whether investment or residential, is that the reaction, which was inevitable, has not been more severe. Much of the buying energy provided by funds accumulated during the war was absorbed in the transactions that made 1919, and 1920 in a lesser degree, remarkable. It may fairly be pointed out that persons who laid out their money in real estate in those years have, on the whole, been saved the mortifying experiences of loss of capital that have been the lot of some who put their money into overcapitalised industrial projects, which were launched in the same period. Just as the 'boom' as some have called it took many people by surprise, so has the contraction.

How well many large landowners, both individual and corporate, did for themselves has been revealed in the results which have from time to time been mentioned of the re-investment of the purchase money in gilt-edged securities. The disclosure last summer in *The Times* of the oppressive nature of the prevalent taxation showed that the large landed proprietor has very great difficulties to contend with and a strong inducement to sell his land. The farmers who tumbled over one another to become their own landlords are in a less fortunate case, it is to be feared, than they had hoped to be. Yet many of them bought at prices that were reasonable enough, and may be regarded as having been better advised than not a few urban buyers of houses. . . .

D. THE PLACE OF AGRICULTURE IN THE STATE

1 The end of England's corn surplus in sight, 1772-3

(William Cobbett, *The Parliamentary history of England from the earliest period to the year 1803* (London, 1812–20), XVII (1771–1774), cols. 475–8)

Debate in the Commons on the Bill to regulate the Importation and Exportation of Corn. April 14.

The House went into a Committee to consider the present State of the Corn Trade, in which Mr Pownall moved the following Resolutions: 1 'That it is the opinion of this Committee, That the importation of wheat and wheat flour, rye and rye meal, into this kingdom, be admitted, for a limited time, free of duty. 2 That the importation of rice, from any of his Majesty's colonies in America, into Great Britain, be admitted, for a limited time, free of duty. 3 That, if the importation and exportation of corn were properly regulated by some permanent law, it would afford encouragement to the farmer, be the means of encreasing the growth of that necessary commodity, of affording a cheaper and more constant supply to the poor, and of preventing abuses in that article of trade.

[*Here followed nine more resolutions that formed the basis of the corn law of 1773.*]

Governor Pownall began by apologizing for his standing forward upon a matter of so much importance; but said that what was intended to be moved was in consquence of several meetings, both last year and the present, of a number of gentlemen of the first interests and abilities in this country, who had maturely considered the business, and had delivered their opinions upon it. He then entered into an explanation of the actual state of the supply and consumption of the kingdom; and shewed, that the present difficulties did not arise from any scarcity; that there was as much, if not more corn grown than formerly; but, from the different circumstances of the country, the consumption was considerably more than the supply and that this disproportion arose from the late immense increase of manufacturers and shop-keepers, the prodigious extent of our commerce, the number of people employed by government as soldiers, sailors, collectors of revenue, etc. etc. and also the prodigious number of people who live upon the interests of the funds; also the great increase of the capital, the manufacturing and sea-port towns; that the surplus which we used to produce, was about one 36th part of the whole growth; and that any one might consider, whether the number of people he had mentioned were not more than one 36th of the whole people; and that therefore the real fact was, we had no longer a surplus. The consequence that he drew was, that if we really meant to have the country well supplied, we must do every thing to encourage the growth, and not discourage the farmer. He spoke much of the nature of the prices of things, and shewed, that though the prices of every thing were nominally risen, yet the price of corn was less so than any other article. He then shewed, from the nature of the market of great towns, that storing of corn must not be discouraged, nor the middle man; for if they were, great towns could never be regularly supplied, but must be in perpetual danger of famine. He concluded with saying, that though the principal end and intention of the resolutions he meant to move were for a permanent Bill, yet such were the present circumstances of the country, that an immediate supply, if it could be got, was absolutely necessary. He therefore moved a temporary Bill for immediately opening the ports for the importation of bread corn; and next moved the resolutions as the foundation of a permanent Bill to take effect when the temporary one expired; and said, that the end proposed by this Bill is that of creating an influx of bread corn for home consumption, in case of internal scarcity; and an aid to our foreign trade in case of our not having a quantity of corn adequate to that important and beneficial commerce. This purpose is conducted under such regulations as shall prevent any interference with the landed interest. In other words, (said he) if I may be permitted to use an allusion to natural operations, it means to introduce into our supply an additional stream, and to fix such a weir at such a height as shall always keep the internal supply equal, and no more than equal, to internal want, yet preserve a constant overflow for all the surplus, so as never on one hand to endanger the depression of the landed interest, nor on the other the loss of our foreign market for corn—by our not being able, as has been the case for several years past, to supply the demands of that foreign market—as it is hoped that this measure will be formed into a permanent law. It is meant by the provisions in the Bill formed for the carrying it into execution— that its operations may go on, as the state of things does actually and really require, not as the interests of designing men may wish and will them to go; that this commercial circulation of subsistence may flow through pools whose gates are to open and shut as the state of the droughts, and floods, and tides may require, not to consist of sluice-doors which are to be locked up and opened by the partial hands and will of men.

2 The debate on the corn law, 1815

(*Parliamentary Debates*, first series, XXIX (1814–15), cols. 798–818)

House of Commons. February 17, 1815. The State of the Corn Laws

The *Hon. Frederick Robinson* immediately rose. ... He had never disguised from himself, and he was not ashamed to confess it, the extreme difficulty, as well as the extreme importance, of this question. He could not, however, but feel that the prejudices on this subject had, from further inquiry, been very much removed. But, above all, he was happy to see that the misrepresentations, for so he thought they were, with respect to the motives of those who supported this measure, and with reference to the effects which it was likely to produce, were done away with. There did not now exist in the public mind the feeling by which it was before influenced. It was not now supposed that the object sought to be accomplished by the alteration of the corn law was the mean and base and paltry one of getting, for a particular class of society, a certain profit at the expense of the rest. 'For my part', said Mr Robinson, 'I declare to God, if I thought this was the motive which actuated any individual who supported the alteration; and, above all, if I conceived that such would be the effect of the measure, no consideration on earth could tempt me to bring it forward'.

*   *   *

... The general result of his reasoning was, in the first place, that it was quite impossible for us safely to rely on a foreign import. If they so did, a necessary result would be a diminution of our own produce, which would become more and more extensive every year, and consequently call for a greater annual supply from foreign countries—a supply which must progressively increase as the agriculture of the kingdom became less encouraged; and that, when the fatal moment arrived, the system of foreign supply would prove completely illusory.

The next point to be considered was the extent to which protection should be given. That was a point on which, undoubtedly, a difference of opinion was most likely to prevail. Some gentlemen would be for going considerably higher than others. Many thought the prohibition ought to be carried to a price considerably above that, without he obtained which it was conceived the agriculturalist could not cultivate. Others would wish that it should be placed much lower; and contend that because a particular species and degree of burden was likely to be removed, the protecting price ought to be much reduced. Now he would be inclined to agree to the first of these propositions, if the necessary effect of it would not be to bring up the price of corn to the highest possible rate, within the limits of the sum at which importation should commence. This certainly might be the case at the first moment, but he believed the ultimate result would not be so. He thought the final effect of the system would be to give such a powerful support to our own agriculture as would greatly increase the general produce of the country. It would excite a strong competition between the different parts of England, and between England and Ireland; so that the growth of corn, if Providence blessed us with favourable seasons, would be sufficiently large to afford an ample supply for the people of this country, and would enable them to be fed at a much cheaper rate, in the long run, than could be effected by the adoption of any other system.

* * *

Mr Philips professed himself equally inclined either to proceed with, or defer the discussion, as might be most agreeable to the wishes of the House. Several members calling out 'Go on', he began by stating his entire concurrence in the opinion of the right hon. gentleman who had moved the resolutions, that this was not a question on which the interests of the commercial and agricultural classes were at variance, but one in which those interests, when fairly and liberally considered, would be found to accord; for no resolution upon it calculated to promote the general prosperity of the country could be adopted without materially benefiting both classes. But if this were not the case, if the question were one in which the interests of two or more descriptions of our fellow-subjects were opposed, he should say that it was the duty of parliament not to legislate for the advantage of one class in contradistinction to, or at the expense of another, but to legislate for the benefit of the whole community. Looking at the question under the influence of this principle, he could not help feeling and expressing some surprise at the occasion of their present deliberations. What was the object of their deliberations? To provide a remedy for the low price of corn. That which all ages and countries had considered as a great national benefit was now discovered to be a great evil, against which we were imperiously called to legislate in self-defence. The real object of the resolutions, however disguised and disavowed, was to raise the price of corn. [Here Mr Robinson expressed his dissent.] Mr Phillips proceeded to say that this not only was their object, but if that object were not attained, the advocates of the resolutions would regard them as nugatory. The right hon. gentleman must at least allow that their object was to raise the present price of grain; but he contended that moderation and uniformity of price would be their ultimate effect. It did seem somewhat inconsistent, on the part of the hon. gentleman, to tell the House that the effectual way to lower price was to acquiesce in a measure expressly intended to raise it. But how are this moderation and uniformity of price to be produced? By contracting the market of supply. Thus, while in all other instances moderation and uniformity of price are to be found in proportion to the extent of the market of supply, in the instance of corn they are to be in proportion to the limitation of it: and in a commodity peculiarly liable to be affected by the variation of seasons, moderation and uniformity of price, and abundance, are to be attained by preventing importations from foreign countries correcting the effect of varieties of climate, and of a scanty harvest in our own. To him it appeared that no measure could be better calculated to produce directly opposite consequences.

$$* \qquad * \qquad *$$

In considering the relation between the price of provisions and of labour, Mr Phillips observed that it was necessary to distinguish the countries and the trades from which examples were taken. In a new country where the value of land is extremely low, and agriculture rapidly progressive, in a new and thriving manufacture, the price of labour may be so high in proportion to that of the necessaries of life as to be little affected by their fluctuations. ... But this state of things cannot exist in old manufactures, such as those generally established in this country, where competition has reduced profits, and that reduction of profit has brought the wages of the labourer to a level with his subsistence in tolerable comfort. In such manufactures if you raise the price of provisions without proportionately raising that of labour, to what privations and evils must you necessarily expose the labourer! He was ready to admit with the noble lord[1] that, *ceteris paribus*, the immediate effect of a high advance of provisions might

---

[1] Lord Lauderdale in evidence before a committee of the House of Lords.

probably be a reduction of the price of labour; because labourers being desirous of obtaining the same comforts that they had been used to, might be stimulated to more diligence. They might work sixteen hours a day instead of ten, and thus the competition for employment being increased among the same number of workmen, without any increase of demand, the price of labour might fall. But will any person contend that this state of affairs can long continue? The labourer must go to the parish, or turn to some more profitable employment, if by chance any can be found, or he must emigrate, or work himself out by overstrained exertion. The proportion being then altered between the demand for labour and the supply, its price will rise. This effect sooner or later must happen, but till it has actually taken place how dreadful must be the situation of the labourer!

*　　*　　*

Having thus shown both by reasoning and by reference to facts, that the price of provisions must ultimately and on the average regulate that of labour, he proceeded to show the effect that an advance of provisions must have on our manufacturing interests. And here Mr Philips said that he wished on such topics, to reduce his reasoning as much as possible to numerical calculation. He would suppose, for the sake of argument, without at all entering into the enquiry, that three-fifths, or 60 per cent of the labourer's wages were spent in provisions, and that provisions were 80 per cent dearer here than they were in France, or any manufacturing country on the continent. By multiplying 60 by 80, and dividing by 100, the committee would see that the excess of the price of labour here above that of France would, from these datas, and according to his reasoning, be 48 per cent. He wished the committee to consider what must be the effect of such an excessive price of labour employed in our manufactures, when compared with the low price of labour employed in the manufactures of France, and what an advantage it must give to the French manufacturers in their attempts to rival us on the continent.

*　　*　　*

[After quoting Malthus] he observed that there were two ways of equalising subsistence and population, one by increasing food, the other by limiting population, and warned the committee against being led into measures whose tendency might be to produce that effect in the latter way. Why (said Mr Philips) should a commercial and manufacturing country like this have such a jealousy and dread of the importation of corn? An importation of corn cannot take place without a corresponding export of commodities on which British industry has been employed. The export will increase your wealth, that wealth will increase your population, and that increased population will produce an increased demand for your agricultural produce. ... Mr Philips observed that no country in the world was so interested as this in establishing the principle of free trade, because no other country could profit equally by the general recognition of that principle. Foreign nations, mistaking, like the advocates of the regulation before the committee, the circumstances which have operated against our wealth for the causes of it, are now following our example. They are prohibiting or imposing restraints on the import of our fabrics, in order to encourage their own manufactures, from which they will receive inferior fabrics at a higher price. Let us convince them, by an example, of their mistake. Let us convince them that by leaving industry and enterprise unfettered, and by allowing capital to take its natural and voluntary direction, we are persuaded that the true interests of this country and of every other will be most effectually promoted.

Mr Philips proceeded to say that Great Britain was geographically a commercial country, that commerce had stimulated her agriculture rather than agriculture had stimulated her commerce. It had given wealth to her people, and diffused fertility over her soil. Take care, said he, that in attempting to change the natural character of your country, you do not stop the progress of national prosperity.

3 Repeal of the corn laws, 1846

(*Parliamentary Debates*, third series, LXXIII (22 January, 13, 20 February 1846), cols. 68, 69–71, 849–50, 1345–7)

Address in Answer to Her Majesty's Speech, January 22nd 1846

Sir Robert Peel

Sir, the immediate cause which led to the dissolution of the Government in the early part of last December, was that great and mysterious calamity which caused a lamentable failure in an article of food on which great numbers of the people in this part of the United Kingdom, and still larger numbers in the sister kingdom, depended mainly for their subsistence. That was the immediate and proximate cause, which led to the dissolution of the Government. But it would be unfair and uncandid on my part, if I attached undue importance to that particular cause. It certainly appeared to me to preclude further delay, and to require immediate decision—decision not only upon the measures which it was necessary at the time to adopt, but also as to the course to be ultimately taken with regard to the laws which govern the importation of grain. I will not assign to that cause too much weight. I will not withhold the homage which is due to the progress of reason and to truth, by denying that my opinions on the subject of protection have undergone a change.

✳   ✳   ✳

Sir, those who contend for the removal of impediments upon the import of a great article of subsistence, such as corn, start with an immense advantage in the argument. The natural presumption is in favour of free and unrestricted importation. It may, indeed, be possible to combat that presumption; it may be possible to meet its advocates in the field of argument, by showing that there are other and greater advantages arising out of the system of prohibition than out of the system of unrestricted intercourse; but even those who so contend will, I think, admit that the natural feelings of mankind are strongly in favour of the absence of all restriction, and that the presumption is so strong, that we must combat it by an avowal of some great public danger to be avoided, or some great public benefit to be obtained by restriction on the importation of food. We all admit that the argument in favour of high protection or prohibition on the ground that it is for the benefit of a particular class, is untenable. The most strenuous advocates for protection have abandoned that argument; they rest, and wisely rest, the defence of protective duties upon higher principles. They have alleged, as I have myself alleged, that there were public reasons for retaining this protection. Sir, circumstances made it absolutely necessary for me, occupying the public station I do, and seeing the duty that must unavoidably devolve on me—it became absolutely necessary for me maturely to consider whether the grounds on which an alteration of the Corn Laws can be resisted are tenable. The arguments in favour of protection must be based either on the principle that protection to domestic industry is in itself sound policy, and that, therefore, agriculture, being a branch of domestic industry, is entitled to share in that protection; or, that in a country like ours, encumbered with an enormous load of debt, and subject to great taxation, it is necessary that domestic industry should be protected from competition with foreigners; or, again—the interests of the great body of the community, the laborious classes, being committed in this question—that the rate of wages varies with the price of provisions, that high prices imply high wages, and that low wages are the concomitants of low prices. Further, it may be said, that the land is entitled to protection on account of some peculiar burdens which it bears. But that is a question of justice rather than of policy; I have always felt and maintained that the land is subject to peculiar burdens; but you have the power of weakening the force of that argument by the removal of the burden, or making compensation. The first three objections to the removal of protection are objections founded on considerations of public policy. The last is a question of justice, which may be determined by giving some counterbalancing advantage. Now, I want not to deprive those who, arguing *a priori*, without the benefit of experience, have come to the conclusion hat protection is objectionable in principle—I want not to deprive them of any of the credit which is fairly their due. Reason, unaided by experience, brought conviction to their minds. My opinions have been modified by the experience of the last three years. I have had the means and opportunity of comparing the results of periods of abundance and low prices with periods of scarcity and high prices. I have carefully watched the effects of the one system, and of the other—first, of the policy we have been steadily pursuing for some years, viz., the removal of protection from domestic industry; and next, of the policy which the friends of protection recommend. I have also had an opportunity of marking from day to day the effect upon great social interests of freedom of trade and comparative abundance. I have not failed to note the results of preceding years, and to contrast them with the results of the last three years; and I am led to the conclusion that the main grounds of public policy on which protection has been defended are not tenable; at least, I cannot maintain them. I do not believe, after the experience of the last three years, that the rate of wages varies with the price of food. I do not believe that with high prices, wages will necessarily rise in the same ratio. I do not believe that a low price of food necessarily implies a low rate of wages. Neither can I maintain that protection to domestic industry is necessarily good.

*   *   *

Adjourned Debate. February 13, 1846
House of Commons
Sir Douglas Howard said:

*   *   *

I have often imagined—and it was for this that I moved for, and obtained the order of this House for, the extensive returns which are now preparing, namely, the various colonial tariffs and commercial relations at present subsisting between all the Colonies of the Empire and the mother country, and between the Colonies themselves—that it might really be possible to treat Colonies like counties of the country, not only in direct trade with the United Kingdom, but in commercial intercourse with each other, by free trade among ourselves, under a reasonable moderate degree of protection from without, and so resolve the United Kingdom, and all her Colonies and possessions, into a commercial union such as might defy all rivalry, and defeat all combinations. Then might colonization proceed on a gigantic scale—then might British capital animate British labour, on British soil, for British objects, throughout the extended dominions of the British Empire. Such an union is the United States of America—a confederation of sovereign States, leagued together for commercial and political purposes, with the most perfect free trade within, and a stringent protection from without; and signally, surely, has that commercial league succeeded and flourished. Such an union, too, is the German Customs League; and it has succeeded to an extent that really is, in so short a time, miraculous. But free trade—the extinction of the protective principle—the repeal of the differential duties—would at once convert all our Colonies, in a commercial sense, into as many independent States. The colonial consumer of British productions would then be released from his part of the compact—that of dealing, in preference, with the British producer; and the British consumer of such articles as the Colonies produce, absolved from his; each party would be free to buy in the cheapest, and sell in the dearest market. I defy any hon. member opposite to say that this would not be a virtual dissolution of the colonial system.

*   *   *

Adjourned Debate. February 20, 1846
Mr B. Disraeli:

*   *   *

I have now nearly concluded the observations which I shall address to the House. I have omitted a great deal which I wished to urge upon the House; and I sincerely wish that what I have said had been urged with more ability; but I have endeavoured not to make a mere Corn Law speech; I have only taken corn as an illustration; but I don't like my friends here to enter upon that Corn Law debate which I suppose is impending, under a mistaken notion of the position in which they stand. I never did rest my defence of the Corn Laws on the burdens to which the land is subject. I believe that there are burdens, heavy burdens, on the land; but the land has great honours, and he who has great honours must have great burdens. But I wish them to bear in mind that their cause must be sustained by great principles. I venture feebly and slightly to indicate those principles, principles of high policy, on which their system ought to be sustained. First, without reference to England, looking at all countries, I say that it is the first duty of the Minister, and the first interest of the State, to maintain a balance between the two great branches of national industry; that is a principle which has been recognised by all great Ministers for the last two hundred years; and the reasons upon which it rests are so obvious, that it can hardly be necessary to mention them. Why we should maintain that balance between the two great branches of national industry, involves political considerations—social considerations, affecting the happiness, prosperity, and morality of the people, as well as the stability of the State. But I go further; I say that in England we are bound to do more—I repeat what I have repeated before, that in this country there are special reasons why we should not only maintain the balance between the two branches of our national industry, but why we should give a preponderance—I do not say a predominance, which was the word ascribed by the hon. member for Manchester to the noble lord the member for London, but which he never used—why we should give a preponderance, for that is the proper and constitutional word, to the agricultural branch; and the reason is, because in England we have a territorial Constitution. We have thrown upon the land the revenues of the Church, the administration of justice, and the estate of the poor; and this has been done, not to gratify the pride, or pamper the luxury of the proprietors of the land, but because, in a territorial Constitution, you, and those whom you have succeeded, have found the only security for self-government—the only barrier against that centralising system which has taken root in other countries. I have always maintained these opinions; my constituents are not landlords; they are not aristocrats; they are not great capitalists; they are the children of industry and toil; and they believe, first, that their material interests are involved in a system which favours native industry, by insuring at the same time real competition; but they believe also that their social and political interests are involved in a system by which their rights and liberties have been guaranteed; and I agree with them—I have these old-fashioned notions. I know that we have been told, and by one who on this subject should be the highest authority, that we shall derive from this great struggle, not merely the repeal of the Corn Laws, but the transfer of power from one class to another—to one distinguished for its intelligence and wealth, the manufacturers of England. My conscience assures me that I have not been slow in doing justice to the intelligence of that class; certain I am, that I am not one of those who envy them their wide and deserved prosperity; but I must confess my deep mortification, that in an age of political regeneration, when all social evils are ascribed to the operation of class interests, it should be suggested that we are to be rescued from the alleged power of one class only to sink under the avowed dominion of another. I, for one, if this is to be the end of all our struggles—if this is to be the great result of this enlightened age—I, for one, protest against the ignominious catastrophe. I believe that the monarchy of England, its sovereignty mitigated by the acknowledged authority of the estates of the realm, has its root in the hearts of the people, and is capable of securing the happiness of the nation and the power of the State. But, Sir, if this be a worn-out dream; if, indeed, there is to be a change, I, for one, anxious as I am to maintain the present polity of this country, ready to make as many sacrifices as any man for that object—if there is to be this great change, I, for one, hope that the foundations of it may be deep, the scheme comprehensive, and that instead of falling under such a thraldom, under the thraldom of Capital—under the thraldom of those who, while they boast of their intelligence, are more proud of their wealth—if we must find a new force to maintain the ancient throne and immemorial monarchy of England, I, for one, hope that we may find that novel power in the invigorating energies of an educated and enfranchised people.

4 Marketing boards, 1931

(*Parliamentary Debates, House of Commons*, fifth series, CCXLVIII (1930–31) (9 February 1931) cols. 53–4, 59–62)

Agricultural Marketing Bill

Order for Second Reading read.

The *Minister of Agriculture* (*Dr Addison*): I beg to move, 'That the Bill be now read a Second time.'

This Bill is concerned with what is undeniably an outstanding weakness in our agricultural system. It relates to conditions which have handicapped and often impoverished producers, which are a burden on retailers and of no value whatever to consumers. If we consider the facts in relation to our home food market we find a very remarkable contradiction, which has been particularly noticeable during the past 25 years. There has been a great improvement in our industrial food market; in fact, it is not too much to say that the food market in the great centres of population in Britain is the best food market in the world, but the production of many foods has been practically stationary in our own country and the increased market is being met very largely with surplus goods from outside this country. We have a very exceptional soil, our countryside for the most part is quite near to the market, and we can produce as good beef and good mutton, as good butter and bacon and fruit and potatoes, as any country in the world.

How is it then that in this country agricultural production has not corresponded to the industrial demand? It is quite a common thing to attribute this to foreign supplies; they are commonly alleged to be cheaper. I notice that that contention is embodied in what, I suppose, will be the official Opposition Motion to-day. But when we come to examine foreign supplies which are particularly competitive with what we ourselves could produce almost in sufficient quantity—I exclude particularly cereals—we find that foreign countries do not send us their worst; they send us their best. If we examine their supplies we see that Denmark sends us her best bacon, New Zealand sends us her best butter, the Dominions send us their best fruit, and so on. As a matter of fact we find that the British market is studied and catered for to receive the best.

If we look a little closer at these competitive foreign imports we find that in nearly every case their collection and standardisation and distribution and supply to our markets are in the hands of extraordinarily capable and highly specialised organisations. It is not the thousand-and-one individual outside producers who have sent to the British market, but it is a highly organised system, with very specialised branches to study the need of our market, to save expense wherever possible, to improve the grade, and to advertise their goods. If anyone goes down to the railway tubes to-day he will see what is at least new to me—a very ingenious and attractive advertisement of foreign bacon. We are exhorted to have two rashers a day, English or Danish, I shall come to the meaning of that in a few minutes. It is a very clever advertisement. But that is not the product of an individual producer; it is the product of a great organisation catering for the British market.

Let us turn to the other side and see what has happened in our own home country. I have here a very interesting table, and I may say that the fact that the two figures I shall quote are almost identical, is an accident: it was not designed by me at all. But as giving an indication of the measure of the opportunity that is before the home producer I asked the Department to get out for me the values of home products and foreign imports in those things in which we could quite well compete, namely, beef, pigmeat, butter, milk, cheese, poultry, eggs, potatoes, and such fruits as we grow at home. It so happens that the home production of these commodities is about £203,000,000 per annum, and that the imported products are almost identical in value. In other words, in this group of commodities, which this countryside is eminently fitted to produce and can produce in the best quality, our home product in mass is about half our supplies and no more.

*   *   *

A marketing scheme must be based on commodities. You cannot deal with an area basis only. It must be something appropriate to the particular commodity you are marketing. The Bill provides that you may have commodity boards of various kinds. You may have them limited to small commodities produced in certain areas, say Cheshire cheese, or you may have them applicable to a commodity produced over the whole countryside, like milk. There is no one board which will require all the powers set out in the Bill. Some powers are appropriate to one class of commodities, and some to another. ...

Let us see what a commodity marketing board would do under the Bill. In the first place, it would help to organise the grading, packing and standardisation of quality of the commodity. I suggest that nothing save large-scale organisation can do that. It is this improvement in quality, this gradual lifting-up of standards, which has been the chief asset of many of our foreign competitors and if we are to compete, say, with Danish bacon, we must see that in grading, quality, reliability the British article is just as good as the foreign article. Another function of the commodity marketing board must be to deal with bulk supplies and to make forward contracts. We could not have a better illustration of the necessity for this kind of thing than what happened last year in connection with our fruit industry. Large numbers of small growers were unable to sell their soft fruit, or some of it. Finally, in despair, as it was not worth picking at the prices then prevailing, they let the fruit drop to the ground and it was wasted.

5 Guaranteed prices, 1947

(The Agriculture Act, 1947, 10 and 11 Geo. VI c. 48)

Be it enacted by the King's most Excellent Majesty, by and with the advice and consent of the Lords Spiritual and Temporal; and Commons, in this present Parliament assembled, and by the authority of the same, as follows:

PART I

Guaranteed Prices and Assured Markets

1—(I) The following provisions of this Part of this Act shall have effect for the purpose of promoting and maintaining, by the provision of guaranteed prices and assured markets for the produce mentioned in the First Schedule to this Act, a stable and efficient agricultural industry capable of producing such part of the nation's food and other agricultural produce as in the national interest it is desirable to produce in the United Kingdom, and of producing it at minimum prices consistently with proper remuneration and living conditions for farmers and workers in agriculture and an adequate return on capital invested in the industry.

(2) This Part of this Act shall extend to Scotland and Northern Ireland.

2—(I) As at such date in each year as the Ministers may determine, they shall review the general economic condition and prospects of the agricultural industry.

(2) If it appears to the Ministers at any time between two annual reviews under the last foregoing subsection that there has been, or there is likely to be, a change in the economic condition of the agri-

cultural industry or any section thereof, arising (otherwise than in the course of a continuous development) from a substantial alteration of costs of production or any other special cause, and that the change is or is likely to be of sufficient importance to require that the Ministers should exercise their powers under this subsection, the Ministers may hold a special review of the matters referred to in the last foregoing subsection, in so far as they are or may be affected by the change.

(3) In holding any review under this section the Ministers shall consult with such bodies of persons as appear to them to represent the interests of producers in the agricultural industry.

*   *   *

FIRST SCHEDULE

Produce to which Part I of Act applies

>Fat cattle
>Fat sheep
>Fat pigs
>Cow's milk (liquid)
>Eggs (hen and duck in shell)
>Wheat
>Barley
>Oats
>Rye
>Potatoes
>Sugar beet

3

The evolution of industry

The growth of manufactures and mining is the basic fact of English economic history since the middle of the eighteenth century. Advantages of geology and geography made modern industry and an extensive foreign trade possible. A protectionist economic policy and a long period of slowly accumulating prosperity from the Restoration onwards provided a favourable background for the technical and organisational advances called the Industrial Revolution. Not least, individual Englishmen (and perhaps even more Scotsmen in proportion to the population) displayed the qualities of leadership and foresight that go to make a successful businessman (*section A*). It is often asserted that enterprise has deserted the shores of England at some date since the 1860s, but whatever may be true for businessmen as a class there can be no doubt that modern times have seen individual businessmen displaying much the same qualities as made Richard Arkwright and James Watt famous.

Technical progress is inseparably connected with the industrial history of the last five hundred years. The process of mechanisation spread from the staple industries—textiles, ironmaking and engineering—to paper-making, the clothing trades and many others too numerous to be documented (*B 1, 2, 4*). Even industries that adopted steam power early continued to rely extensively on hard manual labour into the twentieth century (*B7*) but flow methods of production often leading to sharp reductions in labour costs gradually spread through industry. The most spectacular example is probably to be found in the making of sheet steel (*B8*). Similar changes had however already taken place in papermaking and in the production of the cheaper motorcars, and flow methods were to be widely used in chemicals, soapmaking and brewing. The nineteenth century drew its energy supplies from the coalfields. There is still an abundance of coal underground and the demand for energy steadily increases but coal is a dwindling source of supply. Nuclear power (*B10*) may prove to be the ultimate successor to coal's hegemony though in the shorter run oil and natural gas are the chief competitors. Technical change may provide a motive for the creation of new monopolistic businesses (*C5*) yet it is only one motive among many, ranging from the need to combine against labour (*C1*) to a desire for 'orderly' marketing and the high profits associated with it (*C4*). Combinations do not always succeed (*C3*) and few firms find themselves in the happy position of being without competitors, even in recent times when a handful of giant firms often control most of the output of a particular industry, be it motor cars or brewing.

A. THE ENTREPRENEUR

1 Ploughing back the profits, 1762–4

(Arthur H. John, ed. *The Walker family: iron founders and lead manufacturers, 1741–1893* (Council for the Preservation of Business Archives, 1951), p. 7. The Walkers were Quakers of Rotherham.)

November 1762

This year we made $405\frac{1}{4}$ tons of castings, and supposed the valuation of the nett stock, after taking out £210 for a dividend

... ... £13,500   0   0

This year we had great expences in Parliament and in law with the proprietors of the River Dun [Don]; and great expence, loss, and loss of time at the blast furnace, etc., and therefore added to the supposed capital only £1,000.

This year we built 4 houses in the Kettle Croft; a sand-house and another carpenter's shop at the Holmes. And this year the Meeting House was built so far as to be covered in.

November 1763

This year we made $448\frac{1}{2}$ tons of castings, and supposed the valuation of the nett stock, after taking out £350 for a dividend, and other deductions, to be £15,200   0   0

This year the Works turned out very well, supposed £3,000 but deducted £1,300 for law expences, losses, the above dividend, interest of money, clark's wages, etc., and supposed to have added to the capital £1,700.

This year we built Thribrough Forge, and 4 dwelling-houses, carpenter and smith's shops, charcoal yard, and other necessaries there. Built a stable and 2 large workshops or smitheys in ye Yellands, built a new [an]nealing room at the Holms Mills; built an additional piece to the scrap house at Rotherham Forge.

November 1764

This year we made 472 tons of castings, and supposed the valuation of the nett stock, after taking out £350 for a dividend, and other deductions, to be £17,000 0 0

This year we think better than the year before; suppose, then, we call the improve £3,100, and make the same deductions as last year, then we add to the capital, as above £1,800.

This year we built 4 dwelling-houses, a barn and foldstead, garden walls, and other out-fencing, at Thribergh Forge. Built a small accompting house at the mill at Holmes; 4 houses facing the road, and a large shop for frying-pan makers; 3 small shops, and an engine or turning shop over it, in the Yelands.

2 The entrepreneur as cost accountant, 1832

(C. Babbage, *On the economy of machinery and manufactures* (2nd ed. 1832), pp. 200–1)

The great competition introduced by machinery, and the application of the principle of the subdivision of labour, render it necessary for each producer to be continually on the watch, to discover improved methods by which the cost of the article he manufactures may be reduced; and, with this view, it is of great importance to know the precise expense of every process, as well as of the wear and tear of machinery which is due to it. The same information is desirable for those by whom the manufactured goods are distributed and sold; because it enables them to give reasonable answers or explanations to the objections of inquirers, and also affords them a better chance of suggesting to the manufacturer changes in the fashion of his goods, which may be suitable either to the tastes or to the finances of his customers. To the statesman such knowledge is still more important; for without it he must trust entirely to others, and can form no judgment worthy of confidence, of the effect any tax may produce, or of the injury the manufacturer or the country may suffer by its imposition.

One of the first advantages which suggests itself as likely to arise from a correct analysis of the expense of the several processes of any manufacture, is the indication which it would furnish of the course in which improvement should be directed. If a method could be contrived of diminishing by one fourth the time required for fixing on the heads of pins, the expense of making them would be reduced about thirteen per cent; whilst a reduction of one half the time employed in spinning the coil of wire out of which the heads are cut, would scarcely make any sensible difference in the cost of manufacturing of the whole article. It is therefore obvious, that the attention would be much more advantageously directed to shortening the former than the latter process.

3 Stereotype of a self-made man, 1858

(Anthony Trollope, *Doctor Thorne*, Chapters II and IX)

Roger Scatcherd had also a reputation, but not for beauty or propriety of conduct. He was known for the best stonemason in the four counties, and as the man who could, on occasions, drink the most alcohol in a given time in the same localities. As a workman, indeed, he had higher repute even than this: he was not only a good and a very quick stonemason, but he had also a capacity of turning other men into good stonemasons: he had a gift of knowing what a man could and should do; and, by degrees, he taught himself what five, and ten, and twenty—latterly, what a thousand and two thousand men might accomplish among them: this, also, he did with very little aid from pen and paper, with which he was not, and never became, very conversant.

*　　*　　*

... He had become a contractor, first for little things, such as half a mile or so of a railway embankment, or three or four canal bridges, and then a contractor for great things, such as government hospitals, locks, docks, and quays, and had latterly had in his hands the making of whole lines of railway.

He had been occasionally in partnership with one man for one thing, and then with another for another; but had, on the whole, kept his own interests to himself, and now, at the time of our story, he was a very rich man.

And he had acquired more than wealth. There had been a time when the government wanted the immediate performance of some extraordinary piece of work, and Roger Scatcherd had been the man to do it. There had been some extremely necessary bit of a railway to be made in half the time that such work would properly demand, some speculation to be incurred requiring great means and courage as well, and Roger Scatcherd had been found to be the man for the time. He was then elevated for the moment to the dizzy pinnacle of a newspaper hero, and became one of those 'whom the king delighteth to honour'. He went up one day to Court to kiss her Majesty's hand, and came down to his new grand house at Boxall Hill, Sir Roger Scatcherd, Bart.

'And now, my lady', said he, when he explained to his wife the high state to which she had been called by his exertions and the Queen's prerogative, 'let's have a bit of dinner and a drop of som'at hot'. Now a drop of som'at hot signified a dose of alcohol sufficient to send three ordinary men very drunk to bed.

4 A new broom at the works, 1905

(Robert D. Best, *Brass chandelier: a biography of R. H. Best of Birmingham* (Allen & Unwin, 1940), pp. 152–5. The writer is B. B. Blackburn, the new assistant works manager and later director.)

... I had had fifteen years of experience in the brass trade of Birmingham and came from a well-known Birmingham factory of gas and electric light fittings—nearly all gas at that time—a factory which had made great strides in the production of medium class and cheap goods, where capstan lathes were turning out huge quantities of gas brackets and pendants, innocent of castings, untouched by the file.

Coming from a model modern building consisting of long straight shops which ran on either side of an up-to-date foundry with the minimum of interruption to transport, air and light, the newcomer was amazed at the congeries of rooms and passages, of steps and slopes and stairs, of converted cottages and odd corners which time and Mr Best's genius had gradually compiled and adapted.

That abuses should prevail, and wastefulness run riot in such a place, where the works manager could not see any of the workpeople from his quarters, let alone keep any sort of a check on their goings and comings, their startings and stoppings, where system had slackened into the deputizing of decision and authority to unworthy and incapable hands, was inevitable. How the business could pay at all under the conditions was a mystery to me till I saw the prices which were obtained for a monopoly article of outstanding merit. A large proportion of the profits which should have accrued from this source, the Surprise pendant, were dissipated by old-fashioned processes, an enormous amount of fetching and carrying, of transport back and forth, by loose control.

At that time a horse was kept, a fine upstanding gelding which would have done credit to a brewer's dray. It was generally in its stable. The animal was in fine condition, well fed and impeccably groomed by Tom Freeman *père* and Tom Freeman *fils*. Sometimes it was harnessed to a heavy two wheel cart and sallied forth collecting from suppliers odds and ends of metal, sheet and tube, glass and screws, and other materials. One enquired why it was necessary to keep a horse and cart and man for those odd errands. Why couldn't the makers and suppliers of such requirements deliver them to us? Well, we were rather a long way out, and besides there were occasional deliveries of our own products to be made in town. And the manure was so essential for Mr—'s garden, too. A light broke. So that was it. Well, the horse was sold and the stable became yet another odd corner of the factory, an extension of the glass stores, though the hayloft overhead still disgorged occasionally a missing hand who needed sleep, generally the packer in his wilder days.

A hole in the wall at the back of the dipping yard where the morning beer was pushed through had to be stopped up, and preventive glass stuck on the top of the wall.

The rat trap was set nightly and Mr—'s terrier was fetched down to the works when a likely specimen was found encaged in the morning. It was good practice for the dog and afforded much amusement to the favoured few invited to the rat-baiting. That and other sports in working hours became less popular under the new regime.

As one moved round the works and suddenly opened a door into many of the shops there was sure to be a scuffle and an effort to look 'busy'; the line of closets in the yard was far too full morning and afternoon to be accounted for by any normal percentage of irregularity, but these are universal factory symptoms and though minimized by publicity and supervision will never disappear.

On many soldering hearths there was the sizzling of bacon and the boiling of tea for half an hour or so after the morning arrivals, at night, and again as one o'clock approached, but then the aroma changed to one of chops and steaks.

A plan drawn of the devious course which the product took from inception to dispatch convinced Mr Best there was much room for improvement in that respect, and as facts emerged showing that the working foremen themselves decided upon whether and what stock was to be made, with no record kept of it, that the same men decided on their own to 'work' overtime and chose their own intimates to help them, stayed behind without any supervision and were found reading books and engaged in other less innocent private occupations and amusements till the time came to go home, it was decided after due and careful deliberation to have a clean sweep of some of the 'passengers' and to pull others up sharply. ...

And of course the new manager was unpopular. When I had to visit St James's Church at the top of the hill to see about refinishing the candelabra, the vicar questioned me about the 'young American' who was upsetting all the workmen—and others too—at Cambray Works, putting in time clocks, expecting workpeople to start within three minutes of starting time, registering everything on cards, insisting on 'passes' and 'permits' for over- and under-time, depriving the men of time-honoured allowances and perquisites! That would not last long, opined the reverend gentleman. He had heard a lot about him from some of his flock, and nothing to his credit, said he, and was rather taken aback when his visitor admitted diffidently that he was afraid he himself was the poor unpopular wight!

Things happened in the next five years. Mr Best cooperated whole-heartedly with the 'younger' party, while he who prided himself upon being the brake on the wheel, saw they didn't go too fast. Old shops were altered and improved, new ones were built, more efficient production methods were installed, costing became more accurate, components were machine-made.

Time-recording machines were increased in number and some were moved, bringing them as near the lathes and benches as might be, stocks and labour were properly controlled, delivery promises systematically recorded and kept (a prominent indicator showing how many, if any, were overdue, and by how many days), complaints were investigated and the responsibility for them brought home; there were a hundred reforms.

Payment by individual results was instituted throughout the shops as far as possible, even the toolmakers and toolsetters being paid a bonus on the earnings of the groups they were responsible for. Any 'scrap' was reported, examined, and liability brought home to the offender.

5 Recipe for success in business, 1924

(P. W. S. Andrews and E. Brunner, *The life of Lord Nuffield*: *a study in enterprise and benevolence* (Oxford: Blackwell, 1955), pp. 118–21. From articles contributed by W. R. Morris, later Lord Nuffield, to the journal, *System*.)

We took the right steps to achieve ... a continually improving product ... a reduction of cost so that prices could fall continuously, [and] ... a conservative financial policy which should avoid the risks to which a small, quickly expanding concern is liable. So many small men concentrate on one of these aims—and forget the other two. Yet to fail in one alone is to court disaster. ...

So long ago as 1912, I became convinced that the best way for the small concern to manufacture ... was to get specialists on every separate unit of the job. The work is better and more cheaply done while the cost and worry of more plant is avoided. Money is conserved for better use in other directions. Even at the end of the war, when we could have started to produce the Morris car very largely in our own works, I held to the other policy. There is no point in producing any article yourself which you can buy from a concern specializing in the work. I only buy a concern when they tell me they cannot produce enough of the article in question for our programme. ... Even at the present moment, we have contracts running with at least 200 firms for various parts. At Oxford we merely assemble. However, we now make our own engines at Coventry, and bodies at our own works at Coventry. But our method of using outside services is, I believe, more thorough than is usual. To begin with, we buy our own raw material for the job; we fix up contracts, possibly with four or five firms up and down the country, to maintain that supply of raw material for us. We personally inspect the raw material before it is delivered from contractors; we settle the method of machining; we supply gauges, and in many cases we design the actual fixtures. The whole of the finished parts are delivered to us; they are tested on our own inspection benches, and then issued into our component stores. The firms working for us simply undertake a machining contract; we guarantee the material; we allow a percentage of scrap free; and we place long contracts, so that the financial liability of the concern is limited.

It will be argued that, with our large output, we could do the thing cheaper ourselves; that we are piling up transport costs, and so on. This is largely an illusion. The outside firm that makes perhaps only one important part, is probably making in even larger quantities than we should. It is interested in nothing else; therefore it can keep its governing brains on the problems connected with that unit, in a way impossible to a concern manufacturing a highly complicated article. Even the cost of transport is very little increased. For it costs only a fraction more to send steel from, say, Sheffield to Birmingham to be machined and afterwards to transport the finished unit here, than to pay carriage on the full weight of the raw material direct from

11*

Sheffield to our works. Personally I believe that in a highly organized industrial country like ours there are great opportunities in many lines for the small, keen firm that will devote itself to highly controlled and standardized assembling of many types of goods.

[*Keeping up standards of workmanship*] The secret lies largely in the judgment of the man who places the initial contract. Our production manager invariably gets to know personally the chief of any firm with whom he considers doing business; indeed, he cultivates those responsible right down to the shop foreman. It is his rule never to place a contract unless he has seen the works.

Naturally, economy being our main principle, we frequently ask contracting firms to do work at prices that they believe to be impossibly low. Usually they have not realized the savings which we can help them to effect by standardizing operations, by giving a continual flow of work, and by technical assistance. Many firms have taken on work purely on our assurance that it can be done at a certain price; and none has yet 'gone broke'. Our experience is that the arrangement becomes either semi-permanent—or it does not last five minutes.

[*Prices*] I have said that one of the vital elements in securing success to-day is price. My aim is to keep ahead of the market. We have never waited for the public to ask for a reduction. We get in with the reduction first. Is it quite sufficiently realized in this country that every time you make a reduction, you drop down on what I may call the pyramid of consumption power to a wider base? Even a ten-pound price reduction drops you into an entirely new market. If the man cannot pay the last £10 . . . he cannot buy the car. . . . The one object in life of many makers seems to be to make the thing the public *cannot* buy. The one object in my life has been to make the thing they *can* buy.

There is another current opinion that, if you only make enough of a given article, your price will come down romantically. This is not the fact. To get bed-rock selling price you have to be practising every sort of economy, great and small, all the time. There are no marble halls at Cowley, and there never will be. Not an unnecessary penny is spent on selling or publicity. Every ha'-penny available is put into the production. So far this policy has justified [itself by] results. From 1912 to the end of 1918 I did not spend one shilling in advertising. The goods sold themselves. Even now, I rarely go outside the trade press. But to keep down costs you must have a staff of workers who are interested. My experience is that if you look after your men they will look after you. . . .

[*Financial policies*] I have never gone to the public for ordinary capital. In consequence, all the directors are 'still under one hat'. This ... gives rapidity of action ... I can get things done while [a Board of Directors] would be brooding over them. For instance, I rang up the principal of a certain firm and proposed buying their factory one Wednesday morning. The following Friday afternoon I had bought the concern and completed the financial arrangements. Six weeks afterwards we were into production. ...

My second financial policy is to demand payment promptly and to pay absolutely on the spot. A firm that is supplying us knows that whatever happens on the day which is arranged for our monthly payment our cheque will be in the office. The small manufacturer can budget accurately. He can save himself—and me—an enormous amount of money by being able himself to pay all his bills promptly. We are equally stern with others. As soon as we 'black-list' a customer we cut him off from supplies. I believe in facing financial facts; and the fact is that if you do not pay your bills promptly the other man assumes you have not got the money. He therefore determines to make an extra profit out of you to cover what he generally considers an extra risk. Finally, I believe in the budgeting of finances every week. The Secretary presents me with a weekly statement which actually shows the financial condition of the company at that moment.

[*The sales side*] Personally, I do not think that this country has yet really taken to motoring seriously. Until the worker goes to his factory by car, I shall not believe we have touched more than the fringe of the home market. So I have little fear for the disposal of our 35–50,000 cars this year!

The problem of our sales side is not so much finding the buyer—although it may come to that when our sales are many times their present volume—as discovering means of educating the retailer to our point of view which is—expansion. ...

B. TECHNOLOGY AND THE SIZE OF FIRMS

1 James Watt's steam engine, 1775

(15 Geo. III c. 61)

An Act for Vesting in James Watt, Engineer, his Executors, Administrators, and Assigns, the sole use and property of certain Steam Engines, commonly called Fire Engines, of his Invention, described in the said Act, throughout His Majesty's Dominions, for a limited time.

Whereas His Most Excellent Majesty King George the Third, by His Letters Patent, under the Great Seal of Great Britain, bearing date the fifth day of January, in the ninth year of his reign, did give and grant unto *James Watt*, of the city of Glasgow, Merchant, his executors, administrators, and assigns, the sole benefit and advantage of making and vending certain engines, by him invented, for lessening the consumption of steam and fuel in fire engines within that part of his Majesty's Kingdom of Great Britain called England, the Dominion of Wales, and the Town of Berwick upon Tweed, and also in his Majesty's Colonies and Plantations abroad, for the term of fourteen years; with a Proviso obliging the said *James Watt*, by writing under his hand and seal, to cause a particular description of the nature of the said invention to be inrolled in His Majesty's High Court of Chancery, within four months after the date of the said recited Letters Patent:

And whereas the said *James Watt* did, in pursuance of the said Proviso, cause a particular description of the said engine to be inrolled in the said High Court of Chancery, upon the twenty-ninth day of April, in the year of our Lord one thousand seven hundred and sixty-nine, which description is in the words and form, or to the effect following; that is to say,

My method of lessening the consumption of steam, and consequently fuel, in fire engines, consists of the following principles:

First, That vessel in which the powers of steam are to be employed to work the engine, which is called the Cylinder in common fire engines, and which I call the Steam Vessel, must, during the whole time the engine is at work, be kept as hot as the steam that enters it; first, by enclosing it in a case of wood, or any other materials that transmit heat slowly; secondly, by surrounding it with steam, or other heated bodies; and, thirdly, by suffering neither water nor any other substance colder than the steam, to enter or touch it during that time.

Secondly, In engines that are to be worked wholly or partially by condensation of steam, the steam is to be condensed in vessels distinct from the steam vessels or cylinders, although occasionally communicating with them; these vessels I call Condensers; and, whilst the engines are working, these condensers ought at least to be kept as cold as the air in the neighbourhood of the engines, by application of water, or other cold bodies.

Thirdly, Whatever air or other elastick vapour is not condensed by the cold of the condenser, and may impede the working of the engine, is to be drawn out of the steam vessels or condensers by means of pumps, wrought by the engines themselves, or otherwise.

Fourthly, I intend in many cases to employ the expansive force of steam to press on the pistons, or whatever may be used instead of them, in the same manner as the pressure of the atmosphere is now employed in common fire engines; in cases where cold water cannot be had in plenty, the engines may be wrought by this force of steam only, by discharging the steam into the open air after it has done its office; (which fourth article the said *James Watt* declares, in a note affixed to the specification of the said engine, should not be understood to extend to any engine where the water to be raised enters the steam vessel itself, or any vessel having an open communication with it).

Fifthly, Where motions round an axis are required, I make the steam vessels in form of hollow rings, or circular channels, with proper inlets and outlets for the steam, mounted on horizontal axles, like the wheels of a water-mill; within them are placed a number of valves, that suffer any body to go round the channel in one direction only; in these steam vessels are placed weights, so fitted to them as entirely to fill up a part or portion of their channels, yet rendered capable of moving freely in them, by the means hereinafter mentioned or specified: When the steam is admitted in these engines, between these weights and the valves, it acts equally on both, so as to raise the weight to one side of the wheel, and by the re-action on the valves, successively, to give a circular motion to the wheel, the valves opening in the direction in which the weights are pressed, but not in the contrary; as the steam vessel moves round, it is supplied with steam from the boiler, and that which has performed its office may either be discharged by means of condensers, or into the open air.

Sixthly, I intend, in some cases, to apply a degree of cold, not capable of reducing the steam to water, but of contracting it considerably, so that the engines shall be worked by the alternate expansion and contraction of the steam.

Lastly, Instead of using water to render the piston or other parts of the engines air and steamtight, I employ oils, wax, resinous bodies, fat of animals, quicksilver, and other metals, in their fluid state.

And whereas, the said *James Watt* hath employed many years, and a considerable part of his fortune, in making experiments upon steam, and steam engines, commonly called fire engines, with a view to improve those very useful machines, by which several very considerable advantages over the common steam engines are acquired: but upon account of the many difficulties which always arise in the execution of such large and complex machines, and of the long time requisite to make the necessary trials, he could not complete his intention before the end of the year one thousand seven hundred and seventy-four, when he finished some large engines as specimens of his construction, which have succeeded so as to demonstrate the utility of the said invention:

And whereas, in order to manufacture these engines with the necessary accuracy, and so that they may be sold at moderate prices, a considerable sum of money must be previously expended in erecting mills, and other apparatus; and as several years, and repeated proofs, will be required before any considerable part of the publick can be fully convinced of the utility of the invention, and of their interest to adopt the same, the whole term granted by the said Letters Patent may probably elapse before the said *James Watt* can receive an advantage adequate to his labour and invention:

And whereas, by furnishing mechanical powers at much less expense, and in more convenient forms, than has hitherto been done, his engines may be of great utility in facilitating the operations in many great works and manufactures of this kingdom; yet it will not be in the power of the said *James Watt* to carry his invention into that complete execution which he wishes, and so as to render the same of the highest utility to the publick of which it is capable, unless the term granted by the said Letters Patent be prolonged, and his property in the said invention secured, not only within that part of Great Britain called England, the Dominion of Wales, the Town of Berwick upon Tweed, and his Majesty's Colonies and Plantations abroad, but also within that part of Great Britain called Scotland, for such time as may enable him to obtain an adequate recompence for his labour, time, and expence:

To the end, therefore, that the said *James Watt* may be enabled and encouraged to prosecute and complete his said invention, so that the public may reap all the advantages to be derived therefrom in their fullest extent, may it please Your Most Excellent Majesty (at the humble petition and request of the said *James Watt*) that it may be enacted.

And be it enacted, by the King's Most Excellent Majesty, by and with the advice and consent of the Lords Spiritual and Temporal, and Commons, in this present Parliament assembled, and by the authority of the same, that, from and after the passing of this Act, the sole privilege and advantage of making, constructing, and selling the said engines, herein-before particularly described, within the Kingdom of Great Britain, and his Majesty's Colonies and Plantations abroad, shall be, and are hereby declared to be, vested in the said *James Watt*, his executors, administrators, and assigns, for and during the term of twenty-five years; and that the said *James Watt*, his executors, administrators, and assigns, and every of them, by himself and themselves, or by his and their deputy or deputies, servants or agents, or such others as he the said *James Watt*, his executors, administrators, and assigns, shall at any time agree with, and for no others, from time to time, and at all times, during the term of years herein-before mentioned, shall and lawfully may make, use, exercise, and vend the said engines, within the Kingdom of Great Britain, and in his Majesty's Colonies and Plantations abroad, in such manner as to him the said *James Watt*, his executors, administrators, and assigns, shall in their discretions seem meet; and that the said *James Watt*, his executors, administrators, and assigns, shall and lawfully may have and enjoy the whole profit, benefit, commodity, and advantage, from time to time coming, growing, accruing, and arising, by reason of these his said inventions, for the said term of twenty-five years, to have, hold, receive and enjoy the same, for and during and to the full end and term of twenty-five years as aforesaid; and that no other person or persons within the Kingdom of Great Britain, or any of his Majesty's Colonies or Plantations abroad, shall, at any time, during the said term of twenty-five years, either directly or indirectly, do, make, use, or put in practice, the said inventions, or any part of the same, so attained unto by the said *James Watt* as aforesaid, nor in anywise counterfeit, imitate, or resemble the same; nor shall make, or cause to be made, any addition thereunto, or subtraction from the same, whereby to pretend himself or themselves the inventor or inventors, devisor or devisors thereof, without the licence, consent, or agreement of the said *James Watt*, his executors, administrators, or assigns, in writing under his or their hand and seal, or hands and seals, first had and obtained in that behalf, upon such pains and penalties as can or may be justly inflicted on such offenders.

2 Power-loom weaving in the cotton industry

(i) *William Radcliffe's experiments*, 1785–1807

(William Radcliffe, *Origin of the new system of manufacture commonly called 'power-loom weaving'* (Stockport, 1828), pp. 9–10, 15–16, 20–1, 41, 59–60, 61–2, 65)

The principal estates being gone from the family, my father resorted to the common but never-failing resource for subsistence at that period, viz., the loom for men, and the cards and hand-wheel for women and boys. He married a spinster (in my etymology of the word) and my mother taught me (while too young to weave) to earn my bread by carding and spinning cotton, winding linen or cotton weft for my father and elder brothers at the loom, until I became of sufficient age and strength for my father to put me into a loom. After the practical experience of a few years, any young man who was industrious and careful, might then, from his earnings as a weaver, lay by sufficient to set him up as a manufacturer, and though but few of the great body of weavers had the courage to embark in the attempt, I was one of the few. Availing myself of the improvements that came out while I was in my teens, by the time I was married (at the age of 24, in 1785), with my little savings, and a practical knowledge of every process from the cotton-bag to the piece of cloth, such as carding by hand or by the engine, spinning by the hand-wheel or jenny, winding, warping, sizing, looming the web, and weaving either by hand or fly-shuttle, I was ready to commence business for myself; and by the year 1789, I was well established, and employed many hands both in spinning and weaving, as a master manufacturer.

From 1789 to 1794, my chief business was the sale of muslin warps, sized and ready for the loom (being the first who sold cotton twist in that state, chiefly to Mr Oldknow, the father of the muslin trade in our country). Some warps I sent to Glasgow and Paisley. I also manufactured a few muslins myself, and had a warehouse in Manchester for my general business.

＊　　＊　　＊

At Midsummer, 1801, on taking stock very accurately we[2] found we had upwards of £11,000 in our concern; I had also a landed estate in Mellor, in which was comprehended Podmore, where my father

---

[2] Radcliffe and his partner Ross.

was born, with a rent roll, and good tenants of upwards of £350 per annum, charged with about £1,800 on mortgage. Mr Ross's father was a merchant and magistrate in Montrose, and rich, and, my partner being an only son, could at any time lend us a few thousands, which he afterwards did to the amount of £6,000, including the £2,500 paid down on the formation of our partnership. With this real capital—an unlimited credit (£5,000 with our bankers amongst the rest), an excellent trade, and every prospect of its continuing so for a time, we came to the conclusion of purchasing the premises in the Hillgate, from Mr Oldknow and Mr Arkwright, then standing empty, which I never should have thought of for a moment, but from what had passed at the Castle Inn, for the sole purpose of filling them with looms, etc, on some new plan, and just so much spinning machinery as would supply the looms with weft. But beyond the common warping, sizing, weaving, etc., all was a chaos before me; yet so confident was I, that with such assistance as I could call in, we should succeed, that before I began I laid a trifling wager with my partner, that in two years from the time I commenced, I produced 500 pieces of 7-8ths and 9-8ths printing cambrics, all wove in the building in one week by some new process, which I won easily. And as the price for weaving alone when we began was 17s per piece, and had never been below 16s at any time, we thought we were justified in what we were doing, even if little improvement could be found. And if the goods made abroad from the annually increasing export of twist, and their prohibitions of our goods in consequence, had not gradually reduced this price of weaving from 17s (with a profit of 10 to 20 per cent to the master), to 4s to the weaver (and no profit to the master!), we should have been handsomely rewarded by our trade. But to return from this digression, we concluded our contract about Michaelmas with Messrs Oldknow and Arkwright, for the premises above mentioned: and I brought my family to Stockport in the latter end of December, 1801. I must here observe that we had at that time a large concern in Mellor, that with its various branches for putting out work, employing upwards of 1000 weavers, widely spread over the borders of three counties, in a vast variety of plain and fancy goods, all of which had been raised (like a gathering snowball) from a single spindle, or single loom by myself, and was then upon such a system as apparently might go on without my personal attention.

*   *   *

I shut myself up (as it were) in the mill on the 2nd January, 1802, and with joiners, turners, filers, etc. etc., set to work; my first step was some looms in the common way in every respect, which I knew would produce the cloth so much wanted, and in some degree cover our weekly expenses.

Before the end of the month I began to divide the labour of the weavers, employing one room to dress the whole web, in a small frame for the purpose, ready for the looms in another room, so that the young weaver had nothing to learn but to weave; and we found this a great improvement, for besides the advantage of learning a young weaver in a few days, we found that by weaving the web as it were back again, the weft was driven up by the reed the way the brushes had laid the fibres down with the paste, so that we could make good cloth in the upper rooms with the dressed yarn quite dry, which could not be done in the old way of dressing, when the weft was drove up against the points of the fibres, which shewed us the reason why all weavers are obliged to work in damp cellars, and must weave up their dressing, about a yard long, before the yarn becomes dry, or it spoils.

This accomplished, I told my men I must have some motion attached to either traddles or the lathe, by machinery, that would take up the cloth as it was wove, so that the shed might always be of the same dimensions, and of course the blow of the lathe always moving the same distance, would make the cloth more even than could possibly be done in the old way, except by very skilful and careful weavers.

This motion to the loom being at length accomplished to our satisfaction, I set Johnson to plan for the warping and dressing, suggesting several ideas myself. His uncommon genius led him to propose many things to me, but I pointed out objections to them all, and set him to work again. His mind was so teased with difficulties, that he began to relieve it by drinking for several days together (to which he was too much addicted) but for this I never upbraided him, or deducted his wages for the time, knowing that we were approaching our object; at length we brought out the present plan, only that the undressed yarn was all on one side, and the brush to be applied was first by hand, then by a cylinder, and lastly the crank motion.

*　　*　　*

The partnership being thus dissolved, I proceeded in my business with a double prospect of success; first, by the real business I was doing weekly, of 6 to 700 pieces per week, of printing cambrics, mostly woven in the factory, and the other part in weaving-families in the neighbourhood, on the small looms I had furnished to them, delivering them dressed warps on the beam, and pin-cops for the weft. This system had now become practicable, and was so greatly approved of by the weavers, that, had I weathered the calm, which soon after came upon my credit, I might, in a short time, have had all my looms in the dwellings of the operative weavers on the plan I had been driving at from the first, and from the superior advantage of machine dressing. The evenness produced by this mode of preparation, and the working in my loom, not only rendered these goods of ready sale, but gave me a weekly profit of 90*l* to 100*l*, which, along with the second branch of income that formed my double prospect, viz., the premiums of licenses under patent rights beginning to pour in from the first houses in the trade, to the amount of 1,500*l*, in the eight months from the first of July, 1806, to March, 1807, when my vessel became quite becalmed.

*   *   *

In the year 1770, the land in our township was occupied by between fifty to sixty farmers; rents, to the best of my recollection, did not exceed 10*s* per statute acre, and out of these fifty or sixty farmers, there were only six or seven who raised their rents directly from the produce of their farms; all the rest got their rent partly in some branch of trade, such as spinning and weaving woollen, linen, or cotton. The cottagers were employed entirely in this manner, except for a few weeks in the harvest. Being one of those cottagers, and intimately acquainted with all the rest, as well as every farmer, I am the better able to relate particularly how the change from the old system of hand-labour to the new one of machinery operated in raising the price of land in the sub-division I am speaking of. Cottage rents at that time, with convenient loomshop and a small garden attached, were from one and a half to two guineas per annum. The father of a family would earn from eight shilling to half a guinea at his loom, and his sons, if he had one, two, or three alongside of him, six or eight shillings each per week; but the great sheet anchor of all cottages and small farms was the labour attached to the hand-wheel, and when it is considered that it required six to eight hands to prepare and spin yarn, of any of the three materials I have mentioned, sufficient for the consumption of one weaver,—this shews clearly the inexhaustible source there was for labour for every person from the age of seven to eighty years (who retained their sight and could move their hands) to earn their bread, say one to three shillings per week, without going to the parish.

*   *   *

From the year 1770 to 1788 a complete change had gradually been effected in the spinning of yarns. That of wool had disappeared altogether, and that of linen was also nearly gone; cotton, cotton, cotton, was become the almost universal material for employment. The hand-wheels, with the exception of one establishment, were all thrown into lumber-rooms, the yarn was all spun on common jennies, the carding for all numbers, up to 40 hanks in the pound, was done on carding engines; but the finer numbers of 60 to 80 were still carded by hand, it being a general opinion at that time that machine-carding would never answer for fine numbers. In weaving no great alteration had taken place during these eighteen years, save the introduction of the fly-shuttle, a change in the woollen looms to fustians and calico, and the linen nearly gone, except the few fabrics in which there was a mixture of cotton. To the best of my recollection there was no increase of looms during this period,—but rather a decrease.

*   *   *

I shall confine myself to the families in my own neighbourhood. These families, up to the time I have been speaking of, whether as cottagers or small farmers, had supported themselves by the different occupations I have mentioned in spinning and manufacturing, as their progenitors from the earliest institutions of society had done before them. But the mule-twist now coming into vogue, for the warp, as well as weft, added to the water-twist and common jenny yarns, with an increasing demand for every fabric the loom could produce, put all hands in request of every age and description. The fabrics made from wool or linen vanished, while the old loomshops being insufficient, every lumber-room, even old barns, cart-houses, and outbuildings of any description were repaired, windows broke through the old blank walls, and all fitted up for loom-shops. This source of making room being at length exhausted, new weavers' cottages with loomshops rose up in every direction; all immediately filled, and when in full work the weekly circulation of money, as the price of labour only, rose to five times the amount ever before experienced in this subdivision, every family bringing home weekly 40, 60, 80, 100, or even 120 shillings per week!!!

(ii) The advance of the power loom, 1824

(Select committee on artisans and machinery, *B.P.P.,* 1824, V, p. 302. Evidence given by Mr Thomas Ashton, spinner and power-loom manufacturer of Ashton-under-Lyne.)

(To Mr Ashton) Is there at the present moment a gradual transfer of workmen going on from hand looms to power looms?—Yes.

Do you hold this transfer to be advantageous both to the master and to the workmen?—Yes; that is the general opinion in my neighbourhood.

Does not this transfer of hands from hand looms to power looms, by enabling you to perform the same quantity of work with much fewer hands, throw many workmen out of work?—That has never yet been the case.

Your trade then is advancing in such a rapid degree, as to absorb the whole number of hands who are so thrown out?—Yes, entirely so.

The increase of demand is equal to the number of hands thrown out?—The demand for goods increases as rapidly as we can make machinery to supply it.

Which receive higher wages; the men working at the power loom, or the men working at the hand loom?—The power loom hands get considerably higher wages.

Can you state in what proportion?—Fully one-third more.

What do the men receive at each?—A man engaged in preparing for the looms, which is *dressing,* will earn from twenty-four to thirty shillings a week, clear money. The weaving is principally by boys and women; there are very few men employed as power loom weavers; they are employed in overlooking and other work about the mill; and the overlookers will get about the same wages.

Is the dresser confined to the power or the hand looms?—To the power looms.

How does the dresser prepare; by machinery?—Yes; he dresses the warp for the power loom by machinery.

Have you any men performing a similar duty for hand looms?—No.

That is a new subdivision of labour?—Yes.

What do the hands under him receive?—The boys and women earn from twelve to fourteen shillings.

What do you call them?—Power loom weavers.

One hand takes care of two looms?—Yes.

What wages do they receive in the hand looms?—I cannot speak particularly, for there is no hand weaving in our neighbourhood, in consequence of power looms. I understand it is one third less; but I know the hand-loom weavers are always very anxious to come to the power looms, as soon as they are erected.

3 Size of firms in the cotton industry, 1862

(Report of inspectors of factories, 31 October 1862, *B.P.P.*, 1863, XVII, pp. 18–19)

There is an important point to be taken into consideration when reviewing the part taken by the manufacturers in maintaining their hands, viz., the number of those able to make sacrifices for their hands in a time of general depression.

I find by the last return to Parliament that there were 2,887 cotton factories in the United Kingdom in 1861, 2,109 of them being in my district. I was aware that a very large proportion of the 2,109 factories in my district were small establishments, giving employment to few persons, but I have been surprised to find how large that proportion is. In 392, or 19 per cent, the steam engine or water wheel is under 10 horse power; in 345, or 16 per cent, the horse power is above 10 and under 20; and in 1,372 the power is 20 horses and more. The above are the proportions in my district, and I assume that the proportion for the rest of the kingdom would be as nearly as possible the same. A very large proportion of these small manufacturers—being more than a third of the whole number—were operatives themselves at no distant period; they are men without command of capital, which has been invested in their trade, and are consequently unable, from their resources having been so much crippled, to contribute much to the necessities of others. The brunt of the burden then would have to be borne by the remaining two-thirds, and it is from the methods adopted in several of the larger establishments that we can best learn how far the operatives have been assisted by the manufacturers.

4 Mechanisation of the clothing trade, 1864

(Children's employment commission, Second report, *B.P.P.*, 1864, XXII, p. lxvii)

It is said that hitherto the use of steam as the motive power has not met with much favour, but Mr White and Mr Lord state that this power is now used in some cases, although several employers have adopted and abandoned it, owing to the difficulty of checking the speed and the injury caused to the machines by the constant shaking.

At the Army Clothing Depot, Pimlico, where upwards of 700 women are employed, this difficulty appears to have been overcome, and the same may be said of the very large shirt establishment of Messrs Tillie and Henderson, Londonderry, and of Messrs Tait's army clothing manufactory, Limerick, employing 1,000 to 1,200 hands; it may therefore be anticipated that the application of steam power will extend and become general. The introduction of the machine, joined to the extraordinary and increasing demand in foreign, and especially in the colonial markets, for wearing apparel of English manufacture, is accomplishing quite a revolution in these trades; in fact it is evident that the whole employment is at this time in a state of transition, and is undergoing the same change as that effected in the lace trade, weaving, etc., mechanical power superseding hand labour.

3 Advantages of the Sewing Machine

The history of the sewing machine affords, probably, one of the best illustrations of the benefits conferred upon all classes engaged in industrial pursuits, and especially on the operatives, by the substitution of machinery for hand labour.

It appears from the statement of Mr Tillie, that the machine now performs the work formerly done in London known as the most miserable, and even notorious, of all occupations, under the name of 'slop work', in which grown up women, by working very long hours, could only earn, as in some of the poorest paid branches they still do, from 4s to 6s a week. On comparing the details given further on, it will appear, speaking generally, that the wages of machinists, averaging 14s to 16s a week, are at least one-third higher than those of handworkers in the same department. The economy of production effected by the machine, with the general development of trade in late years, has also led to a great increase in the number of hands. The result of these two conditions combined has, in the aggregate, greatly added to the national wealth. Thus, in the Londonderry district, where the machine shirt business was only introduced 14 years ago by the firm of Messrs Tillie and Henderson, it is estimated by the first-named gentleman 'that the whole sum paid for labour in this branch of manufacture now amounts to nearly a quarter of a million yearly, circulating in cash for the general benefit of all'. Mr Tillie may therefore well say, 'the benefit conferred on this part of Ireland by the introduction of this branch of manufacture is enormous'.

But, in addition to the pecuniary gain, another great boon has been conferred on the operative class by the reduction of the protracted hours of work formerly exacted by the system of hand labour. It will subsequently appear that in the shirt and clothing factories, and especially in Ireland, where the greatest change has taken place, the hours for the most part do not exceed, in the case of young persons and adults, those of the Factory Acts, in fact they are often considerably below these, being at ordinary times only nine or 10 hours. ...

The introduction of the machine has necessitated the employment, on the whole, of older children and girls, the usual age for commencing being about 14, one consequence of which is that in these factories the great majority of the employed being above 13 are either adults or 'young persons', as defined by the Factory Act, and therefore entitled to work full time, thus facilitating the introduction of legislative measures.

5 The Gilchrist–Thomas process, 1879

(Richard Meade, *The coal and iron industries of the United Kingdom* (London, 1882), pp. 407–8, quoting an address by E. W. Richards to the Institution of Cleveland Engineers in November 1880)

... A short history and description of a process which has created so much interest in the metallurgical world during the last two years will no doubt be of interest to you. Messrs Thomas and Gilchrist made numerous experiments on a small scale at the Blaenavon Ironworks, where they were assisted by the manager, Mr Edward P. Martin, and they tried also a couple of casts in a large converter at Dowlais. They prepared a paper, giving very fully the results of their experiments, with analyses, which was intended to be read at the autumn meeting of the Iron and Steel Institute in Paris in 1878; but so little importance was attached to it, and so little was it believed in, that the paper was scarcely noticed, and it was left unread till the spring meeting in London in 1879. Mr Sidney Thomas first drew my particular attention to the subject at Creusot, and we had a meeting a few days later in Paris to discuss it, when I resolved to take up the matter, provided I received the consent of my directors. The consent was given, and on the 2nd of October, 1878, accompanied by Mr Stead, of Middlesborough, I went with Mr Thomas to Blaenavon. Arrived there, Mr Gilchrist and Mr Martin showed me three casts in a miniature cupola, and I saw sufficient to convince me that iron could be dephosphorised at a high temperature. I visited the Dowlais Works, where Mr Menelaus informed me that the experiments with the large converter had failed, owing to the lining being washed out. We very quickly erected a pair of 30 cwt converters at Middlesborough, but were unable for a long time to try the process, owing to the difficulties experienced in making basic bricks for lining the converter and making the basic bottom. The difficulties arose principally from the enormous shrinkage of the magnesian limestone when being burnt in a kiln with an up-draught, and of the failure of the ordinary bricks of the kiln to withstand the very high temperature necessary for efficient burning. The difficulties were, however, one by one surmounted, and at last we lined up the converters with basic bricks, when, after much labour, many failures, disappointments, and discouragements, we were able to show some of the leading gentlemen of Middlesborough two successful operations on Friday, April 4th, 1879. The news of this success spread rapidly far and wide, and Middlesborough was soon besieged by the combined forces of Belgium, France, Prussia, Austria, and America. We then lined up one of the 6-ton converters at Eston, and had fair success. The next meeting of the Iron and Steel Institute in London, under the presidency of Mr Edward Williams, was perhaps the most brilliant and interesting ever held by the Institute. Messrs Thomas and Gilchrist's paper was read, and the explanations and discussions by other members of the Institute were listened to with marked attention. Directly the meeting was over, Middlesborough was again besieged by a large array of Continental metallurgists, and a few hundredweights of samples of basic bricks, molten metal used, and steel produced were taken away for searching analysis at home. Our Continental friends were of an inquisitive turn of mind, and, like many other practical men who saw the process in operation, only believed in what they saw with their own eyes and felt with their own hands— and were not quite sure then, and some are not quite sure even now. We gave them samples of the metal out of the very nose of the converter. Our method of working at that time was to charge the additions of oxide of iron and lime at the same time into the converter, and pour the molten metal upon them. The quantity of additions varied from 15 to 25 per cent on the metal charged, according to the amount of silicon in the pig-iron used. We soon found that the oxide of iron was unnecessary; besides, it cooled the bath of metal, and we afterwards used lime additions only. After about three minutes under blow, a sample of metal was taken from the converter, quickly flattened down under a steam hammer, and cooled in water. The fracture gave clear indications of the malleability of the iron. When the bath was sufficiently dephosphorised to give a soft ductile metal, the spiegel was added.

12*

6 Slow progress with the steam turbine, 1903

(Royal Commission on coal supplies, final report part X, *B.P.P.*, 1905, XVI, qq. 18036–54, 18059–66, 18176–7. Evidence of the Hon. C. A. Parsons, proprietor of the Heaton Works, Newcastle-on-Tyne and managing director of Parson's Marine Steam Turbine Co.)

Will you tell us whether the steam turbine has made much progress yet?—At the present time in England and on the Continent there are about 250,000 horse-power working on land for generating electricity, chiefly in central stations for lighting and for power purposes.

Are they confined to the generation of electricity and land services? —Those are almost exclusively confined to the generation of electricity: the other applications are very small and are quite at their beginning at present: electricity is the chief and practically the only present application of them at the present time on land.

Are they making any progress on the Continent?—Yes. There is about a half to two-thirds of the horse-power on the Continent that there is in England.

There is more than one kind, is there not?—Yes, there are three chief kinds at present. The oldest is mine and Dr de Laval's, which were almost simultaneous. The Laval turbine is only applied to powers up to about 200 horse-power at the present time. It requires gearing which limits its scope and size.

Is yours applied to purposes to any extent below 200 horse-power? —Not many below 200.

Then may we take it that that forms a sort of division?—That is so.

For higher powers, yours, apparently, is the best adapted?—Yes, for higher power purposes it is. Then there is another turbine of later date—the Curtis turbine, which has been brought out lately in America by the General Electric Company.

Is there much difference in the principle or the construction?—It is different. They are both operated on the turbine principle by the velocity of steam as opposed to the reciprocating engine; but up to the present time there have been no authentic tests published of the Curtis turbine by independent authorities.

Then it has not made much progress yet in practical work?—Well, orders are said to have been taken in the States to the extent of 200,000 horse-power. About three, I believe, were placed in England.

Could you tell us quite shortly in what respect the turbine has advantages over the reciprocating engine?—At the present time large turbines, turbines of electrical generating sets of 2,000 horse-power and upwards, are about from one-half to two-thirds the capital cost of ordinary engines.

Therefore, that is a less capital outlay?—Yes.

Is the cost and maintenance more or less?—A great deal less. The cost of oil is avoided; it is practically nil. The cost of steam is less, according to the circumstances of the case, from 10 to 25 per cent.

May we take it that the saving is greatest in the cases of the larger sizes?—It is greater in the larger sizes.

And there is a less total capital outlay and a less cost of running and maintenance?—That is so.

(*Sir William Lewis*) That is to say, the same amount of steam and the same pressure would do 25 per cent more work with the turbine than with the reciprocating engine?—Yes. There was a test made three weeks ago by the Cunard Company at Wallsend; the turbine at full load took 25 per cent less steam per horse-power or per kilowatt generated than the reciprocating engine.

Under precisely similar circumstances?—Yes.

(*Chairman*) Do you claim that you can, with advantage, use higher pressure or superheated steam to greater advantage by the turbine? —Yes, there is no limit at all as regards the turbine; the boiler or the superheater is the only limit. The turbine will stand any reasonable heat or any pressure.

I see you speak of condensers. Have you found a plan by which these turbines can be used in that way—with condensers?—Yes, they are generally, almost universally, used with condensers in conjunction, as condensing turbines.

And is the advantage relatively as great as the difference between an ordinary high-pressure and a condensing engine?—The advantage increases with condensing more than it does in the case of a reciprocating engine. A turbine benefits more by a condenser—by a vacuum— than a reciprocating engine. Those comparative tests of consumption which I gave were with condensers. Without a condenser the turbine takes rather more steam than the reciprocating engine. . . .

Is the application of the turbine to ships increasing?—At the present time there is about 40,000 horse-power actually in commission under the British flag.

Is that for the Admiralty?—It is both for the Admiralty and the Mercantile Marine? There are only three Admiralty ships, one completed, one under trials and one in course of having the machinery put on board.

As to the completed one, has that been tested in actual ordinary work?—Yes.

I mean have you got beyond the stage of completion and got it into actual practice?—It is, I believe, handed over to the Admiralty, but not in commission yet, I think.

Therefore, we can get no information as to its performance in actual work?—No. The one undergoing trials now is one of the latest class of Destroyers.

(*Sir William Lewis*) Is the turbine in that case used for subsidiary engines on board ship or for the main driving engines?—The main propelling engines.

For the propellers?—Yes, 7,000 horse-power, $25\frac{1}{2}$ knots.

(*Chairman*) Is it anticipated that a higher speed will be obtained by that means?—So far as trials have gone, they seem to indicate that the consumption of coal will be as good as with the sister vessels fitted with reciprocating engines.

*   *   *

I understand from your evidence that the United Kingdom has taken the lead in this matter of turbines as against Germany and America. We have heard so much about the lead they are taking in everything that I am glad to hear we are first in something?—Yes, the United Kingdom has been first in turbines. The Westinghouse Company bought the rights for our turbines about eight years ago for the States, but did not do anything practically with them until recently. Now they are doing a very large amount of business: they have over 200,000 h.p. on order at the Westinghouse Company's works at Pittsburg—turbines made from our plan. Then, in addition to that, the General Electric Company have the Curtis turbine, which they also have taken about the same amount of orders for. Then, on the Continent, Messrs Brown, Boveri & Company, who are working with us, have taken orders for, or executed, over 100,000 h.p.

That is all emanating from your firm?—Yes, from our plans, excepting, of course, the Curtis turbine. Then there is the Laval turbine, which is only made up to 200 h.p.; that is the Swedish turbine which is made by Messrs Greenwood & Company, of Leeds, for England, and by Messrs M. Brequet & Cie., Paris, for France, and several other firms on the Continent; but it is only for comparatively small sizes.

7 Iron-making a labour-intensive process, 1907

(Lady Hugh Bell, *At the works* (Arnold, 1907; Newton Abbot: David
& Charles, 1970), pp. 31–8)

The material tipped by the gantrymen into the kilns is gradually, as
it sinks down, drawn off at the bottom through the 'hoppers', which
can be opened or closed as required by the huge iron shutters before
described. A hopper, be it said, is practically any vessel or receptacle
with a hole at the bottom and sloping sides. Every now and again,
if the kiln is too hot, the ironstone begins to fuse, and the pieces stick
together, making a lump too big to get through the inside of the kiln
into the hopper. The lump is called a 'scarr', and it then has to be
broken into bits with an iron rod from the outside by a man called the
'scarrer', who stands on a movable wooden trestle, about the height of
the bottom of the hopper, and breaks up the lump by thrusting his
bar into the opening until the lump is small enough to come through.
The opening into the kiln above the hopper is called the 'eye'. The
scarrer earns from 30*s* to £2 a week; he spends most of the eight-
hours day standing at the bottom of the kiln, the iron rod in his
hand, ready to thrust it into the kiln whenever the obstacle shows
itself. By him is standing another man ready to add his weight to the
thrust if the strength of the first one is not enough to deal with the
obstacle. For whatever operation is being carried on at the ironworks,
there are always a number of men standing round in a state of watchful
concentration, their attention on the alert, ready to lend a hand in a
case of emergency. The spectator receives an overpowering impression
of what that watchfulness needs to be, of what sudden necessities may
arise, of what may be the deadly effect of some swift, dangerous variation, some unexpected development in the formidable material which
the men are handling.

The 'scarrs' having been broken, and the calcined ironstone, now
of a dull red colour, drawn out from the hopper into shallow iron
barrows, it is wheeled from the kiln to the lift of the blastfurnace,
20 to 30 yards off, by a man called a 'mine-filler', who does this ten to a
dozen times during the hour. On the way to the furnace he stops to
have the weight of his load checked outside the 'weigh cabin', a little
dark, dusty shed, outside the window of which there is a platform
like that used for weighing luggage at a station. The men get so well
accustomed to judging the 'burden' required—that is, the weight of
ironstone or limestone that ought to be in the barrow—that they
generally hit off almost exactly before it is weighed the required quan-

tity of rough pieces of stone piled up in it. The mine-filler is paid 42s a week.

The weight being ascertained to be correct, the weighman puts down a straight stroke for each barrow that passes him, and after every four draws a horizontal line through the four strokes, each of these groups of lines therefore representing five barrows. The man in the weigh-cabin receives about £1 per week. He is generally one of the older men, obliged to accept as his strength declines a lighter and less remunerative form of work. His job is to sit for the eight hours in which he is on duty in his little cabin, sheltered, at any rate, from the weather, but not from the smells, the noise, and the dust of the blast-furnaces close to which his cabin stands.

The mine-filler, the weight of his barrow being checked and found correct, then wheels it to the lift, a square, unenclosed platform which goes up and down to the top of the furnace and back again between four huge supports.

In the place we are describing there are first two furnaces and then a lift, then two furnaces again, then a lift, and so on; the same lift always serves the same pair of furnaces. The six barrows of ironstone, three of limestone, and six of coke, constituting the 'round' already described, are placed on the lift and hoisted to the top of the furnace, where they are taken off by the chargers, and wheeled close to the big round aperture at the top of the blast-furnace. These men are standing on a platform about 10 feet wide that runs round this aperture, and is guarded on the outside by a railing. That aperture is closed by a huge lid, a bell and hopper in this shape: the centre edges of a hopper (which in this case is the shape of a deep saucer or shallow bowl with its middle out) rest upon the sides of a shallow cone called the 'bell'. The charge is lowered on to the top of the bell, on which the inner edge of the hopper is resting; the bell is then lowered, leaving, therefore, an opening, and the charge, no longer kept in place by the sides of the bell, slides down them into the furnace below. The bell is then rehoisted into position. The sudden jet of vivid flame from the top of the furnace, so familiar to those who live in the neighbourhood of ironworks, is produced at the moment of charging. The chargers have from 30s to £2 per week. The work of the charger is arduous and trying to the health. Men with susceptible lungs are apt to be much affected by the combination of the rapid breathing necessitated by handling the heavy barrows and the fumes inhaled with every panting step.

The temperature inside the furnace is about 3,000° F. This great mass of combustion is being perpetually fanned by the hot blast, heated, as has been before described, by passing through the stoves made hot by the passage of the gas, and incessantly driven into the furnaces at the top of the 'hearth', just above the slag opening, by blowing-engines moved by machinery. The iron, being heavier, gradually sinks to the bottom, and the slag (which may be called an artificial lava, and is practically the scum or dross of the iron) rises, floating on the top of the iron, just below the place where the blast is blown in. The materials put into the furnace gradually sink down as those which preceded them are consumed, and the metal is separated from the dross with which it is associated in the stone. As they sink more material is put in above, so that the level of them at the top is always within a few feet of the bell. The heavier iron falls molten to the bottom, while the dross floats molten on its surface. As the process continues, the mass of liquid iron rises, as well as the slag or dross, until the latter flows out of the furnace through a hole provided for the purpose. Four times in the twenty-four hours the furnace is 'tapped'; a hole is made at the bottom of the furnace, and the iron is allowed to run out. This is an important moment in the process of manufacture. To say that a hole is made in the furnace for the iron to run out sounds simple enough: but a stream of molten iron cannot be drawn off like water, and time after time the tapping of the furnace, accomplished by breaking by main force through the piece of fireclay, which, having been thrust cold into the red-hot aperture after the last casting, has baked and hardened in the opening to a solid mass, is a strenuous encounter with a potent and deadly enemy. To be a 'furnace-keeper', and responsible for the furnace being in absolute working order, is one of the most responsible posts at the works. The keeper gets from £2 10s to £3 per week.

A great square platform, the 'pig-bed' of firm moist sand, dug from the river-bed, extends 10 feet from the ground in front of each furnace, at a level of about 1 foot below the bottom of it. In this, before each casting, the channels are prepared into which the molten stream is to run, by the helpers of the furnace-keeper. These channels consist of one long main channel, 16 inches wide and 10 inches deep, at right angles to which are other channels varying in number, generally about 16 feet long; between these, parallel to the main channel, are rows of shorter channels, also varying in number according to the size of the pig-bed; there may be twenty-four, or there may be even as many as thirty-six. These shorter channels are for the pieces of iron which when cast will be known as 'pigs'; the transverse and longer channels are called the 'sows.' The area of the sandy platform having been made tolerably smooth and level again after the last casting, the men take first of all a long piece of wood the size that the sows are to be, with an iron ring at each end of it for facility of handling, and put it across the platform where the first sow is to be. They then take a number of short oblong blocks of wood, the shape of the pigs, and drop them rapidly one after another all along at right angles to the sow, with about 3 inches of sand between them; other men standing ready with spades then throw sand into the interstices between the wooden blocks, forming a partition between each. When the row is finished, therefore, each block of wood is lying practically in a little rectangular sandy hollow. The long transverse piece, the sow, is then lifted up by two men holding the iron ring at either end and thrown across to others, who put it across at the bottom of the shorter ones, and repeat the operation. Then all the shorter wooden pieces are taken up by the ring at the end of them in the same way (except that they can be lifted by one man) and thrown across to the others standing below the next sow, who repeat the operation, as with the sow, of dropping them into position and filling up the interstices with sand. Into the first row of moulds, now left empty, a man holding a wooden instrument, a thin piece of wood fastened transversely at the bottom of a long handle, like a broom ending in wood instead of bristles, goes rapidly along the channels, flattening them at the bottom; after which he is followed by another man with an instrument which has at the end of its handle a cross-piece of iron, with raised letters on it, with which he in his turn goes along the channel of pigs, stamping the name of the brand at the bottom of them. And so admirably adapted is the river sand for the purpose of a mould, especially after several castings have been run over it, thoroughly drying it and burning the lime out, that this stamping by hand by a mould with raised letters, firmly pressed into each channel in turn, is enough.

To the outsider, indeed, part of the absorbing interest of watching the manufacture of iron is that in this country, at any rate, it is all done by human hands, and not by machinery. From the moment when the ironstone is lifted off the trucks, then dropped into the kilns, afterwards taken to the furnace, and then drawn out of it, it has not been handled by any other means than the arms of powerful men, whose strength and vigilance are constantly strained almost to breaking-point.

8 Flow production—a modern technique, 1937

(i) The case for a continuous strip mill

(*The Times*, 26 January 1937. This is an extract from a prospectus of Richard Thomas & Company, which was raising £7 million by the issue of 4 per cent debentures to finance the modernisation of its steelworks at Ebbw Vale.)

Development. The Company's development programme includes the erection at Ebbw Vale of the latest American type of:

(a) Combination Billet Blooming and Slubbing Mill, with a capacity of 900,000 tons per annum.

(b) Continuous Hot Roll Strip Mill, with a capacity of 600,000 tons per annum, capable of making steel strip 48 inches wide from 2 inches thick down to 18 gauge [about 0·05 in].

(c) Cross Rolling Mill capable of making sheets up to 72 inches wide.

(d) Cold Reduction Plant, plus a finishing plant capable of producing 370,000–400,000 tons per annum of highest grade sheets and tinplates. . . .

The product of these modern continuous mills is vastly superior in quality to and lower in production cost than the product at present produced by any existing plant in this country. Hence, although existing sheet and tinplate capacity is in excess of present demand it is expected that the product of the new mills, for which there is already an urgent call, will itself increase the demand for flat steel for building, furniture and motor car bodies. . . .

(ii) The threat to batch-production methods

('Report of the Import Duties Advisory Committee on the present position and future prospects of the iron and steel industry', *B.P.P.*, 1936–37, XII, p. 39)

. . . In a statement which we have received from the Sheet Trade Board, a body representative of employers and operatives engaged in the manufacture of black and galvanised sheets, it is pointed out that methods of manufacture are changing rapidly, hand-operated mills being replaced by mechanical and continuous strip mills, of enormously greater capacity of output. There are at present 25 sheet works employing some 18,000 operatives, and some 80 tinplate works employing over 22,000 operatives: their whole output of plates and sheets could be replaced by a very few mills of the magnitude of that now in process of erection at Ebbw Vale, with a very large reduction in the numbers employed.

9 Man-made fibres — Terylene, 1952

(*The Economist*, 24 May 1952)

The directors of Imperial Chemical Industries devote a good deal of attention in their report to new research projects. ... Meanwhile development sales of Terylene have strengthened earlier impressions that there should be a market for this textile fibre capable of absorbing the output of the large-scale plant now being completed at Wilton; the enthusiasm expressed in the report about the prospects for Terylene seems to be in some contrast with the directors' lack-lustre comments about Ardil.

10 A new source of power, 1955

(A programme of nuclear power, *B.P.P.*, 1954–5, XIII, pp. 1, 4–6)

1 An important stage has been reached in the development of nuclear energy for peaceful purposes. Hitherto the work in this country has consisted of a military programme, a broadly based research and development programme, and the production and use of radioisotopes. The military programme continues to be of great importance but the peaceful applications of nuclear energy now demand attention. Nuclear energy is the energy of the future. Although we are still only at the edge of knowledge of its peaceful uses, we know enough to assess some of its possibilities.

2 Our future as an industrial country depends both on the ability of our scientists to discover the secrets of nature and on our speed in applying the new techniques that science places within our grasp. The exact lines of future development in nuclear energy are uncertain, but this must not deter us from pressing on with its practical application wherever it appears promising. It is only by coming to grips with the problems of the design and building of nuclear plant that British industry will acquire the experience necessary for the full exploitation of this new technology.

3 The application that now appears practicable on a commercial scale is the use of nuclear fission as a source of heat to drive electric generating plant. This comes moreover at a time when the country's

great and growing demand for energy, and especially electric power, is placing an increasing strain on our supplies of coal and makes the search for supplementary sources of energy a matter of urgency. Technical developments in nuclear energy are taking place so fast that no firm long-term programme can yet be drawn up. But if progress is to be made some indication must be given of the probable lines of development so that the necessary preparations can be made in good time. A large power station may take five or more years to complete, including finding the site, designing the station and building it.

*   *   *

15 It is expected that it will be possible to extract as much as 3,000 megawatt-days of heat from every ton of fuel. This is the equivalent of the heat from 10,000 tons of coal. There is as yet no practical experience of this level of irradiation at high temperatures and the metallurgical behaviour of the fuel elements is uncertain. But there are many lines of development which should overcome such metallurgical defects as may appear.

*   *   *

19 ... Taking what appears to be a reasonable value for the plutonium, the cost of electricity from the first commercial nuclear station comes to about 0·6d a unit. This is about the same as the probable future cost of electricity generated by new coal-fired power stations.... If no credit were allowed for the plutonium the cost of nuclear power would be substantially more than 0·6d a unit. Later stations should show a great improvement in efficiency, but the value of plutonium would probably fall considerably during their lifetime. Even so their higher efficiency should enable them to remain competitive with other power stations.

20 These estimates assume that all the plutonium is used for civil purposes, as would be most desirable. No allowance has been made for any military credits.

PART III

A Provisional Programme

21 Her Majesty's Government consider that the development of nuclear power has reached a stage where it is vital that we should apply it commercially with all speed if we are to keep our position as a leading

industrial nation and reap the benefits that it offers. The programme outlined below is provisional and must be considered only as the best indication that can now be given of the probable line of development. Types of stations, numbers and dates are all subject to change.

22 Although the decision to go ahead with a nuclear power programme does not depend on precise comparisons of cost, the outline given above has shown that the cost of nuclear power should not be greatly different from the cost of power from coal-fired power stations. This country has a rapidly growing demand for energy, particularly in the form of electric power, and increasing difficulty in producing the necessary quantities of coal. These facts by themselves would justify a great effort to build up a nuclear power system.

23 The stations will be built in the normal way by private industry for the Electricity Authorities, who will own and operate them. The Atomic Energy Authority, as the only body with the necessary experience, will be responsible for giving technical advice on the nuclear plant. British industry and consulting engineers have as yet no comprehensive experience of nuclear technology. They will be faced with a major task in training staff, in creating the necessary organisation and in designing the stations. This work has already begun. Owing to its complexity and diversity teams drawn from several firms may have to be formed. The preparatory work will call for great efforts from all concerned, and even so it will not be practicable to start building any commercial stations before 1957.

24 It is intended that the Electricity Authorities and private industry should obtain as quickly as possible the practical experience in designing and building nuclear power stations that will be the necessary foundation for a big expansion in the later stages of the programme. The Atomic Energy Authority, while giving as much assistance and advice to industry as possible will remain primarily a research and development organisation and will continue to design, build and operate pioneering types of power reactor. They will also be responsible for buying uranium, fabricating the fuel elements, processing the used fuel and extracting the plutonium from it. There will therefore have to be a continuous process of co-operation and of financial adjustment between the Electricity and Atomic Energy Authorities. The exact arrangements to be made are at present being discussed with them.

C. INDUSTRIAL COMBINATION

1 Disciplining the domestic worker, 1777

(Herbert Heaton, *The Yorkshire woollen and worsted industry* (Oxford University Press, 1920), p. 427, quoting *Leeds Mercury*, 19 August 1777)

The Committee of manufacturers of combing wool have nominated [seven persons here named] to be inspectors for preventing frauds and abuses committed by persons employed in the manufactures of combing wool, worsted yarn, etc, ... and do hereby give notice, by virtue of an Act passed last Session, and forewarn all spinners who shall be guilty of reeling false or upon false reel that they will be prosecuted and punished by the said inspectors, as the law directs, without any favour or partiality. They likewise give notice to all agents or persons hired or employed to put out wool to be spun into worsted, that by the said Act such agents are liable to pay a penalty of five shillings for every parcel of yarn made up which is short weight, and which is false or short reeled, unless they produce and do give in evidence what person was the reeler of such yarn, so that he or she may be lawfully convicted; for which purpose it will be expected that the putters-out ticket their yarn.

2 An ineffective combination to raise iron prices, 1818

(Madeleine Elsas, ed. *Iron in the making: Dowlais Iron Company letters, 1782–1860* (Cardiff: Glamorgan County Records Committee and Guest Keen Iron & Steel Company Limited, 1960), p. 5)

W. Pinton, Chairman of Staffordshire Ironmasters, to Guest & Co.

Dudley [*Worcestershire*], *March 2nd 1818*

On the 24th October last the Ironmasters in Staffordshire declared an advance of 15/–per Ton on Bar, 25/–per Ton on Rod Iron; as no alteration was then or has since been made in the Cardiff price, and as a reduction has lately been made in London, it must be obvious that the advance cannot be established in Staffordshire & it has therefore been determined by a general meeting held this day to abandon this attempt. The price at our respective Works will therefore be Bars, 6 months—£11 15*s*, 3 months—£11 5*s*, Rods, 6 months—£13, 3 months—£12 10*s*, which will correspond with the proportion as heretofore settled between the Staffordshire & Cardiff terms. As it is extremely probable that the result of this meeting may be misrepresented, it has been thought necessary to acquaint you with these particulars.

3 The Newcastle coal vend, 1771–1830

(Reports from Committees on the coal trade, *B.P.P.*, 1800, X, p. 540, and *B.P.P.*, 1830, VIII, pp. 6, 254–5)

(*i*) *Evidence of Francis Thompson* (formerly manager of Washington colliery), 1800

Is there any regulation or limit as to price they[1] may give to the coal-owners? In August, September, and October, 1771, I found great irregularities in the Coal Trade, particularly with respect to the measure. I communicated my sentiments to two of the most respectable agents of the owners ...; upon which it was agreed that a meeting should be had of the coal owners belonging to Sunderland, to be convened by me, and the coal owners at Newcastle, to be convened by a Mr Gibson and Mr Morrison, which was done; and we had three or four meetings, and I was appointed Secretary. ... Since that time, according to the best enquiries I have been able to make, the coal owners have had frequent meetings for the purpose of stipulating the vends[2]; that is, that five of the collieries of the best coals, viz., Walls End, Walker, Wellington, Hebburn, and Heyton, are permitted to vend the greatest proportion, and at the best price; after that there is a second class, which sells one shilling per chaldron lower, being coals of an inferior quality, and also less in proportion as to quantity; there is likewise a third class, at a shilling less than the second, and who are allowed to sell a still less proportion as to quantity.

By what means do you understand those vends have been limited? —By the meetings of the coal owners frequently for the purpose of ascertaining the vends.

Was there any positive agreement for that purpose?—That cannot be well known, being contrary to Act of Parliament.

---

[1] The fitters or agents between coal-owners and ship-owners.
[2] The name by which the agreements as to output were known.

(ii) Evidence of Robert William Brandling, 1830

The proprietors of the best coals are called upon to name the price at which they intend to sell their coals for the succeeding twelve months; according to this price, the remaining proprietors fix their prices; this being accomplished, each colliery is requested to send in a statement of the different sorts of coals they raise, and the powers of the colliery; that is, the quantity that each particular colliery could raise at full work; and upon these statements, the committee, assuming an imaginary basis, fix the relative proportions, as to quantity, between all the collieries, which proportions are observed, whatever quantity the markets may demand. The committee then meet once a month, and according to the probable demand of the ensuing month, they issue so much per 1000 to the different collieries; that is, if they give me an imaginary basis of 30,000 and my neighbour 20,000, according to the quality of our coal and our power of raising them in the monthly quantity; if they issue 100 to 1000, I raise and sell 3,000 during the month, and my neighbour 2,000; but in fixing the relative quantities, if we take 800,000 chaldrons as the probable demand of the different markets for the year; if the markets should require more, an increased quantity would be given out monthly, so as to raise the annual quantity to meet that demand, were it double the original quantity assumed.

*   *   *

What means have been resorted to in the north of England, with a view to keep the price of coal at such a rate as should compensate the owners of those collieries in which the expense of raising is the greatest?—We have entered into a regulation at different times, which regulation is in existence now, and which has for its object to secure us a fair uniform remunerating price, and enables us to sell our coals at the port of shipment under our immediate inspection, instead of being driven, by a fighting trade, to become the carrier of our coals, and to sell them by third persons in the markets to which they are consigned; thereby trusting our interests to those over whom we have no direct control whatever.

So that practically the real quantity to be sold is fixed with reference to each colliery each month?—Yes. ...

The basis originally fixed, is the proportion taken between all the collieries?—It is merely an imaginary quantity to fix the relative proportions. ...

13   "E-532 Documents"

Has the scale of prices now in operation been varied materially from that which was adopted when the regulation of the vend was last on?—I have already stated in my evidence that ours is a competition price, that we endeavour to get the best price we can, which is a little below what the consumer can get the same article for elsewhere. In the regulation in 1828 we found we had fixed our prices too high; the consequence was, it created an immediate influx of coals from Scotland, Wales and Yorkshire, and more especially from Stockton; so that when the coal owners met together, to enter into another arrangement last year, we were obliged to fix our prices a little lower.

4 Restrictive practices in the sale of wall-paper, 1906

(Henry W. Macrosty, *The trust movement in British industry: a study of business organisation* (Longmans, 1907), pp. 310–11. The existence of this circular had first been made known to the public in the *Daily Mail*, 23 October 1906.)

THE WALL-PAPER MANUFACTURERS, LTD.

125 High Holborn
London, W.C., 17th October, 1906

Gentlemen,—In view of the fact that certain manufacturers are pressing for orders, permit us respectfully to remind you that by the terms of your agreement with this company you have engaged not to 'stock nor cut up patterns, nor issue in your pattern books, nor sell for stock any paper-hangings or any raised materials other than those manufactured by the company'.

We have reason to believe that some of our customers, either from negligence or under the advice of interested parties, have been induced to commit small breaches of Clause 18 of the agreement.

In those cases which have come to our knowledge, we have, in bringing the matter to our customers' notice, obtained from them formal recognition of their obligations, and a promise to comply with them in the future; and we have instructed our solicitors to commence proceedings and enforce the payment of damages against any of those who commit breaches of the agreement.

In the general interest, and at the request of a large body of agreement customers, we are writing this letter to all, and take the liberty of bringing the matter to your attention, so as to avoid any chance of misapprehension.

Yours truly,
The Wall-Paper Manufacturers, Ltd.

5 Advantages of bigness in the motor industry, 1966

THE BRITISH MOTOR CORPORATION LIMITED

Longbridge,
Birmingham.

To: The Shareholders   *29th July, 1966*
Dear Sir (or Madam),

Merger with Jaguar Cars Limited

It was announced on 11th July, 1966, that the Boards of The British Motor Corporation Limited ('B.M.C.') and Jaguar Cars Limited ('Jaguar') had reached agreement in principle for the two Companies to join forces by means of a merger to be effected by B.M.C. making an offer to acquire the entire issued Ordinary and 'A' Ordinary Share capital of Jaguar as a first step towards the setting up of a joint holding company to be called British Motor Holdings Limited. . . .

This development is in line with world trends towards larger and more comprehensive units possessing the greatest possible resources for the development, manufacture and marketing of a complete range of products. The Boards of B.M.C. and Jaguar share the opinion that, if the British motor industry is to remain truly competitive, a closer integration of its various units must be achieved. The ranges of vehicles produced by the B.M.C. and Jaguar Groups are basically complementary rather than competitive and benefits should result from the merger. . . .

Yours faithfully,
G. W. Harriman,
Executive Chairman

6 . . . And in brewing, 1967

(*The Guardian*, 21 July 1967)

The biggest brewery group in the world will be created if a proposed merger between Bass, Mitchells and Butlers, and Charrington United Breweries goes through.

The combined group, to be called Bass Charrington, would be responsible for $18\frac{1}{2}$ per cent of the total beer production in this country, would own more than 11,200 public-houses and other outlets, and would be capitalised at some £200 millions in the stock market.

13*

The unexpected announcement of the merger proposals last night brings to a peak the recent rash of takeovers and amalgamations in the industry. A little while ago, Bass itself succeeded in acquiring Bent's Brewery after a hard-fought battle with Watney Mann. The merger, which is supported by the directors of both companies, is to be achieved by an offer for the shares of the two companies which will result in the ordinary capital of the new group being equally divided between the present shareholders. An effective increase in the dividend is forecast.

President and chairman

The new company is to be headed by Mr John Charrington, the C.U.B. chairman, as president, and Mr Alan Walker, now chairman of Bass, as chairman. The object of the merger is twofold. In the first place, it will create a completely national distribution for the two companies' beers. Bass has its main strength at present in the Midlands, while Charrington is well represented in both the South and the North. Beyond this, however, Mr Walker and his colleagues are looking for greater strength of the combined company in attacking the overseas markets, particularly in Europe and North America, where they see great opportunities for expansion.

The directors said yesterday: 'This is a merger of truly complementary groups, and, if approved, will give the new organisation a completely national coverage. The directors of Bass and of C.U.B. are of the opinion that the merger of the two groups is desirable and advantageous and the terms of the merger of the two companies are fair and reasonable to the shareholders of both companies'.

D. THE FORTUNES OF INDUSTRY

1 Trade depression, 1775

(Historical Manuscripts Commission, *Dartmouth MSS*, III, 220. The writer was John Wesley.)

I aver that in every part of England where I have been (and I have been east, west, north, and south within these two years) trade in general is exceedingly decayed, and thousands of people are quite unemployed. Some I know to have perished for want of bread; others I have seen walking up and down like shadows. I except three or four manufacturing towns which have suffered less than others. . . .

Even where I was last, in the West Riding of Yorkshire, a tenant of Lord Dartmouth was telling me 'Sir, our tradesmen are breaking all around me, so that I know not what the end will be'. Even in Leeds I had appointed to dine at a merchant's, but before I came the bailiffs were in possession of the house. Upon my saying 'I thought Mr ... had been in good circumstances', I was answered 'He *was* so, but the American war has ruined him'.

2 Lead mining and smelting, 1798 and 1849

(Myrddyn J. Bevan-Evans, *Early industry in Flintshire* (Flintshire Record Office, 1966), pp. 16–17, quoting (*i*) Richard Warner, *A second walk through Wales* (Bath, 1799) pp. 249 *et seq*. and (*ii*) Edward Parry, *Parry's railway companion from Chester to Holyhead, together with some account of the stupendous railway works* (London, 1848), pp. 41–2.)

(i) Lead mining, 1798

... Another hour (from the Allelujah monument at the Rhual) brought us to the great object of our day's ramble, Llyn-y-Pandu mine, the most considerable lead-mining speculation in England. The scenery of this place is wonderfully wild and romantic; a deep valley, rude and rocky, shut in by abrupt banks, clothed with the darkest shade of wood. Straggling through the bottom of this dale, is seen the little river Allen, which, having pursued a subterraneous course for nearly three miles, makes its second appearance close to the lower engine belonging to these stupendous works.

Llyn-y-Pandu mine is the property of John Wilkinson, esq; the great iron-master, who has, with infinite spirit and perseverance, encountered obstacles in bringing it to its present state, that would have exhausted the patience and resolution, as well as the coffers, of most other men. With all his exertions, however, he has not been able to render it complete; the mine even now contains so much water, that he has been under the necessity of erecting four vast engines (of Messrs Boulton and Watt's construction) upon the premises to drain it. The steam cylinder of the lower one is forty-eight inches diameter, and works an eight-feet stroke in a pump twenty-one inches diameter, to a depth of forty-four yards; the steam cylinder of the mountain engine is fifty-two inches diameter, and works an eight-feet stroke in a pump twenty-one inches diameter, to a depth of sixty yards; the steam cylinder of Perrins's engine is twenty-seven inches diameter (double) and works a six-feet stroke in a pump twelve inches diameter, to a depth of seventy yards; and the steam cylinder of Andrew's engine is thirty-eight inches diameter, and works an eight-feet stroke in a pump twelve inches diameter, to a depth of sixty yards. The mountain engine has been lately erected, in consequence of a lease of ground of upwards of a third of a mile in length upon the range of the Llyn-y-Pandu vein, called Cefn-Kilken, granted by Earl Grosvenor to Mr Wilkinson. Many thousand tons of lead ore are now in stock upon these premises waiting for a market, the war having almost suspended the demand for lead, and lessened the price to nearly one half of what it formerly sold for. The engines also are quiet, and the works at a stand. When the bottom level, intended to communicate all the engines, is finished, great expectations are entertained with respect to the produce of this mine; as it contains one head of solid ore upwards of six feet wide, another of four feet, and about two feet upon the average for ninety yards in length upon the bottoms. The ore is of two kinds, the one *blue*, which yields sixteen cwt of lead per ton, and the other *white*, which yields thirteen cwt. They are both gotten in the same vein; the white lying in general on the south, and the blue on the north side. When peace shall again have opened a market for lead, these ores are intended to be smelted at works now erecting by Mr Wilkinson on Buckley Mountain, near to the road which leads from Mold to Chester.

(ii) Lead smelting, 1849

. . . The dark columns of smoke, rising high into the heavens and then spreading a sable cloud through a vast expanse, indicate the large trade transacted here [Bagillt]. Nearly all the lead ore produced in North Wales, and considerable quantities from South Wales, Isle of Man, and Ireland is brought here for smelting, and is manufactured into different useful articles of trade, and shipped to various parts of the kingdom. The Flintshire lead ore markets are held alternately at Flint and Holywell, every fortnight, and whole cargoes are invariably sold a day or two after their arrival in the Dee. These lead markets are considered the largest in Great Britain, and it is acknowledged that the Flintshire smelters manufacture more than one-fourth of the lead made in the United Kingdom, the average of which is 50,000 tons annually. These works give employment to several thousands of industrious labourers and artizans. The following are the principal firms that carry on these works: Sir E. S. Walker & Co., Thomas Mather and Co., Newton, Lyon & Co., J. P. Eyton, Esq. &c.

3 Handicaps of the West-Country woollen industry, 1845

(Alfred Plummer, *The Witney blanket industry* (Routledge, 1934), p. 108. The railway reached Witney in 1861)

Quantity of Wool used in Witney would average 150 packs weekly, which would be about 16 Tons, equal to 800 Tons annually. Averaging this at £10 p. pk. would be £1,500 weekly, or £75,100 annually: when manufactured into goods would be about £150,000 annually. About three-fifths of this amount goes to London at an expense of 35*s* per ton, the rest to all parts of England and Wales. One-third of the wool consumed here comes from London. One-third West of England. One-third from Worcestershire and adjacent counties, Liverpool and Wales—more would be brought from Liverpool but for the expensive carriage which, costing 20*s* p. ton from Lpool to Birmingham, costs in addition 50*s* p. ton from Birmingham to Witney. The high price of coals here prevents all idea of steam being introduced and we are obliged to get water Power in the neighbourhood at great inconvenience, loss of time and labour, and no chance of increasing the trade of the district as the available water power is all engaged. About 100 tons of oil used yearly worth £4,000. The present price of coal to consumers is for—

Staffordshire	31*s* 6*d*	p.	ton
Moira	29*s*		ton
Limboing (?)	21*s*		ton

Tenby coal for Malting is costing about 8*s* p. ton in Wales—costs here 45*s* p. ton. We consider by this branch line into Worcester and Wolverhampton, all should (in addition to the saving by Railway Communication) have the advantage of Competition—the Welch Coals on the one hand, and those now used on the other, and would lead to an enlarged Consumption beyond the present demand, which is estimated for the town and neighbourhood at about 8 to 10,000 tons annually. We are also deprived the chance of getting Waste and Noils from the North which would be much used in the town but for the expense of carriage. All our machinery, cards, and other apparatus is made in the North of England and the carriage is very expensive, putting us to a difficulty in competing with the Yorkshire Manufacturers; in this respect the same objection applies to Oils and Dye Wood which we should be able to buy from Lpool but for carriage.

4 The cotton famine, 1862–4

(Reports of inspectors of factories, 31 October 1862, *B.P.P.*, 1863, XVII, pp. 18 and 23)

K. L.—Mills entirely closed. Gives his workpeople soup and bread four times, and bacon and bread twice a week.

M. N.—Mills entirely closed. Distributes nearly 100*l* each week to the hands employed at these mills, as their necessities require it.

O. P.— Working three days a week. Has been selling bacon at wholesale price to his hands, and is giving them soup every Saturday.

Q. R.— Working about two days a week. Are selling bacon at wholesale price to their hands, and giving it away to the most necessitous. They pay the school wage of the children employed by them, and have remitted the rents of their cottage property inhabited by their workpeople.

S. T.—Mills entirely closed. Are giving relief to their workpeople and families on the following scale:

	s	d
Heads of families	2	0
Single man or woman (or married without children)	2	6
Children under five years of age	1	0
Do. over five years of age	1	6

In addition to this they give soup and bread weekly to the most necessitous, and are selling potatoes, bacon, and meal at wholesale prices. They have opened sewing classes for the females and schools for the young persons and children of both sexes. They give all their overlookers four days a week wages.

U. V.— Mills entirely stopped.

The most perfect system of relief that I have seen at any mill in my district is that in practice at these mills, which is so excellent that it deserves especial notice. I have therefore drawn out for your information a brief account of the method they pursue, as witnessed by me at a visit I have just paid to the mills.

These extensive mills are now entirely stopped, but the proprietors are giving each week at the rate of a day and half wages to their hands, and in addition to this have opened provision stores within their mills, where their hands, if they please, can purchase provisions at wholesale prices. This advantage, as will be seen by the table given below, increasing the value of the day and half wages to that of about two and a half days.

Articles Sold	Prices Sold at	Usual Retail Price
	d	d
Bread	4 per 4 lb loaf	6
Bacon	4 per lb	$7\frac{1}{2}$
Tea	2 per oz	$2\frac{1}{2}$
Coffee	$1\frac{1}{2}$ per 2 oz	2
Meal	$7\frac{1}{2}$ per 5 lbs	9
Rice	$1\frac{1}{2}$ per lb	3
Soup	1 a quart	—

5 A glimpse of the Black Country, 1874

(Samuel Griffiths, *Guide to the iron trade of Great Britain* (London, 1873 [1874]; Newton Abbot: David & Charles, 1967), pp. 204–5)

Willenhall, the real seat of the lock, door-bolt, and latch manufacturers for the world, is a township in the parish of Wolverhampton, and is connected with it by two lines of railways, viz. the old Grand Junction line, and one recently opened, the Wolverhampton and Wallsall line. Willenhall being just three miles from either town, now contains about 20,000 inhabitants, a complete hive of industry. We believe there are some five to six hundred separate manufactories (of course some only small concerns) of rim, mortice, drawback, dead, cupboard, drawer, box, and pad locks; all kinds of latches and door bolts; currycombs, gridirons, box-iron stands, and skewers; horse scrapers and singers; carpet-bag frames and locks; box corners and clips; keys of all descriptions; and stampers of an endless variety of articles for the gun, steel, toy, and other trades carried on in neighbouring towns. There are also Ironfounders, brassfounders, and wrought Iron works, blast furnaces and collieries in abundance. To these mainly must be attributed the rapid growth of this industrious town. The writer can well remember when three to four thousand was the extent of its population. In those happy old days of the past there was one church, with a blaspheming drunken parson, who spent six times more of his time in the public-house than in the church, the only one the place possessed. In those times there was no Methodist or Dissenting resident minister; and what is more, no magistrate, no lawyer, no police, and not an inhabitant (except the parson) but what was engaged in some kind of business. At the present time the township of Willenhall contains four churches, five Wesleyan chapels, four Baptist chapels, five Methodist chapels of various denominations, and one Roman Catholic chapel, which represents one place of worship for every thousand of the population, a fact few towns can boast of; with good school accommodation, British, National, and Wesleyan, and a literary institute of no mean pretensions, having its reading, recreation, and class rooms, a good lecture hall, and a well-furnished library.

On visiting some of the manufactories of Willenhall, we found the Albion works one of the most prominent, employing some hundreds of work-people. The business carried on here was established in the last century by the father of one of the present proprietors; and one of the principal branches of the trade, that of door bolts, was extensively carried on by the grandfather of the other more than eighty years ago. At these works we find manufactured rim, dead, and mortice locks; spring, rim, night, Norfolk, Suffolk, and Lancashire thumb-latches, in various ornamental designs; door-bolts in prodigious quantities.

6 Survey of the woollen and worsted industries, 1886

(Royal Commission on the depression of trade and industry, Second Report, *B.P.P.*, 1886, XXI, qq. 6699–719; evidence of Sir Jacob Behrens)

... First I would ask what are the branches of industry in which you are specially interested, and what is the area of which you are going to speak in the evidence which we hope you will give?—Perhaps you will allow me to make a difference between the industry and the trade, for Bradford is the centre of a large industry and also of a much larger trade. The district in which the worsted industry is carried on, and of which Bradford is the centre, extends on the one side to Halifax, Sowerby Bridge, Elland, Stainland, and Luddendenfoot; and on the other side to Shipley, Saltaire, Bingley, Keighley, Skipton, and up to Colne in Lancashire. But Bradford besides being the principal seat of the worsted *industry* of the West Riding, contains probably the largest silk mill in the world, that of Messrs Lister and Co., the well known iron works of Low Moor and Bowling in the immediate neighbourhood, while the combing machines, spinning frames, and power looms made in the district are used at home and abroad, wherever worsted yarns and tissues are manufactured. The *trade* of Bradford, and especially the export trade, deals with every article of wool manufactures, and I believe I am within the mark in assuming that fully two-thirds of the woollen exports of the West Riding go through the English and foreign houses established in the town. I am not, however, prepared to deal with anything but the wool industry in general, and shall, with your permission, make the attempt to give an estimate of the value of the worsted, as distinct from the woollen branch of it, though I am well aware of the insufficiency of the data upon which alone I can found my calculations.

What is the area of Bradford?—The area of the borough of Bradford itself is 10,776 acres, and the number of inhabitants in the borough is 214,431; and the annual value of the property assessed to the poor rate is 919,231*l*.

In what proportion do the trades and industries of which you speak find their market at home or in foreign countries, and with regard to the latter in which countries chiefly?—I estimate the value of the whole wool industry in 1884 at 60,400,000*l*; and the value of the worsted branch I estimate at 33,000,000*l*. Of the whole of the woollens and worsted goods and yarns I estimate that we export 21,400,000*l* to foreign countries, or 35·4 per cent; to colonial possessions 6,000,000*l*, or 10 per cent; and therefore we leave for home consumption 33,000,000*l* sterling, or 54·6 per cent. That is the value of the whole woollen and worsted industries arrived at by calculation based on the consumption of material in 1884.

Then distinctly the largest proportion of your production is for the home market?—Rather better than half. Then of the 33,000,000*l*, which I estimate the worsted industry to come to, we export somewhere about the same proportion, namely, 40·5 per cent to foreign countries; 6·08 per cent to colonies and possessions; and 53.43 per cent is left for home consumption. Therefore the home consumption is about the same in both cases, or rather better than one-half. Of course this is a bold guess—it cannot be anything else—but I believe I have taken it upon the best available data.

Taking the last 20 years or so, should you think that the proportions of the trade to foreign countries to the home market have altered at all?—That depends very much upon whether you take the volume or the value.

I mean as regards the proportions. I do not mean the total amounts. Do you suppose that 53 per cent would probably be about the ratio of the home sales 20 years ago as it is now, or that the home trade was larger and the foreign trade less?—I am not exactly prepared to answer that question, because it would have involved an enormous amount of work; but to speak just at random, I believe that we have exported more from our district in 1884 than in any previous year, and perhaps quite as large a proportion at any former time.

Have you any statistics as to the progress of the volume and the value of trade in the period from 1865 to 1870?—Yes; but the only reliable data that I have is the consumption of the raw material annually worked up during certain quinquennial periods. What we had retained of foreign wool, alpaca, mohair, and weaving yarn in the period from—   *lb*

1860 to 1864 was 126,234,031
1865 to 1869 was 158,810,934
1870 to 1874 was 202,878,901
1875 to 1879 was 220,595,954
1880 to 1884 was 247,791,814
and in 1884 was 279,335,151

Of English wool the consumption has decreased; less was produced, and more exported in these quinquennial periods. The average annual weight retained for home consumption in—

lb
1860 to 1864 was 135,000,000
1865 to 1869 was 140,636,174
1870 to 1874 was 147,319,598
1875 to 1879 was 144,499,885
1880 to 1884 was 118,879,866
and in 1884 was 114,324,420

So that the total consumption of material has only increased by 50 per cent since the period of 1860 to 1864, whilst the foreign material alone has increased by more than 120 per cent.

(*Professor Bonamy Price*) The raw material that you spoke of first, I presume, is of domestic growth?—Yes; English and Irish wool.

(*Chairman*) These are the figures of the volume of trade; how is it as to its value?—With regard to the value of the trade, I can only come to that by taking the value of the exports each year and estimating them either at the low value of the year 1884, or by calculating the declared value of each of the said quinquennial periods on the basis of the value of our exported yarns which in—

d
1860 to 1864 was 35·23 per lb
1865 to 1869 was 37·84 per lb
1870 to 1874 was 36·54 per lb
1875 to 1879 was 32·95 per lb
1880 to 1884 was 25·90 per lb
and in 1884 it was just 24*d* per lb

The total amount of wool yarns and tissues exported in 1884 was 25,292,000*l*, when yarns were worth 24*d* per lb, which at the prices ruling in these quinquennial periods would have given:

£		£
37,165,000 in 1860 to 1864		18,443,000
39,898,000 in 1865 to 1869		26,481,000
38,507,000 in 1870 to 1874	instead of the	31,950,000
34,723,000 in 1875 to 1879	actually de-	22,915,000
27,294,000 in 1880 to 1884	clared value of	22,961,000

Thus the volume of our exports in 1884 must have been 100 per cent greater than it was in 1860–64; 50 per cent more than in 1865–69; 20 per cent more than in 1870–74; 50 per cent greater than in 1875–79; and even 23 per cent greater than in 1880–84. I must be allowed to observe that the comparison with the declared values of 1870 to 1874 is vitiated by the glaring inaccuracies in the returns for the early part of that period, which render them 'not quite so suitable for comparison with later years', as Mr Giffen expresses it in *official* language in his evidence (page 19).

That is only another way of calculating the volume?—Yes, and I do not see how I can get the data in any other way.

Have you any means of giving or calculating what was the value? —In the preceding answers I have attempted to fix the volume of the wool industry by two methods, viz., the ascertained weight of the material worked up, and the value of our exports calculated upon the basis of the prices ruling in each quinquennial period. It was satisfactory to me to find that both methods gave nearly identical results. To compare the value of each period's exports, it may perhaps make my meaning clearer if I reverse the method, and calculate the exports on the prices of 1881, when the exports of 1860–64 would have been only 12,592,000*l* in lieu of 18,443,000*l* as they were at the then prices.

(*Mr Ecroyd*) The volume is best ascertained by what you can get absolute information upon, namely, the home and foreign grown wool retained for consumption each year?—Yes, I could not get anything better, and I shall be very glad to show Mr Ecroyd the way in which I estimate the volume and the value of it; there is nobody better able to criticise and judge of it. But I have brought it all together in a tabular form which I shall be very happy to hand in.

(*Chairman*) Can you tell us anything about the fluctuations in the profit of the trades or industries? —That is a question which I thought it was impossible for me to answer, but having been furnished with returns under Schedule D for the last 25 years for the district of Bradford from Somerset House, the Commission may not object to let me put them in.

Have you not any of the figures by you?—Yes, and I find that they average per year in—

	£
1860 to 1864	1,414,049
1865 to 1869	2,028,602
1870 to 1874	2,943,783
1875 to 1879	3,001,446
1880 to 1884	2,405,346
and in 1885	2,502,663

In 1870–77 all incomes up to 150*l* were made free from income tax, instead of only up to 100*l* as formerly. I have no means of ascertaining the influence of this change upon the amount assessed.

Can you give us any information as to the amount of capital invested or the quantity of labour employed in those periods?—No. I was in hopes of giving you an estimate of the value of the capital employed in trade, and I only got this morning a letter from a gentleman whom probably Mr Ecroyd knows, Mr Prince Smith, of whom I asked the question, what would be the cost of a mill, complete in all its appliances, erected here and in France? and he says he has not yet been able to make it out, but he has promised it, and perhaps, with your permission, when I get it I will hand it in later on. I have got another estimate for a similar mill, but I think it will be better to get others first, so as to compare two or three. I do not know what the plant, what the building, and what the whole cost would be. I am not very familiar with that; but by adding the same amount for the floating capital to work it, I think that that is as near as you can get it, and I should hope to be able to give you an estimate of that kind.

(*Mr Ecroyd*) I should think that an amount of floating capital equal to that sunk in buildings and machinery would be very inadequate for the satisfactory conduct of the business, would it not?—Not of the machinery only, but of the building and the land, and everything connected with it. For instance, for a mill such as the one that I have got an estimate for making at the cost of from 70,000*l* to 71,000*l*, I do not think with the present appliances it would require more than

70,000*l* of floating capital, and have been so informed by large mill owners.

(*Mr A. O'Connor*) Could you say how much out of that 71,000*l* would represent the charge on account of land?—I have a letter here from an architect who has been employed in the building of a great many mills, and he gives it me without the details, which I shall get afterwards to compare with those that I get from other gentlemen.

What would you think would be about the general average, taking the whole of Yorkshire, between the cost of a mill and the total cost payable on account of land?—I have had a letter on that point to-day, which I have not had time to read. This is a letter from Mr Prince Smith:—'We duly received your esteemed favour of the 17th, and have been very carefully looking into the question of cost of mills, but must confess the work of collecting the necessary particulars increases in magnitude as we pursue our investigations, so that it is quite impossible for us to furnish you with full details by the time fixed in your last letter. Fortunately, however, we are in a position to state from actual personal experience that the Roubaix worsted spinner has to pay 20% more for his new machinery than the Bradford spinner pays'—(I wanted to ask him what is the cost of putting down the same machinery here as at Roubaix, as I thought it might be interesting to the Commission)—'in both instances the machines being delivered and erected by us in their mills ready for working. A short time ago we sent out to our agent at Roubaix an experimental plant of our machinery, and were most particularly careful to note every item of cost in connexion with same; hence we have the soundest possible basis to work upon in giving you this statement. We might add for your guidance that the estimate affording this per-centage information includes cost of machines, packing cases, carriage by rail from Keighley to Goole, insurance, shipping charges or freight from Goole to Calais, customs duties, carriage from Calais to Roubaix by rail, cartage to mill, statistiques, also our erectors' time and expenses in setting the machinery up ready for work same as we do in Bradford. Since we instituted our inquiries upon the subject we have had the enclosed book "Chamber of Commerce Report" for year 1879 brought under our notice, and we beg to refer you to page 50 for Messrs Godwin and Illingworth's statement upon the same subject in their special report upon English *v.* French mills. Of course you will notice their statement is 25% to $33\frac{1}{3}$% as against our 20%'. I have a letter here from another gentleman, an architect, who says: 'As nearly as I can tell, in the absence of definite plans and special requirements, an establishment of 500 broad and 500 narrow looms for weaving goods (textile fabrics) which are now made in Bradford, including the necessary combing, carding, preparing, and spinning, and also including the necessary buildings, engines, boilers, shafting and accessories, and the land, would not cost less at present prices than 70,000*l* or 71,000*l*'. The representative of an eminent Lancashire firm writes as follows: 'In Lancashire, at the present time, a good fire-proof mill, say of 70,000 to 80,000 mule spindles (for spinning only), costs 21*s* 6*d* to 22*s* 6*d* per spindle. Of course, this includes everything. Twenty years ago the price was the same as to-day. Ten years ago the price was 28*s* to 30*s*. A mill in America costs 60 to 80 per cent more than in England. I cannot give you the price of a mill in France or Germany, but should say it would be at least 20 to 30 per cent more than in England'.

You have not got the proportion of the total cost?—No, but I shall get the details later on.

(*Chairman*) During what periods in the last 20 years should you say that trade had been at its normal level or above or below it?—That has been a question which has bothered the Chambers of Commerce more than anything else, as to what is the normal level. Where we have been progressing every five years we could not find the normal level, and therefore, at all events, no normal depression can be found in the volume of the raw material used.

Then we may consider that the trade and industry of the district that you represent, or at all events the woollen and worsted trades, are not to be described as depressed?—They have not been so progressive, and were perhaps at a stand-still, from 1874 to 1879, but they never declined with regard to volume.

7 Heavy industry in the trough of depression, 1886

(*The Times*, 10 January 1887)

Messrs Matheson and Grant, in their Engineering Trades' Report for the second half of 1886 make the following observations:

The year just closed has been one of the worst ever experienced in the engineering trades, but in several branches a slight improvement has set in which bids fair to continue, and the prospects for the coming spring are brighter than they have been for the last two years. Since the date of our July report there have been large investments of capital in undertakings which give employment to engineers, and an increasing expenditure among manufacturers may be looked for.

The revival of railway enterprise in the United States has given great impetus to the iron and steel trades of that country, and though free exchange with Great Britain is restricted by their fiscal system, the intimate connexion between the two countries has always led to a corresponding improvement here.

Coal has been cheaper than ever during the last year and in South Wales, as in the North of England the reduced output of the collieries has told severely on the owners whose charges for royalties and maintenance cannot be reduced in proportion....

Iron. The continued fall in the prices of pig iron during the first half of 1886 was arrested by the natural remedy of a reduction in the output, and prices advanced in consequence. The prices of rolled iron have fluctuated during the last six months, and are now slightly higher than during the summer. They would be still higher but for the competition of steel, not only in shipbuilding, but in boilers, bridges and other structures. At many of the leading rolling mills Siemens and other steel-making plant has been established to meet the altered demand, and at some of the works favourably situated for suitable ore and fuel the puddling furnaces and other appliances for making wrought iron are likely to be abandoned altogether.

Steel, which in the spring and summer of 1886 fell even more rapidly than iron, has during the last few months recovered from 5s to 10s per ton. Owing to the collapse of the English and Continental rail-makers combination, prices of heavy steel rails fell as low as £3 12s 6d per ton, but the price is now from £4 to £4 5s. Although this recovery is assisted by the considerable manufacture of ship and bridge steel the immediate cause has been the revival of the American demand.

8 Coal's fluctuating fortunes, 1872–1901

(Royal Commission on coal supplies, First Report, *B.P.P.*, 1903, XVI, qq. 2890–4, 2920–8, 2939–41. Evidence of Mr Alfred Hewlett)

I am the managing director of the Wigan Coal and Iron Company, Limited, whose collieries are in and around Wigan, in the county of Lancaster, and who are developing mineral ground in Nottinghamshire. Their output is about two and a quarter million tons per annum. They also carry on largely the making of iron, steel, and coke. As to minimum thickness of workable seams:—I have had considerable experience in working a thin seam, and I have had taken out, for the last 30 years, the cost of working four separate seams by the company.

First, the Cannel Seam, which is very thin; secondly, the Yard coal, which is rather thin; thirdly, the Arley Mine, which is of medium thickness; and fourthly, the West Leigh Seven Feet, which is a thick seam. I can give you the thickness of those for comparison. The Cannel is about 18in, the Yard 2ft 8in, the Arley Mine 3ft 10 in, and the West Leigh Seven Feet 6ft 8in. I have taken those four seams to try to show, so far as possible, how the cost varies between very thin and thick, and how it has been affected during those 30 years.

And what does the cost include?—Everything except exhaustion.

(*Mr Young*) Any interest on capital?—No, nothing in respect of capital. In the year 1872, taking the seams in the same rotation, the Cannel cost 18*s* 4·5*d* per ton. It ought to be remarked there that the year 1872 was an abnormal year; it was the year that used to be called the 'coal famine' year, and, therefore, everything was somewhat abnormal at that time. But I included that year in order to make up 30 years. As a matter of fact, we ended with a year which was abnormal also. In 1878, which I have called a more normal year, the same seam was 9*s* 4·3*d*. In the year 1889 it was 10*s* 10·8*d*, and last year (1901), 21*s* 4·7*d*. The cost of the Yard coal in the year 1872 was 6*s* 0·9*d*; in 1878 it was 5*s* 1·7*d*; in 1889, it was 5*s* 1·2*d*, and last year 7*s* 11·5*d*. The cost of the Arley Mine for the year 1872 was 4*s* 8·9*d*; in the year 1878, 5*s* 0·1*d*; in the year 1889, 5*s* 3·2*d*; and last year 7*s* 4·2*d*. The Seven Feet was not at work in the first two periods which are set down; it had commenced working in 1889, and the cost was 5*s* 6·8*d*, while last year it was 7*s* 7·2*d*. So that if we make a comparison with 1872, the percentages are: in Cannel, 16½ per cent higher; in Yard coal, 31 per cent; in Arley, 55 per cent; and in the Seven Feet 44 per cent; but bear in mind that the latter was only worked a portion of the time.

But these figures of cost apply to the whole coal, large and small? —Everything brought to the surface.

In other words, these figures are the cost per ton of all the coal of each seam which is brought to the surface?—That is so; and you may take it that in Lancashire it is practically all brought to the surface. No small coal is left in Lancashire. As we have to pay for it at per ton per acre, we take care to bring up every bit we possibly can; so that it is the absolute produce of the mine.

I want to ask a question with regard to the variation in the cost. The figures you have given us dealing with the period of 30 years show that the first year was abnormal, and the last year abnormal. Take the year 1878 and the year 1889, which you give us, the cost is rather less in 1878 than it was in 1889?—That is so.

Would you say that either of these years, or the average of those two years, would be what you would call a normal year?—Not quite so. I should like to start with last year, 1901. It was an abnormal year, but it was abnormal more on one side than on the other. In the year 1872 both the cost and the selling price were very abnormal. In the year 1901 it was more that the selling price was abnormal than that the cost was abnormal. It was higher, as you see by the figures, a great deal, but it had kept gradually getting up; it increased to nothing like as much as it did in 1872; so that even the two abnormal years are not quite on all fours with one another.

(*Mr Young*) 1878, I think, was a very bad year?—Yes.

(*Chairman*) That is the lowest point you have given?—Yes.

(*Mr Young*) In 1889 prices were high?—Yes.

(*Chairman*) Can you tell us what the variation is due to? Is it a question of wages plus materials, or changes in rent, or what is the main factor?—Practically it is wages. I may say at once that rents do not come in at all; nearly all our leases have been going on for a good long time, and I do not think the tendency of rents has been upwards in our district; there is something in material, not a great deal; there is a great deal in rates and taxes. Of course, it is known very well to the Commission that the important factor in all coal-mining is the question of wages. I suppose at the present moment probably 70 per cent of the total cost is wages. You quite understand I am not making a tirade against wages, but that is really the important factor.

(*Mr Sopwith*) Would you say that the physical conditions of the mine have not contributed to the variation in the cost?—Yes. The physical conditions of the mine have remained very much the same. Possibly the gradually increasing depth has caused a little greater outlay, but that does not affect the point really.

(*Mr Young*) Supposing you took the first 10 years of the 30 and the last 10 years of the 30, including, as you yourself have pointed out, the abnormal year, which was the first of the period, and the abnormal year, which is the last of the period, would there be much variation, and, if so, could you say how much variation between the first 10 years and the last 10 years?—I think I can give you year by year, but I have not summed them up so as to average them in that way; I have only averaged them in the way I have already placed before you. They are as follows:

14*

Total Cost of Production exclusive of Depreciation

	Cannel		Yard		Arley		Seven feet	
	s	d	s	d	s	d	s	d
1872	18	4·5	6	0·9	4	8·9	—	
1873	—		8	0·1	5	11·5	—	
1874	22	8·3	8	1·5	6	0·5	—	
1875	18	6·0	7	1·0	6	3·0	—	
1876	14	10·9	5	9·8	5	3·4	—	
1877	10	0·9	5	9·1	5	5·7	—	
1878	9	4·3	5	1·7	5	0·1	—	
1879	—		4	10·0	4	8·1	5	3·3
1880	—		4	8·4	4	10·6	4	10·9
1881	—		4	10·4	5	1·0	4	6·5
1882	—		5	0·5	4	9·8	4	4·7
1883	—		5	2·1	4	10·8	4	9·6
1884	—		5	0·7	4	7·6	5	1·1
1885	—		5	2·5	4	6·0	5	1·9
1886	—		4	9·9	4	9·0	5	4·3
1887	—		4	9·2	4	9·6	5	2·3
1888	—		4	9·7	4	9·9	5	2·4
1889	10	10·8	5	1·2	5	3·2	5	6·8
1890	15	2·1	5	11·3	6	0·9	6	4·7
1891	11	5·0	6	1·6	5	10·4	6	8·8
1892	11	9·3	6	2·4	6	0·3	6	8·5
1893	12	7·3	6	8·4	6	4·4	6	6·9
1894	11	9·4	6	2·5	5	9·4	6	3·2
1895	9	11·2	5	8·5	5	5·5	5	11·7
1896	10	0·7	5	4·9	5	4·4	5	7·4
1897	10	10·0	5	5·6	5	4·3	6	2·4
1898	11	6·6	5	6·1	5	3·1	5	11·2
1899	11	8·7	5	9·9	5	5·9	6	2·3
1900	14	11·1	7	0·5	6	5·2	6	9·6
1901	21	4·7	7	11·5	7	4·2	7	7·2

(*Professor Hull*) If it had not been for the special valuable character of the Cannel seam, do you think it would have been worked at all, considering its thickness?—Oh no, it certainly would not. It is a specific article, and I am sorry to say it is of diminishing value.

Are coal cutters employed to any large extent in these seams?—We have very few, I am sorry to say. We are trying to increase the number greatly. I suppose we shall. There are a great many difficulties in introducing them, and in some parts of our collieries we could not introduce them at all, because the ground is so faulty. In one set of our collieries we have so many faults that it costs £1,000 a week for dead work year in and year out, and has done for the last 30 years, I should think. Coal cutters there would not be any good at all. Where you have a good open road without any faults, then, coal cutters should be used as much as possible.

(*Mr Bell*) With reference to your answer that perhaps in the future the quantity of coal got by coal-cutting machines would increase greatly or otherwise, is that opinion based upon any knowledge of your own, or upon any statistics which show that the use of coal-cutting machines is gradually and continuously increasing?—It is gathered from both sources, both my own knowledge and also from the statistics which are published as regards mines, giving the number, and the numbers are increasing. So far as my company is concerned, I certainly intend to put in more during the next two or three years. I have pointed out one of the difficulties, and no doubt we shall have plenty, but there is no doubt whatever in my own mind that there will be a great many more used within the next two or three years.

(*Sir George Armytage*) Can you tell us what the comparative cost of working by means of the machine and in the ordinary way is?—I should not like to say, because personally I have not had enough experience. I believe it is a little less, but is not going to be as much less as many people thought it would be. It is going to be less, and I think the proportion of round to small will be greater by using the coal cutter than by hand work. That is almost the most important factor, but I could not say more.

9 State of the engineering industry, 1907

(*Report of the tariff commission*, Vol. 4, *The engineering industries* (P. S. King, 1909), section II, paragraphs 15–32. The tariff commission was an unofficial body set up to investigate the effect of foreign tariffs on British industry. Although biassed in favour of tariff reform the commission took evidence from protectionists and free traders alike and published its findings (and the evidence on which they were based) in the manner of royal commissions. The passages printed below are from the commission's analysis and summary of evidence and statistics 'without comment of any kind by the Commission.')

The Engineering industry is more complicated than any other industry which has formed the subject of inquiry by the Commission. It is in fact a vast group of industries which continually extends as the different trades become revolutionised by the application of machinery and new motive forces to processes hitherto carried on by older methods. For the detailed examination of the movements which are taking place in the industry the evidence of witnesses and the replies to Forms of Inquiry are especially important because of the inadequacy of the classification adopted in the official returns. Whole classes of engineering products which should be dealt with separately are grouped together and the extent or direction of movements affecting important branches of the engineering trade thus escape statistical observation. It is only a few years since the exports of ships were included at all in the Board of Trade returns, and even now it is not possible to compare effectively the statements of witnesses with official figures.

As will be seen from the detailed figures which follow, out of £5$\frac{1}{4}$ millions of engineering imports in 1907 not less than £3 millions worth remain unclassified, while of the £31$\frac{3}{4}$ millions of exports there are not details of £11 millions worth. The most important omissions relate to the various classes of industrial machinery such as machine tools, machinery for milling printing, papermaking, woodworking, sugar, leather, brickmaking,' and cement. Hydraulic and heating, lighting and ventilating machinery is also unclassified and other important engineering products are omitted. Thus the items of one-half of the electrical importations are unenumerated. The statistics of 1908 introduce material changes in the classification and these will permit of a more detailed examination of that one year's trade, but in many of the new groups there are no earlier figures with which to make comparison, and even in their revised form the official returns of 1908 show practically the same amount of unclassified engineering products. The only effect of the revision is to improve the classification of those goods which were previously classified.

The general tenour of the evidence is that, despite labour and other difficulties, the engineering industry on the whole has progressed during recent years, but not so rapidly as the demand has increased for engineering products, or as rapidly as the engineering industries of other countries, particularly the United States and Germany. The increase which has taken place appears from the evidence to have been stimulated by the growth of the Navy and armaments generally; and especially by the widespread protection of the home industry secured by the exclusion of foreign competition in Admiralty and War Office contracts; the development of traffic facilities by municipalities and other bodies; the growing popularity of electric lighting and the increasing application of electrical and other power; the development of new branches, such for example as the motor car industry; changes in methods in industries in which machinery is applied, such for example as the milling industry; and the growth of oversea commerce which has had the effect of increasing and at the same time making competition more keen in the shipbuilding industry. There is however overwhelming evidence that in practically all branches foreign competition in the home market has increased in recent years, and there can be no doubt that as compared with a generation ago the control of the home market by the British manufacturer is relatively less extensive and less secure. The evidence shows that efforts are made from time to time to mitigate what are regarded as unfair conditions of competition by combinations both international and domestic among manufacturers of particular engineering products but it is not clear how far this movement has gone or how far it is of a permanent character. On this point a manufacturer says:—'During the boom (1904–1907) the whole world has been busy and has left us pretty much to our own market. However we find a considerable amount dumped, and the prices quoted here are unremunerative as the German dumping price fixes ours. One very bad effect of the previous low prices arising from dumping here led the iron and steel manufacturers to form syndicates to keep up to a remunerative level the prices of their product to English buyers. There is hardly a branch of these trades which is not combined for this object. Of course this will ultimately result in foreigners dumping more than ever as they need not dump so low and then may be able to reduce their price to their home consumers which will give them a better chance of competing with us in the English grades of manufactures'.

Table 1—*United Kingdom. Imports of Machinery and Millwork from Principal Countries (in thousand £)*

From	1897	1902	1907	Increase, 1897—1907	
				Amount	Per cent
United States	1,620	2,984	3,117	1,497	92
Germany	303	763	906	603	199
Belgium	88	313	378	290	330
Holland	85	252	333	248	290
France	119	223	190	71	60
Canada	30	62	105	75	250
Total from all countries	2,371	4,761	5,312	2,941	125

The imports of machinery products in 1907 were $2\frac{1}{4}$ times those of 1897, which was the first year in which machinery and mill work was separated from iron and steel manufactures in the import returns.

It will be seen that the increase has been large and practically continuous. Correcting the import figures as given above so as to indicate the countries from which the goods were consigned instead of shipped, it appears that the German importation into the United Kingdom is considerably larger and the importation from Holland and Belgium considerably smaller than is shown in the foregoing table. In 1907, for example, the importation from Germany was £906,000 worth whereas the actual German consignments were of the value of £1,355,000. On the other hand the imports from Holland were £265,000 and from Belgium £193,000 larger than the consignments.

The United States holds 59 per cent of the import trade of the United Kingdom; eleven years ago it held 68 per cent. On the other hand Germany, Holland and Belgium together accounted for 20 per cent in 1897 as compared with 30 per cent in 1907. Canada is the only British Dominion substantially represented in this trade; the importations from Canada are chiefly agricultural machinery.

There are insuperable difficulties in making any adequate analysis of the course of the import trade by groups. In the first place the classification in use up to the end of 1907 was not adopted until 1901; prior to that year Sewing Machines only were returned separately from Machinery. Hence no comparative figures can be carried further back than seven years. In the second place, as is pointed out above, no less than £3 millions still remain unclassified, out of a total of £5$\frac{1}{4}$ millions in 1907; in other words only 43 per cent of the imports are grouped in sufficient detail to permit of a survey of the progress of the leading branches.

Analysing the £2,258,000 worth of imports in 1907 which are classified (and including electrical machinery) we get the following table:

Table 2—United Kingdom. Imports of Machinery and Mill Work in Principal Groups (in thousand £)

	1897	1902	1907	Change, 1902–07	
				Amount	Per cent
Steam Engines—					
Locomotive	*	15·5	5·6	−9·9	−64
Agricultural	*	6·6	·8	−5·8	−88

Other descriptions ...	*	479	82	−397	−83
Not being Steam Engines—				
Agricultural	*	397	770	+373	+94
Sewing machines	291	378	486	+108	+29
Mining	*	27	81	+54	+200
Textile	*	119	230	+111	+93
Other descriptions				
(including electrical)	2,080	3,338	3,657	+319	+10
Total	2,371	4,761	5,312	+551	+12[1]

* Included in 'Other Descriptions'.   [1] Increase since 1897, £2,941,000 or 124 per cent.

It will be seen that the most important classified group is non-steam agricultural machinery which accounts for about one-third of the classified total of £2,258,000. This group has increased by 94 per cent since 1902. Sewing machines account for more than one-fifth of this total and textile machinery one-tenth. The former increased by 29 per cent and the latter by 93 per cent in the last five years.

The witnesses and firms whose evidence is contained in this volume give many illustrations of the effects of these large and growing importations of engineering products. ... The varied character of the importations is fully indicated. Over 500 items, many of them comprehensive groups in themselves, are contained in the list of competing imports. Many specific illustrations of dumping are quoted in later sections of this summary, and speaking generally it may be said on the authority of manufacturers that in most cases these imports have displaced British manufactures and that practically all of them could have been made with equal advantage in this country under more equitable administrative and fiscal conditions.

Seventy-seven engineering firms state that they do not suffer from foreign competition in the home market. These are chiefly shipbuilders and marine engineers, manufacturers of steam hydraulic machines, large steam engines, textile machines, etc. In most cases they explain that they do not manufacture the type of machines in which foreign competition exists, or that theirs is a local trade or that they are protected by Government prohibition of foreign materials or by patents.

Speaking generally it may be said that foreign competition in engineering products in the United Kingdom is increasing both in area and in severity. This competition is dependent upon the free market maintained in the United Kingdom and is often carried on by methods which British manufacturers consider most unfair. Thus a Continental company will establish an agency here under an English name, put up an English nameplate on a small office, and while paying little or no income tax or other contribution to British taxation, sell foreign goods as British at prices which undercut British manufacturers. 'Duties', said one witness, 'would unmask these houses that adopt British names. There is a certain sentiment which is not always shared by corporate bodies, I am sorry to say, that actuates our fellow countrymen towards home-made stuff and it must be in order to get round that that these firms pose under British names'. The Merchandise Marks Act is spoken of as being quite unsatisfactory. It is said to deceive the public rather than assist, in view of the absence of any provision that the place of origin shall be placed on the actual goods manufactured. . . .

The increase in the total exports of machinery and millwork during the last thirty years has been large and continuous, and in 1907 they were four times as large as in the five years ending 1875–9. It will be seen that the export figures cover a much longer period than those of the imports, and the classification has been more detailed, though constant alterations and sub-divisions make it impossible to compare for the whole period of thirty years the growths in the groups at present differentiated.

Taking fifteen years only, the growth of exports to principal countries in five-year periods and in groups (in thousand £) has been as follows:

Table 3—United Kingdom. Exports of Machinery and Mill Work to Principal Countries (in thousand £)

	1893–97	1898–02	1903–07	Increase, 1893/7–1903/7	%
Russia	1,903	2,592	1,924	21	1
Germany	1,651	1,846	2,016	365	22
Belgium	720	816	808	88	12
Italy	470	622	1,225	755	161
Argentina	416	446	1,727	1,311	315
Japan	716	572	1,020	304	42
India	2,063	2,435	3,896	1,833	89
Australasia	719	1,435	1,452	733	102
France	1,205	1,473	1,614	409	34

India has always been the chief export market for British machinery and millwork. One sixth of the total exports (£5,364,000 in 1907) now goes there, or more than double the export to any other country.

The most important items are textile machinery and locomotives. The other leading markets on the basis of the 1907 figures are Russia (£2,230,000), Germany (£2,366,000), France (£2,065,000), Argentina (£2,458,000), Japan (£1,828,000), Australia (£1,358,000), Italy (£1,795,000), Belgium (£1,156,000).

One-third of the engineering exports which were of the total value of £31¼ millions in 1907 remain unclassified in the official returns, that is to say allowing for about £700,000 worth of electric machinery which is now separately returned, there remain about £8 millions worth of annual exports during 1903–7 of which no details are available. The exports in all defined groups have increased and the increase is specially marked in locomotives and sewing machines. In textile machinery which is the largest of the groups the increase is very small and is probably entirely accounted for by increased prices.

Table 4—United Kingdom. Exports of Machinery and Mill Work in Principal Groups (in thousand £)

	1893–7	1898–02	1903–7	Increase (first to last period) Amount	Per cent
Steam Engines:					
Agricultural	705	692	1,013	308	44
Locomotive	895	1,732	2,600	1,705	191
Others	1,487	1,704	2,554	1,067	72
Other Sorts:					
Sewing Machines ...	886	1,442	1,936	1,050	119
Agricultural Machinery	815	843	1,082	267	33
Mining Machinery ..	686	613	819	133	19
Textile Machinery ...	5,867	5,777	6,017	150	3
Others (including Electrical Machinery)	3,967	6,045	8,558	4,591	116
Total Machinery (including Sewing Machines)	15,309	18,846	24,580	9,271	61

In regard to what are now foreign protected markets, the evidence both from witnesses and firms is practically unanimous that the development of native industries in these countries and the tariffs which have been imposed in the last thirty years have greatly restricted the British export trade in many classes of engineering goods. The progressive development of the German tariff for example has affected important branches of the British industry, and the continuous effects of the last revision of the tariff, with the greater specialisation of duties and the increase of the rates of duty, is regarded with apprehension by some witnesses.

This movement extends to practically all European countries and to the United States. It is recognised that the engineering development of foreign countries was inevitable and no complaint is made by manufacturers of the legitimate ambition of these countries to establish great engineering industries. This foreign development has been aided even if it has not been mainly created by tariffs, and it is believed that no great expansion of British trade in engineering products in these foreign protected markets can be anticipated, though such is the diversity of the engineering industries that by negotiation a useful exchange of certain products may be arranged and a substantial trade guaranteed for a long period of time.

The injury caused to various branches by foreign tariffs is explained in the evidence in various ways. For example some manufacturers state that the market of the country imposing the tariff is closed to them except in so far as the trade consists of specialities or of engineering products which that country does not as yet make. The limits of trade upon this basis are becoming every year more restricted. Even textile machinery, in which in certain lines the United Kingdom has for years had an undoubted pre-eminence, is now in most countries subjected to a heavy tariff. Moreover, foreign tariffs become more burdensome at every revision, not only because the actual rate of duty tends to increase, but because the classification is worked out with greater precision, with a view to the development of national industries. In all the group of central European countries, bound together by treaties, the engineering classification has been completely revolutionised, while the duties have been considerably raised, and although the United Kingdom enjoys most-favoured-nation treatment, that treatment affords no real protection to British interests. Mutual concessions made by foreign protected countries in negotiation with one another are arranged upon the basis of the mutual interests of the contracting countries without the smallest regard to the interests of the United Kingdom which are entirely unrepresented in any of these tariff negotiations. By means of minute classification the advantages which should be secured by the most-favoured-nation clause are made illusory, and British products remain to all intents and purposes subject to the general tariff of the foreign country and not to the lower tariff based upon conventions.

10 Motoring—a new industry, 1914

(*The Times*, 3 February 1914)

The Motoring Industry
An enormous annual expenditure
Growth of employment

It has been estimated that on January 1 over 440,000 motors of all kinds were in use in the British Isles. Of these over 254,000 are touring, public service and commercial motor-cars, and the rest motor-cycles. Nearly 1,700 additional vehicles are being added to this substantial total every week. It is further calculated that these motor vehicles cover in the aggregate over 3,100,000,000 miles of roads in a year.

The Home Trade

As may be surmised from these figures, motoring has been the means of building up a vast industry. The Board of Trade returns show how erroneous is the popular idea that foreign cars are in the majority, and that our foreign trade is small in comparison with that of other nations. Although we import 4,000 more cars than we send abroad, we export more than 16,500 motor cycles, against less than 1,400 imported; and we are over half a million sterling to the good on gross turnover of the whole trade. This speaks well for the soundness of the home industry, especially when it is remembered that it is but a decade since our export trade was insignificant and that the bulk of the cars in use at home were foreign. In point of fact, last year, in spite of the very large number of cheap cars which are now being imported from America, 73·8 per cent of the cars and 96·8 per cent of the motor-cycles bought by British motorists were of British manufacture, and if a comparison were taken on values the percentage would be higher still. It can be safely computed that the cars and cycles at present running in Britain and Ireland aggregate a present value of £55,000,000 while their first cost, the money spent on motors during, say, the last ten years cannot have been far short of £120,000,000. The purchase price of motor-cycles averages about £50 and the price of cars ranges from as low as £75 for a cycle-car to £1,200 or £1,500 for a six-cylinder limousine. Close on £20 millions sterling—to be exact £19,912,428—was the sum spent on the purchase of new motor-cars by British motorists and cyclists last year, and to this has to be added the value of the 'spares' and accessories—the supply of which forms a very substantial branch of the motor trade—amounting to £5,773,396.

Running Costs

But the annual expenditure on new purchases, large as it is, is completely dwarfed by the cost of running the motor-vehicles which are now upon the road. Motoring has built up an industry in the manufacture of tires which is almost as large as that of the car-building trade itself. The annual tire consumption approximates £14,521,000. Next to tires the cost of petrol and lubricating oil bulks heaviest in the motorist's expenditure. The fuel and oil bill works out at close on £8,500,000. Moreover the cost of repairs, renovations and periodical overhauls amounts to over six millions; while motoring brings into the Exchequer over one and a half million sterling in registration fees and licences, and the insurance companies benefit to nearly the same amount.

... The number of motor-car drivers is about 150,000 and their wages at the rate of 35s per week together with the wages of other 'hands' employed about a car amount to the sum of £13,562,000. These several items together reach to over £47,994,000 as the running expenses of the motorist and when that sum is added to the cost of new cars we have the enormous aggregate of close on £74,000,000. ...

Taking all of the heads of the expenditure of owners of motor-cars the benefit, directly and indirectly, to British labour aggregates £37,550,000. If the average of the wages and salaries of all engaged in the motor trade and its allied industries be taken as high as £100 per annum, this means that 375,500 people obtain their employment and that something like 1,000,000 of the population are supported by the industry of motoring. ...

It is certain that at the present rate of increase, the gross expenditure upon all branches of motoring will, before the present year is out, reach the enormous annual figure of £100,000,000.

11 Contraction of the staple trades of Victorian England, 1932

(G. C. Allen, *British industries and their organisation* (Longmans, 1932), pp. 27–8)

It is not surprising that observers have tended to base their generalizations concerning British industry on data drawn from these great [staple] trades. With the rise of these industries Great Britain had become rich and powerful. Problems of large-scale production and marketing, of industrial relations, of unemployment and of trade fluctuations, were all presented here in their most clear-cut form.

Fluctuations in the exports of iron and steel, coal, textiles and engineering products, could serve, it seemed, as a measure of British economic prosperity and of British efficiency in relation to that of other nations. Since these industries were engaged, for the most part, on the bulk production of homogeneous commodities, it was easy to obtain figures of output, and it was thus possible to judge, with some exactitude, of any change in fortune. The very concentration of the trades brought them to the notice of the public; while the existence of strong trade unions and employers' federations, often organized on a national scale, made conflicts in them as spectacular as they were, in many instances, protracted. For these reasons it became usual to regard the state of the staple industries as indicative of the economic condition of Great Britain as a whole. Consequently, the outlook has seemed increasingly disquieting as the depression, which has affected these trades throughout the post-war period, has failed to lift. To some observers it has seemed that when trades which for a century have been regarded as the foundation of the economic system decay, then a general economic decline is inevitable unless their resuscitation is achieved. As time has gone on, it has become evident that these staples are not likely to resume their pre-war rate of expansion, and, indeed, many of them must be content with a permanently smaller production than that of 1913. But it is not necessary to accept the conclusion that British industry as a whole must decay. It has been argued that this is a period of change in the direction of British economic development, and that the solution of Great Britain's troubles will come from a redistribution of her resources among new industries. This, indeed, has been occurring during the last ten years and, in the long run, it is likely to solve that part of the unemployment problem which has been created by the decline of the staples.

4

Transport and internal trade

The capital needed to stock a farm or set up in manufacturing or trade used to be a sum well within the reach of many men—and often still is. The same rule holds good for the carrier whether he owns a wagon and horses, a barge or a lorry. But the provision of the track—road, canal, railway or pipeline—has always demanded much greater resources, namely capital in abundance and extended legal powers. The provision of better transport facilities therefore has the character of a public improvement rather than of a private venture. Turnpike roads (trusts) raised capital by the issue of bonds and if the road ran at a profit the surplus went to reduce the county rate. Canals and railways were owned by shareholders but like turnpikes, depended on special legal powers for their existence (*1*). At first the new ventures attracted local capitalists; later 'blind' capital flowed freely into railway securities (*8*).

The big railway companies were much more than mere transport undertakings and touched economic life at many points (*12*). In the late nineteenth century their only rivals as carriers were the small ships engaged in the coasting trade: long-distance road traffic had vanished and the canals were little used (*17*). Suburban railways, buses and trams made possible one of the characteristic features of modern civilisation—living at a distance from one's work with its attendant advantages and disadvantages.

The use of city centres largely for business and not for residence had begun by the middle of the nineteenth century and by its close had produced the City of London that we know today—crowded by day, deserted by night (*9, 16*).

The railways ceased to run at a profit during the inter-war years and were saved from speedy ruin by nationalisation after the Second World War. Even nationalisation, however, could not disguise the advantages of road transport for short distances—speed, cheapness, convenience. The social costs of transition from a rail to a road-based transport system are hard to calculate but firms and passengers scarcely bothering themselves with such considerations have opted for road traffic to such an extent that the railway is fast becoming a subsidiary form of transport. (*19, 21, 23*). The aeroplane and the pipeline have not so far proved such damaging competitors to the railway but they may be expected to take a further slice of its traffic in due course.

1 The Trent and Mersey canal, 1765

(Trent and Mersey Canal Act 1765, reprinted 1830; copy in Goldsmiths' Library)

An Act for making a Navigable Cut or Canal from the River *Trent*, at or near *Wilden Ferry* in the County of *Derby*, to the River *Mersey* at or near *Runcorn Gap*.

Whereas the making a Cut or Canal from the River *Trent*, near *Wilden Bridge*, below an ancient Ferry called *Wilden Ferry*, in the County of *Derby*, through or near *Swarkstone* and *Willington* in the said County, *Wichnor*, *Rudgley*, *Stone*, and *Burslem*, in the County of *Stafford*, and through or near *Lawton* and *Middlewich*, and near *Northwich*, in the County of *Chester*, and to the River *Mersey* at or near a certain Place called *Runcorn Gap*, for the Navigation of Boats and other Vessels with heavy Burdens, will open an easy Communication between the interior Parts of the Kingdom and the Ports of *Hull* and *Liverpool*, which will be of great Advantage, not only to the Trade carried on to and from the said Two Ports, but to several different Manufactories which abound in many Towns or Places through or near which the said Canal or Cut is proposed to be made, and will also tend to the Improvement of the adjacent Lands, the Relief of the Poor, and the Preservation of the public Roads, and moreover be of great public Utility: And whereas the several Persons herein-after particularly named are desirous, at their own proper Costs and Charges, to begin, carry on, and complete the said Navigable Cut or Canal: [the persons named] shall for that Purpose be One Body Politic and Corporate, by the Name of 'The Company of Proprietors of the Navigation from the *Trent* to the *Mersey*', and by that Name shall have perpetual Succession, and shall have a Common Seal, and by that Name shall and may sue and be sued, and also shall and may have Power and Authority to purchase Lands to them and their Successors and Assigns, for the Use of the said Navigation, without incurring any of the Penalties of Forfeitures of the Statue of Mortmain. ... [Power to survey lands. Commissioners appointed to settle differences between the Company and landowners.]

$$* \qquad * \qquad *$$

XXI And, to the end that the said Company of Proprietors may be further enabled to carry on so useful an Undertaking, be it enacted by the Authority aforesaid, That it shall and may be lawful to and for the said Company of Proprietors, their Successors and Assigns, to raise and contribute amongst themselves, and in such Proportions as to them shall seem meet and convenient, a competent Sum of Money for making and completing the said Navigable Cut or Canal, provided that the said Sum do not exceed the Sum of One hundred and thirty thousand Pounds in the whole, except as herein-after mentioned, and that the same be divided into such Number of Shares as hereafter directed, at a Price not exceeding Two hundred Pounds *per* Share; and that no Person subscribing thereunto, or becoming a Proprietor in such Navigation, do become a Proprietor of less than One Share or more than Twenty Shares, either in his own Name or in the Name of any other Person or Persons in Trust for him, (except the same shall come to him by Will or Act in Law,) upon pain of forfeiting to the said Company of Proprietors, their Successors and Assigns, all such Shares exceeding Twenty Shares aforesaid; and the Money so to be raised is hereby enacted and appointed to be laid out and applied, in the first place, for and towards the Payment, Discharge, and Satisfaction of all Fees and Disbursements for obtaining and passing this present Act of Parliament, and all other necessary Expences relating thereunto, and all the Residue and Remainder of such Money for and towards making, completing, and maintaining the said Navigable Cut or Canal, and other the Purposes of this Act, and to no other Use, Intent, or Purpose whatsoever.

XXII. And be it further enacted by the Authority aforesaid, That the said Sum, or such Part thereof as shall be raised by the several Persons herein-before named, shall be divided and distinguished into Six hundred and fifty equal Parts or Shares, at a Price not exceeding Two hundred Pounds *per* Share; and that the said Six hundred and fifty Shares shall be and are hereby vested in the several Persons before mentioned, and their several and respective Heirs and Assigns, to their and every of their proper Use and Behoof, proportionably to the Sum they and each of them shall severally subscribe and pay thereunto ... [Power to raise a further £20,000 if necessary.]

*   *   *

XXXIX And be it further enacted by the Authority aforesaid, That in consideration of the great Charges and Expences the said Company of Proprietors, their Successors and Assigns, will be at in making, maintaining, and supplying with Water the said Cut or Canal, and in making and maintaining all the other Works hereby authorized to be made and erected, it shall and may be lawful to and for the said Company of Proprietors, their Successors and Assigns, from Time to Time and at all Times hereafter to ask, demand, take, and recover, to and for their own proper Use and Behoof, for Tonnage and Wharfage for all Coal, Stones, Timber, and other Goods, Wares, Merchandize, and Commodities whatsoever which shall be navigated, carried, or conveyed upon or through the said Cut and Canal, (except such Part thereof as is hereby authorized and required to be made by the Duke of *Bridgewater*, his Heirs or Assigns,) such Rates and Duties as the said Company of Proprietors, their Successors and Assigns, shall think fit, not exceeding the Sum of One Penny Halfpenny *per* Mile for every Ton of Coal, Timber, Stone, and other Goods, Wares, Merchandize, and Commodities which shall be navigated, carried, or conveyed upon or through the said Cut or Canal, and so in proportion for a greater or less Quantity than a Ton;

$$* \qquad * \qquad *$$

XLVI Provided always, and be it further enacted by the Authority aforesaid, That all Persons whatsoever shall have free Liberty to use, with Horses, Cattle, and Carriages, the private Roads and Ways, and with Boats or other Vessels the Navigable Cuts, Canals, or Sluices, to be made by virtue of this Act, for the Purpose of conveying Coal, Stone, Timber, and other Goods, Wares, Merchandize, and Commodities whatsoever, to or from the said Cut or Canal, Trenches or Passages, and also to navigate upon the said Cut, Canal, Trenches, Sluices, or Passages, with any Boats or Vessels not exceeding Seven Feet in Breadth, and to use the said Wharfs or Quays for loading and unloading Coals and other Goods, and the said Towing Paths for haling and drawing such Boats and Vessels, upon Payment of such Rates and Duties as shall be demanded by the said Company of Proprietors, their Successors and Assigns, not exceeding the Rate herein-before mentioned.

2 Bristol and the Midlands, 1775

(Walter E. Minchinton, ed. *The trade of Bristol in the eighteenth century* (Bristol Record Society, XXII, 1957), p. 134, quoting a petition to the House of Commons)

That they observe that leave is given to bring in a Bill for making

15*

and maintaining a navigable canal from or near the town of Stourbridge in the county of Worcester to communicate with the Staffordshire and Worcestershire canal at or near Stourton in the county
of Stafford and for making and maintaining two collateral cuts to
communicate with the intended canal.

That in the towns of Stourbridge and Dudley and their neighbourhood are large manufactorys of nails and other wares with which
the merchants of this city are supplied for their foreign trade, also
extensive mines of coal, particularly usefull in sundry manufactorys
carried on in this city, and other mines of a peculiar kind of clay with
which fire bricks are made which bricks are absolutely necessary in the
brass and iron founderys glass and other manufactures of this city,
and are from hence forwarded to like manufactorys in different parts
of the kingdom.

That from the port of Bristol raw materials of various kinds are
likewise sent up to the places above mentioned to supply their manufactorys and a very great intercourse of trade is maintained between
them.

That the proposed canal and cuts will not only be greatly beneficial
to the trade of this city but be of public utility.

Pray that the Bill may pass into a law.

3 An early-nineteenth-century draper

(Handbill in the possession of the editor)

Joseph Fary

LINNEN DRAPER

at the Hen and Chickens against the Church at

GREENWICH

Sells all sorts of Holland and Irish Linnens—Cambricks, Lawns,
Muslins, Printed and Strip'd Cottons and Linnens, and all other kinds
of Linnen Drapery Goods,
Also very good Black and Colour'd Silks, Perians, Poplings, Bomba-
zenes and Norwich Crapes, Camlets, Yard Wide Stuffs, etc.
Very good Worsted Thread and Cotton Hose, and Gloves, Bath and
other Flannels, fine white Linceys, Bays, etc.
All sorts of Haberdashery wares, Kid and Lamb Gloves, Velvet
Hoods, Short Cloaks and Quilted Petticoats, etc.,
at the very lowest prices.

4 A Frenchman admires England's highways, 1825

(Baron François P. C. Dupin, *The commercial power of Great Britain exhibiting a compleat view of the public works of this country under the several heads of streets, roads, canals, aquaducts, bridges, coasts and maritime ports* (2 vols, London, 1825), I, 181–2, 184–5)

During my travels in England, I everywhere admired the superior excellence of the roads, as compared with the generality of our own. Is this superiority to be attributed to Nature or to Art? Undoubtedly Nature has, in this respect, done much in favour of Great Britain. The soil of most of the counties produces materials well adapted to the construction of roads. The ground over which the roads are traced is, in many parts, naturally very firm, from being composed of a mixture of sand, gravel, and flint, which, at the same time, enables the water to filter easily through it, and thus leaves the road dry, almost immediately after rain. The climate of England too, though habitually damp, is not subject to those heavy torrents of rain which occasion such a rapid destruction of the roads in more southern countries.

These causes, however, are not sufficient to account for the excellence of the roads in Great Britain; for, in many parts of the North of England, and in Wales, where heavy rains are frequent, and where the waters run in rapid torrents, public roads have been constructed of a perfectly good quality. Indeed, even on marshy and clayey soils, roads have been formed remarkable for their solidity, durability, and dryness.

One circumstance which clearly proves, in opposition to the generally-received opinion, that the superiority of the English roads is not to be attributed to the excellence of the materials used in making them, is, that they are to be found in equal perfection, even when materials of the most different nature are employed. In some counties, as for instance, in Essex, Sussex, Shropshire, and Staffordshire, they make use of flints mixed with sand. In Somersetshire, Gloucestershire, and Wiltshire, limestone is chiefly used. This substance certainly offers but little resistance; yet when properly prepared and well laid down, it produces a compact and solid road, and binds more readily than any other material. Its only defect, therefore, is its want of durability.

It is very remarkable that the great high roads which run into London, and which from their beauty are the admiration of foreigners, are formed of the most defective materials, and on this account are, perhaps, the worst roads in all England. This defect is rendered still more serious from the immense traffic carried on upon these roads.

* * *

The roads in the neighbourhood of London being extremely smooth, the traveller, in driving along them, free of all inconvenience from jolting, concludes without further examination, that they are in the best possible condition; but, drivers and postmasters form a different opinion of them. These roads have the fault of being extremely soft, particularly first after their construction and repair. For this reason the stage coaches, etc., must be drawn by horses of very superior strength, to enable them to proceed as rapidly as they do at a greater distance from the capital. Notwithstanding their superior strength, the fatigue endured by these poor animals is so excessive, that they are rendered useless in the short space of three years! The foreigner justly admires the beauty of the horses attached to the public vehicles in the neighbourhood of London; but he is far from suspecting that the choice of these animals is occasioned by the very defects of the road which is so magnificent in appearance, and so pleasant to the traveller.

5 Canal versus railroad, 1825

(*Quarterly Review*, XXXI (March 1825), pp. 356–8, 362–4, 377)

But it is high time that we should advert to the more immediate object of this Article—a discussion of the merits of an old invention, newly revived, which is become a subject of almost as eager and feverish speculation as the mines or the loans. It is one, however, in which the commercial, manufacturing, agricultural, and indeed every class of the community, are most deeply concerned—we need hardly say that we allude to the projected improvement in the internal communications of the country, by which a more speedy, certain, safe and economical conveyance of persons and property is expected to be accomplished. It would be a waste of time to point out, what is so obvious, the vast importance of such a result, which must be felt and understood by all;—by the producer and consumer, by him that travels, and by him that remains quietly at home. It is true that we, who, in this age, are accustomed to roll along our hard and even roads at the rate of eight or nine miles an hour, can hardly imagine the inconveniences which beset our great-grandfathers when they had to undertake a journey—forcing their way through deep miry lanes; fording swollen rivers; obliged to halt for days together when the 'waters were out'; and then crawling along at a pace of two or three miles an hour, in constant fear of being set fast in some deep quagmire, of being overturned, breaking down, or swept away by a sudden inundation.

Such was the travelling condition of our ancestors, until the several turnpike acts effected a gradual and most favourable change, not only in the state of the roads, but the whole appearance of the country; by increasing the facility of communication, and the transport of many weighty and bulky articles which, before that period, no effort could move from one part of the country to another. The packhorse was now yoked to the waggon, and stage-coaches and post-chaises usurped the place of saddle-horses. Imperfectly as most of these turnpike roads were constructed, and greatly as their repairs were neglected, they were still a prodigious improvement; yet for the conveyance of heavy merchandize, the progress of waggons was slow and their capacity limited. This defect was at length remedied by the opening of canals, an improvement which became, with regard to turnpike roads and waggons, what these had been to deep lanes and packhorses. But we may apply to projectors the observation of Sheridan, 'give these fellows a good thing and they never know when to have done with it', for so vehement became the rage for canal-making that, in a few years, the whole surface of the country was intersected by these inland navigations, and frequently in parts of the island, where there was little or no traffic to be conveyed. The consequence was, that a large proportion of them scarcely paid an interest of one per cent and many nothing at all; while others, judiciously conducted over populous, commercial and manufacturing districts, have not only remunerated the parties concerned, but have contributed in no small degree to the wealth and prosperity of the nation.

Yet these expensive establishments for facilitating the conveyance of the commercial, manufacturing, and agricultural products of the country to their several destinations, excellent and useful as all must acknowledge them to be, are now likely, in their turn, to give way to the old invention of *Rail-roads*. Nothing now is heard of but rail-roads; the daily papers teem with notices of new lines of them in every direction, and pamphlets and paragraphs are thrown before the public eye, recommending nothing short of making them general throughout the kingdom. Yet till within these few months past, this old invention, in use a full century before canals, has been suffered, with few exceptions, to act the part only of an auxiliary to canals, in the conveyance of goods to and from the wharfs, and of iron, coals, limestone and other products of the mines to the nearest place of shipment.

*   *   *

Returning, however, to that important question, which will speedily be brought to issue between canals and rail-roads, and perceiving how much the public mind is abroad upon the subject, we think it may not be useless or uninteresting to take a general view of the comparative merits of the two. No accurate estimate can be made of their comparative expense, because both must depend on circumstances constantly varying, and which can seldom be common to either: but it may be stated in general terms, with regard to canals, that the deep cuttings and high embankments to preserve the levels, or, in default thereof, the substitution of numerous bridges and locks; the high price paid for the best land through which they are generally carried; the reservoirs necessary for collecting and preserving water; the repairs required for their locks and banks, the latter of which are constantly subject to injury from floods or frosts; the cost and feed of horses; the building and repairs of boats;—these and other incidental charges occasion a much larger expenditure in a canal than a rail-road, mile for mile, supposing them to run to and from the same places; to say nothing of the excess of length which, in the canal, will be generally about one-third greater than the rail-road. As to the original cost, we have before us a list of the estimates for no less than *seventy-five* canals, including those of the greatest and those of the least expense, and the general average is £7,946 per mile; and as the estimated expense is generally very much exceeded, we may fairly set down the real cost as £9,000 per mile. We have also a list of rail-roads, (some tram-rails, others edge-rails, some of cast and others of wrought iron,) containing upwards of 500 miles, and the general average (allowing them a double set of tracks) is as near as possible £4,000; but, from the imperfections of these old roads, we may extend the average to £5,000 per mile. The estimate for the Liverpool and Manchester rail-road we have understood to be taken at £12,000 per mile, but that road is meant to be executed on a magnificent scale; to be sixty-six feet wide; the rails to be laid down in the best possible manner: and the purchase of land at the two extremities must be paid for at an enormous price; this estimate also includes the cost of engines, waggons, and warehouses. The Union canal, however, is stated to have cost just as much; the Forth and Clyde twice as much, the Regent's canal we are afraid to say how many times as much, and the Caledonian more than four times as much. We observe also that Mr Jessop, after a minute survey of the proposed Peak Forest railway, patronized by the Duke of Devonshire, states its estimated cost at £149,206; and that a canal, to form the same connection as is proposed by the railway, was estimated in October, 1810, by the late Mr Rennie, at £650,000, being more than four times the cost of the former. The disadvantages of a canal are numerous. The frost at one season of the year entirely puts a stop to all conveyance of goods; and the drought at another renders it necessary to proceed with half cargoes. A rail-road is exempt from both these serious drawbacks; and even if snow-blocked, nothing can be so easy as to send forward a scraper at the front of the steam-carriage to clear it as it proceeds.

The speed, by which goods can be conveyed on a rail-road, can be so regulated as to be certain and constant, while boats are frequently delayed for hours at the lockages of a canal. This speed besides is limited on canals, as we shall presently shew, but unlimited, as far as the power of steam can be made to exceed the power of friction, on rail-roads. To what extent, with safety and convenience, this advantage is capable of being carried, nothing but experience can determine, Rail-roads may be made to branch out *in every direction* to accommodate the traffic of the country, whatever be the nature of the surface; the possibility of carrying branches from a canal *in any direction* must depend entirely on the surface, and a supply of water.

✳   ✳   ✳

It has been said that an opposition to rail-roads will be made on the part of the landed proprietors, but the absurdity of this is so glaring, that it must defeat itself. Country gentlemen may not at first see their own interest, but their tenants will find it out for them; they will discover immediately the advantage, which a rail-road will confer along the whole line of country through which it passes, by the increased facility of sending their produce to market, and of receiving the objects of their wants in return. The two great landed proprietors along the projected line of the Liverpool and Manchester rail-road, are the Lords Derby and Sefton, and with regard to them it appears by the plan, that the road does not reach within a mile and a half of the residence of the Earl of Sefton, and that it traverses the Earl of Derby's property over the barren moss of Knowsley, passing about two miles distant from the Hall. 'I would defer to my Lord Derby, my Lord Sefton, or to Lord Stanley', says Mr Sandars, 'on all points affecting their substantial comforts and convenience; and I am convinced that they possess feelings of a character too liberal and patriotic, to urge speculative, frivolous or fanciful objections'. With regard to Lord Sefton, the coal waggons and other traffic now pass on the public road, within a hundred yards of his door; he, therefore, would be a gainer by the rail-road; and so far from its trespassing on the quiet and peaceful enjoyment of Knowsley, the line is meant to pass over a barren heath, far in its rear, and will be separated from the park by a public road; as little likely, therefore, is Lord Derby to be annoyed by the smoke, as the Duke of Wellington is by the smoking chimnies of Kilburn. Thus while the road will disturb neither of these noble Lords, both of them must be aware how immensely their estates will be improved by it.

We have purposely abstained from that part of the question which regards the conveyance of *passengers*. There is no doubt that a diminished weight may be conveyed with increased speed and with equal safety, as far as the strength and stability of the engine are concerned; but we think it would be expedient to wave all thoughts of this part of the subject for the present, until the roads and the engines have acquired that degree of perfection of which they are capable, and such as will remove all apprehension of danger.

6 Public improvements, 1829

(*The British almanac for 1830 of the Society for the Diffusion of Useful Knowledge* (1829), pp. 244–7, 250–3)

Canals and Railways

Birmingham Canal Improvements—The important portion of the system of inland navigation called the Birmingham Canal, was originally made by authority of Parliament in the years 1768 and 1769, for the purpose of forming an easy communication between the mineral districts of Staffordshire and the town of Birmingham on the southern side, and the town of Wolverhampton on the northern side. Its original length was a little more than twenty-two miles from Birmingham to its junction with the Staffordshire and Worcestershire Canal below Wolverhampton, and the facilities which were thus afforded to the active energies of a large manufacturing population, created upon the original line such an important trade as to induce the proprietors from time to time, under sanction of the legislature, to make branches and extensions of their canal, in various directions through the coal and iron districts in its immediate vicinity, and also to make an additional line of canal from Birmingham to Fazeley, in order to open a new road to London through the Coventry and Oxford Canals, so that the whole canal has now spread itself over a distance of more than seventy miles. Within the last three years, the improvements upon this canal have been such as to excite the admiration of scientific engineers, and to command the approbation of all persons who are accustomed to take pleasure in contemplating works of science and national improvement. The original line of this canal, as laid out by the celebrated Mr Brindley, was very circuitous, and one great object of improvement has been to shorten the mileage between Wolverhampton and Birmingham; this has been already effected, under the superintendence of Mr Telford, to a considerable extent, and when the works now in active progress for this purpose shall be completed, the distance by canal between these two towns will be about fourteen miles, being scarcely one mile farther than the turnpike road; many deep cuttings have been made, and numerous embankments have been raised in order to effect this object, which is of great importance to a thoroughfare trade. But what is of still more consequence to the immediate district, the communication between the town of Birmingham and a great portion of the collieries will be expedited by the avoiding of three ascending and three descending locks, which have heretofore existed at Smethwick and Spon Lane. The quantities of earth which have been moved, in order particularly to effect the avoidance of these six locks, are ascertained to be 1,697,414 cubic yards; the whole of which have been excavated and removed, in a length of two miles, and in a period of two years and a half, between March 1827, and September 1829. On this part of the work the slopes are one and a half horizontal to one perpendicular. The greatest depth of cutting is seventy-one feet; the water-way of the canal is made forty feet wide, and five feet six inches deep; it is walled with stone on each side; and the trade is very much facilitated by the existence of a towing path, twelve feet wide on each bank. Over this chasm are thrown, at various places, bridges of brick and stone, in the construction of which considerable ingenuity has been displayed, as several of them are much askew. At the place of greatest excavation is erected the largest canal bridge in the world; it is made of iron: the arch is one hundred and fifty feet span, and over it passes a public roadway twenty-six feet wide; each of the other bridges is fifty-two feet span, and all the bridges are so wide as to admit the water-way and towing-path without being subject to contraction.

Near Spon Lane is constructed an aqueduct of two arches, by means of which the original line of canal is carried across the new works, without the water-way or towing-paths of either the old or new lines being in any way contracted; and at Smethwick has been made an elegant iron-aqueduct, through which any surplus water is conveyed from the upper level of the old canal, across the new works, into a feeder which has been formed to carry water to a magnificent reservoir recently constructed by the Canal Company, covering upwards of eighty acres of ground, and of an average depth of more than thirty feet. This feeder also has been made to convey water from the reservoir into the upper level, or original line of the canal; while the new works, or Birmingham level, can be supplied with water from the same reservoir at any moment, if the quantity provided by the descending lockage from the upper level to the lower, should be at any time insufficient to keep up the canal to its proper level.

Throughout the whole line of improvements are erected, at convenient stations, small accommodation bridges, to enable traders to cross off the line, and thus prevent any inconvenience arising from the regulation, which is rigidly enforced, for the boats passing towards Birmingham to use one towing-path, while those passing towards Wolverhampton and the collieries use the opposite one. But what is particularly to be remarked is, that each of these smaller bridges is constructed of such span, as to render unnecessary the slightest contraction of the towing-paths or water-way.

At Smethwick the upper level decends into the improved new line of canal, and between this point and the town of Birmingham, the line of canal has been so much shortened that in a distance of four miles and three-eighths, nearly two miles have been saved; in order to effect this, there have been formed three considerable embankments, and a depth of cutting has been effected, which, prior to the works completed at Smethwick, were thought to be of extraordinary magnitude. The quantity of earth removed to complete these three embankments, and the intervening deep cutting, was 370,000 cubic yards, which was moved in the space of two years and a half, and throughout a distance of two miles and upwards, as between the different points of improvement lay portions of the original circuitous line; this part of the canal is principally walled on each side, has double towingpaths, is in all parts forty feet wide, and for some distance the water-way is more than fifty feet wide. Where the canal is not walled the water-way is effectually fenced with wood-work. Across this part of the canal are thrown several elegant iron-bridges; but there are two public road-bridges in this district, which were built of common brick in 1826, and are so much askew as to deserve mention, one being fifty-two feet span and fifty feet askew, and the other being fifty-two feet span and forty-six feet askew; the road-way of the former being sixty-six feet, and of the latter thirty feet wide. The same plan of making the bridges of sufficient dimensions to avoid any contraction of the waterway or towing-paths, prevails throughout this portion of the work, and is intended to be carried into effect throughout the line towards Wolverhampton.

At a place called Bloomfield, in the parish of Tipton, an excavation is being made, extending towards Deepfield, by means of which, though the new works will be in extent only one mile, the trade of the country will be able to avoid an old circuitous route of more than five miles. At this part of the improvements the greatest depth of cutting is ninety feet, the canal is formed twenty-four feet wide, and five feet deep; having the sides walled, and a towing-path on each side. The quantity of rock and earth already removed and intended to be removed in effecting this important shortening, may be fairly estimated at not less than one million of cubic yards.

The foregoing statements relate to the works which are immediately connected with the improvements on the main line of the canal; but in forming an idea of the important works which have been undertaken by the Proprietors of this canal, notice should be drawn to the various collateral branches which have been made at various times, and particularly within the last three years, during which new lines have been opened into various estates, and in the formation of which no less a quantity than 500,000 cubic yards of earth have been moved. Great and important as these improvements must be to the commercial interests of the country, it is highly gratifying to know that they are carried into effect without the imposition of the slightest additional charge on parties who trade on the line of this canal.

*   *   *

Liverpool and Manchester Railway.—The attention of the public has been recently directed to the remarkable competition of Locomotive Steam engines upon this Railway.* The success of these experiments has confirmed the value of this great undertaking. We are enabled, from the obliging communication of the Engineer, to continue our notice of the progress of the works.

Dear Sir,

Since my last communication to you on the subject of our works, we have made considerable and rapid advances towards completion.

---

* The Rocket actually accomplished one mile in one minute and twenty seconds being at the rate of forty-five miles an hour.

Commencing at the yard at Wapping, near the docks, which is intended as the grand depôt for the Liverpool end of the line, we have very nearly completed the whole of the earth work; the side walls are already in a very forward state, and the flooring and roof, for an extensive warehouse, capable of receiving a large bulk of merchandise, are in progress. Under this warehouse, four distinct lines of railway are intended to be laid, in order to render as much facility as possible in loading and discharging the goods to and from the waggons.

The tunnel, since last year, has been completed. A double line of railway has been laid throughout, and a row of gas lights has been suspended from the centre of the arched roof at a distance of twenty-five yards from each other. The effect, when lighted up, is strikingly beautiful, for the rays of light from each lamp throw a distinct luminous arch on the roof, and the series diminishing according to the laws of perspective, gives the appearance of a number of distinct arches instead of one continued vault.

The tunnel is lighted up every Friday for public inspection, and many ladies have descended in a carriage at the rate of twenty-five miles an hour, (performing the whole distance in three minutes), without experiencing any alarm or disagreeable sensation.

Since last year another tunnel has been made, quite distinct from the larger one, and which extends from the deep cutting at Edge-hill, to near the Botanic Garden. It is about 300 yards long, fifteen feet wide, and twelve feet high. This tunnel is chiefly intended for the conveyance of passengers and coals to the higher part of Liverpool; it is quite completed, and, like the large one, is lighted with gas.

The excavation for two lines of rails at the top of the tunnel, called Edge-hill, is completed, and the rails are laid down within 100 yards of the tunnel: at this place are making four distinct lines of railway, in order to arrange the trains previous to their starting for Manchester.

Olive Mount Excavation has been completed through the rock, and there now only remains a small portion of clay cutting, which will be removed in three or four weeks.

Two bridges have been thrown over this excavation, and two others are now building.

Broad Green Embankment has also been completed, and nothing now remains to be done but to lay down the permanent rails, which will be commenced immediately.

From Huyton, through Whiston, Rainhill, and Sutton, the excavations are finished; and for a distance of four miles the panerment rails are laid down, and finished off, excepting for about a mile, where a single line is only laid.

It is at Rainhill where the recent trials of the powers of locomotive engines have been made, and which have so strikingly shewn the advantages which may be expected from this rapid and economical mode of conveyance.

I trust I shall not be considered digressing from the subject, when I add, that in contemplating a speed of thirty miles an hour with passengers, and from fifteen to twenty miles an hour with a load of merchandise, at a cost of almost nothing, comparatively speaking, I can scarcely set a limit to the advantages which this country has a right to expect from this improved mode of conveyance and means of inter-course; and even should no further improvements be made;—and I doubt not but many and important ones will follow,—there has been sufficient to shew, that locomotive engines are capable of pro-ducing and maintaining a speed beyond any other means at present known.

The London road at Rainhill has been carried over the railway by means of an oblique or skew arch, the direction of the railway and that of the road varying only 34°.

The arch and abutments are of solid stone-work, and the arched stones are placed at right angles to the face of the bridge.

The great viaduct across the Sankey canal and valley, which consists of *nine* arches of fifty-feet span each, is nearly completed; the arches are all turned, and the centres removed, and the cornice and parapet walls are rapidly advancing.

The Kenyon Excavation has been proceeding without much inter-ruption,—the materials from which have gone to form the embank-ment eastward towards Chat Moss, and westward toward the Sankey Valley. The eastward embankment will be completed in about two months, and the westward in about three months; 80,000 cubic yards yet remain to be removed.

Chat Moss.—The permanent rails have been laid nearly the whole of the way across the moss. A single line is laid for four miles, (one mile and a half of which has rails laid for the double line, but not finished off,) and *horses with loaded waggons, each weighing five tons, are constantly moving over those parts of the moss, which originally would scarcely bear a person walking over it.*

The Barton Embankment, on the east of the Moss, is nearly done, and the Eccles Excavation requires about 20,000 cubic yards to be removed for its completion—both these portions of the work will be finished in little more than a month: from Chat Moss to Manchester, the permanent way is laid for more than one half the distance. The bridge across the river Irwell has commenced: the foundations are in, and the abutments are now building.

There are forty-four Bridges on the line already built—nine are now building, and a few others of minor importance remain to be erected.

Two Locomotive Engines have been in use (one near Liverpool, the other at Eccles) during the last summer, dragging the marl and rock from the excavations, by which a great saving has been effected. The one near Liverpool has saved near 50*l* per month, when compared with horses; and the one at Eccles in a still greater proportion.

The 'Rocket', locomotive engine, which gained the premium of 500*l*, is about to be put on Chat Moss, to drag the gravel for finishing the permanent way, and there is no doubt but a proportionate reduction will take place—besides doing away with the wear and tear of the horse-track, which on all new-made roads is so considerable.

The immense quantity of rain which has fallen during the last six months has had considerable influence in retarding the progress of the works—particularly in the marl excavations. It was expected that the line might be opened by the 1st January, 1830; and had the season been at all favourable there is no doubt but this might have been effected; but, from the nature of such a work as this, it is almost impossible to calculate on its completion with much certainty; but if we have a moderate winter, I have no doubt but the line may be opened in five months from this time, and the line as far as the collieries may be opened in two months.

I am, dear Sir,

Your obedient and faithful servant,

Liverpool, Oct. 25th, 1829.   George Stephenson.

P.S. Nov. 19. Since the date of this Report, the works have made such progress as to justify the confident expectation that one line will be opened on the 1st of January, 1830.

7 New methods in retailing, 1833

(Select Committee on Manufactures, Commerce and Shipping, *B.P.P.*, 1833, VI, qq. 1418–21, 1437–42)

In passing along the streets of London as a common observer, is it often not obvious to you, that the practice of ticketing the prices of goods in shops has very much increased within the last few years?

—Certainly; it is now much more general than it was 10 or 15 years ago, but I do not know that it has increased within the last three, or four, or five years. [The witness was Thomas James, wholesale draper.]

Is it not the fact, that within the last 10 or 15 years the practice of ticketing the prices of goods has grown up altogether—I think that within the last 10 or 15 years the practice has increased very considerably.

What do you mean by ticketing—I mean affixing to goods in shop windows the prices at which they are to be sold.

Does not the increase of that practice arise from an entire change in the mode of carrying on the retail business, that is, that it has now become more casual and less dependant upon regular constant customers to particular shops than it used to be formerly?—I think it does.

Is there any disadvantage arising from the ticketing system; does it betray a desire to sacrifice goods at a less cost, or is it only an endeavour to tempt customers to purchase?—I do not think it necessarily infers a desire to sacrifice the goods, because we find some of that class of men as respectable and as regular in their payments as others.

Is it not the case that the more respectable retail dealers do not have recourse to that practice if they can avoid it?—Certainly; it is a practice that is resorted to in particular trades, and the object is to attract a particular class of customer; it is not a practice that would be resorted to by those who would seek their customers from the higher class of the community, but in populous neighbourhoods it is more resorted to, where they seek principally customers among the lower and middling classes of people.

Taking, for example, two great thoroughfares in London, Oxford-street and Regent-street, do not you think that three out of four retail shops in both of those streets resort to the practice of ticketing? —I am not so frequently near those streets that I can answer that question accurately, but I should say it is more general in Oxford-street than it is in Regent-street, on account of the different class of customers that are more likely to resort to the one than the other.

And more frequently still in Ratcliffe Highway and Shoreditch?— Yes, and in Whitechapel.

Do you think the practice of ticketing is any evidence of a person's poverty?—Not at all.

Would a house setting up in your line of business have any chance of success unless they had a very large capital?—I should think not. I have said that it is the practice of wholesale houses generally to pay monthly for four-fifths of their purchases, and to hold a stock and to give credit of three months, and it will be manifest that that requires a large capital.

8 Railway shareholders, 1837

(Minutes of evidence before the committee on the London and Birmingham Railway Bill, *B.P.P.*, 1839, XIII, appendix No. 32.)

Name	Residence	Description	£100[†]	£25
Abbott, Benjamin	Bolton	basketmaker	2	2
Abbott, Robert	Liverpool	tea-dealer	3	3
Aberdein, Francis C.	Old Broad-street	esquire	5	5
Abbey, Mary, wife of Edmund Abbey	Coton	gentleman	1	1
Abbott, John	Portland-place	ditto	2	2
Abram, Ralph	Liverpool	merchant	2	2
Adams, Samuel	Brewer-street	gentleman	1	1
Adams, William	Capesthorne	ditto	1	1
Adams, William	Liverpool	ditto	2	2
Addison, John, the younger	Preston	barrister	20	20
Aikin, Grace	Liverpool	spinster	1	1
Alcock, John	Gatley, Cheshire	merchant	70	70
Alcock, Samuel	Manchester	manufacturer	95	95
Addenbrook, Edward	Kingscomford	gentleman	2	2
Allright, William	Charlbury	ditto	1	1
Alderson, Harrison	Blackburn	tea-dealer	3	3
Alleyne, Sarah Anne	Liverpool	widow	20	20
Alexander, William	Ditto	surveyor for Lloyd's	10	10
Aldred, Joseph	Nottingham	gentleman	1	1
Alanson, John, and William Brandreth	Liverpool	wine-merchants	10	10
Alanson, John	Ditto	wine-merchant	13	13
Allen, Peter	Ditto	gentleman	5	5
Alison, Richard Edward	Charnock Chorley	ditto	10	10
Anderson, Joseph	College-street	merchant	10	10
Archmard, Jean Samuel	Geneva	esquire	6	6

[†] This table is to be read: Benjamin Abbott holds two £ 100 shares and two £ 25 shares.

Archmard, Samuel Cesar	Ditto	ditto	5	5
Armstrong, Elizabeth	Salford	widow	5	5
Armstrong, Richard Baynes	Staple-inn	gentleman	15	15
Armstrong, Thomas	Manchester	merchant	65	65
Armand, Elias	Liverpool	esquire	10	10
Arthington, Robert Morley, and George Crossfield	Ditto	gentlemen	1	1
Ashurst, Margaret	Grimsargh, Preston	widow	2	2
Ashley, John	Lymme, Cheshire	gentleman	1	1
Ashton, John	Manchester	ditto	30	30
Ashton, Samuel	Irwell House, Bury	esquire	20	20
Ashton, Thomas	Manchester	manufacturer	20	20
Atherton, Eleanora	Hersall Cell	spinster	20	20
Atherton, Elizabeth	Winwick	ditto	1	1
Atherton, William	New Brighton, Cheshire	gentleman	1	1
Atkins, Joseph	Chipping Norton	ditto	6	6
Atkinson, Edward	Drogheda		5	5
Atkinson, Ellen, now wife of Edward Patten, of Liverpool, gentleman			2	2
Axford, Richard	Liverpool	major, East India Company's service	3	3
			508	508

9 Growth of suburbs

(*i*) *For the rich*, 1848

(T. B. Macaulay, *The history of England from the accession of James II* (Everyman ed. 1906), I, p. 272)

The whole character of the City has, since that time, undergone a complete change. At present the bankers, the merchants, and the chief shopkeepers repair thither on six mornings of every week for the transaction of business: but they reside in other quarters of the metropolis, or at suburban country seats surrounded by shrubberies and flower gardens. This revolution in private habits has produced a political revolution of no small importance. The City is no longer regarded by

16*

the wealthiest traders with that attachment which every man naturally feels for his home. It is no longer associated in their minds with domestic affections and endearments. The fireside, the nursery, the social table, the quiet bed are not there. Lombard Street and Thread-needle Street are merely places where men toil and accumulate. They go elsewhere to enjoy and to expend. On a Sunday, or in an evening after the hours of business, some courts and alleys, which a few hours before had been alive with hurrying feet and anxious faces, are as silent as the glades of a forest. The chiefs of the mercantile interest are no longer citizens. They avoid, they almost contemn, municipal honours and duties. Those honours and duties are abandoned to men who, though useful and highly respectable, seldom belong to the princely commercial houses of which the names are renowned through-out the world.

In the seventeenth century the City was the merchant's residence. Those mansions of the great old burghers which still exist have been turned into counting houses and warehouses: but it is evident that they were originally not inferior in magnificence to the dwellings which were then inhabited by the nobility. They sometimes stand in retired and gloomy courts, and are accessible only by inconvenient passages: but their dimensions are ample, and their aspect stately.

(*ii*) *For all*, 1891

(Local government and taxation committee of the corporation of the city of London, *Ten years' growth of the City of London, 1881–1891: report of the day-census, 1891*, 1891), pp. 13–17)

Considerations Making a Day-Census of the City Desirable

First—The Returns as to the population of the City made under the Imperial Enumeration are *absolutely worthless* as forming any indica-tion of its commercial importance.

Indeed, to a superficial reader, these Returns can only convey the impression that the City is hopelessly *on the wane*.

Thus the population of the City according to the *Imperial Census* was:

In 1861 112,063
In 1871 74,897
In 1881 50,652
In 1891.......... 37,694

showing a *decrease* in the number of inhabitants:

> From 1861 to 1871 of 37,166
> From 1871 to 1881 of 24,245
> From 1881 to 1891 of 12,958

while the Mercantile and Commercial population, as ascertained by the *Day-Census*, was found to be:

> In 1866 170,133
> In 1881 261,061
> In 1891 301,384

showing an *increase*:

> From 1866 to 1881 of 90,928
> From 1881 to 1891 of 40,323

No comment is necessary to point out the unfairness to the City of the Imperial Enumeration.

The extraordinary discrepancy between the results arrived at by the two enumerations is assignable to the fact that the Imperial Census is a registration of the *night* population of the City, while the Census taken at the instance of the Court of Common Council registers its *day* population.

Surely no system of estimating the number of a city's inhabitants can possibly be more unsatisfactory than a computation made at a time when the bulk of the citizens are absent.

It may not be out of place to show that the circumstances which, unquestionably, conduce to the City of London occupying the forefront of commercial importance inevitably tend to make it a *non-residential* city.

The circumstances are these:

(*a*) The great and ever-increasing demand for business premises.

Sites in the City upon which, a few years ago, churches and dwelling-houses stood, are now covered by offices and warehouses, and one has but to note the comparatively few of these latter which are unlet or untenanted to form a very just estimate of a great city's requirements for business accommodation.

(*b*) The abnormally high value of property in the City.

This is so great that, were other considerations excluded, the larger number of business men would be unable to afford the rent which premises made suitable for family residences would command.

Owners of City property under the existing state of things are able to realise far more by the conversion of buildings into offices than by letting them as dwelling-houses. Even were this not so, the circumscribed area of the City would be totally inadequate for its citizens becoming resident.

(c) The unsatisfactory method of imposing the Inhabited House Duty.

If an owner or tenant resides in any portion of his building, no matter how extensive the premises may be, or how small his requirement for sleeping purpose, he renders the whole building liable to Inhabited House Duty. The Inland Revenue allows a caretaker, but insists that such person shall be a 'menial'. While I must agree that the occupier should be charged a Tax on the *portion* of the building he uses, I consider that the present unfairly imposed taxation must have the effect of causing many engaged in the City to be non-resident.

(d) The facilities in the way of Railway travelling.

There is, undoubtedly, when the day's business is done, a relief in dissociating oneself, for a time, from the scene and surroundings of commercial pursuits. The desire for quietude, unattainable in the busy thoroughfares of a city; for a sight of trees and fields; for a sense of freedom altogether incompatible with the necessarily circumscribed limits of a city dwelling-house, invests a suburban or country residence with great attraction.

This desire, coupled with the ease and expedition with which, under improved Railway locomotion, citizens can now travel daily to and from their place of business has, in no inconsiderable degree, resulted in the bulk of the City's population residing outside its borders.

10 Railways and the London meat trade, 1850

(Royal Commission on Smithfield Market, *B.P.P.*, 1850, XXXI, qq. 1859-72)

Mr Robert Moseley examined

Are you the traffic manager of the Eastern Counties Railway? —I am.

Has the transport of live stock to London by railway increased lately? —Very considerably.

Is it principally cattle or sheep?—There is an increase in sheep, and there is an increase in oxen.

What railways do you speak of particularly?—The Eastern Counties Railway and the Norfolk Railway.

Is there greater facility for sending live stock by railway than in any other manner?—Yes.

Do they come up in better condition if sent by railway than if they walk?—Yes.

Is any dead meat sent by railway?—Yes, very considerable quantites.

At what times of the year does it principally come?—The heavy season for dead meat commences in December, and continues till the end of February. We are receiving dead meat throughout the year, but not in such considerable quantities as in the three months which I have mentioned.

Does the transport of dead meat increase?—Yes, it does.

Is the chief part of the dead meat beef or mutton?—From Norfolk, it is principally beef; from Lincolnshire, principally mutton.

What are the principal counties which send up both live stock and dead meat?—The great supplies come from Norfolk, Suffolk, Essex, Cambridgeshire, and Lincolnshire.

Has the effect of the railways been to obtain supplies from a greater distance than formerly?—Yes, it has.

Is that effect likely to increase?—I think it is; as in the year 1845 the quantity of dead meat carried on the Eastern Counties Railway was about 100 tons per week. The quantity of dead meat which we are now carrying is about 600 tons per week, from the facilities which we give to the trade. We find what are called hampers or peds, and cloths, and small butchers are induced to hire them, who make it quite a business to feed a little stock in the country and send it up to London, paying us a moderate rate for using the peds. They could not afford to find peds themselves, for in the event of loss, it would be more than equal to any little profit that could be obtained in the sale of the meat. The consequence has been that by the construction of the railway through those producing districts, we have enabled men in the country to connect themselves with the London meat markets. Formerly, they were completely shut out by expensive road conveyances, they wanted cheaper communication; the establishment of railways has benefited them considerably, and it is bringing upon our railway an enormous increase of traffic.

Do all those cattle go to Smithfield, or do any go directly to the butchers?—In a general way they are landed at Tottenham, that is the Eastern Counties great landing place for live stock; it is six miles below London, from whence they are walked thence to the lairs, where they rest about 24 hours and then taken to Smithfield at a very early hour on Monday morning, so that they may get into the market by 2 or 3 o'clock. The dead meat comes to our goods' station in London, from whence we remove it to Newgate and Leadenhall markets by our vans and waggons; we pitch the meat at the various salesmen's stalls.

11 A capital-saving invention—the electric telegraph, 1857

(Samuel Smiles, *The life of George Stephenson: railway engineer* (2nd ed. John Murray, 1857), pp. 518–19, quoting Robert Stephenson's presidential address to the Institution of Civil Engineers, 1856)

The automatic working of the telegraph shows the officers at every station, that for a considerable number of miles in advance of the station, whether up or down, the line of way is clear. This knowledge, imparted instantaneously and comprehended by a glance, enables the officers to augment very materially the traffic over the portion of the line to which their duty may apply. The telegraph, in fact, does the work of an additional line of rails to every company that uses it, and does it at a cost perfectly infinitesimal in comparison with the cost of constructing another line.

At one period of its history, the North Western Railway appeared to be so overcrowded with traffic, that additional lines for its relief were believed to be indispensable; but at the very moment when the demands upon the system were beginning to outgrow the machinery for safety, this remarkable invention came to its relief, and the capacity of the line for traffic has consequently been immensely increased. The very first use made of the telegraph was to enable the Company to meet the difficulty of a strike among the artizans. During the Great Exhibition of 1851, when 750,000 passengers were conveyed to London by the North Western excursion trains alone, the whole of the extraordinary traffic of the line was conducted by means of the electric telegraph. At the present moment the ordinary traffic is double what it was when the telegraph was invented, and there is a greater capacity for increase than at any period since the line was opened.

Moreover, it must be observed, that great as is this saving to a Railway Company, it is not the only economy effected by the use of the electric telegraph. On every line where it is thoroughly employed, it effects a very material saving in the expensive element of rolling stock. The officers of a Company are enabled, the first thing every morning, to consider the wants and requirements of the day. They find, that on one portion of their line there is likely to be extra traffic, whilst at some other station, during the previous day, or night, there has been an accumulation of passenger carriages or vans. By the use of the electric telegraph, nothing is so easy as to supply the wants of one station from the surplus stock at the other; whilst the probabilities are, that without the facility afforded by the telegraph, the stock at one place would have been lying idle, athough it was urgently needed at another. Probably most lines would require fully 20 per cent more carriage stock than they now possess, if it were not for the telegraph.

12 The London and North-Western railway—a many-sided business, 1867

(Harold Pollins, 'Railway auditing—a report of 1867', *Accounting Research*, VIII (1957), pp. 17–20)

A considerable amount of attention is bestowed upon the large sums credited under the head of Traffic, and a careful adjustment is periodically made in the Accountant's Office, which demonstrates that the Company's Agents and others have fully accounted for the monies earned at each of the 565 Stations, the accuracy of the details of such earnings being duly tested by the staff in the Audit Office, who also certify that all settlements of Traffic interchanged with other Railway Companies have been correctly made.

The Rents due to the Company, amounting to nearly £80,000 a year, also claim attention, the monies actually received being compared with the Rent Roll, and the accounts of the working of the Euston and Victoria and Holyhead Hotels are also examined. Care is also taken to see that proper dividends are received on all the shares held by the Company in other undertakings.

The Auditors also visit half-yearly, and oftener when needful, the large manufacturing and other centres of the Company, at Crewe, Wolverton, Earlestown, Holyhead, Birmingham, Watford, and Stafford; they go through the books kept at these places, to satisfy themselves that the large sums there disbursed, are properly accounted for, and that a due amount of repair and renewal is regularly executed in order that the rolling stock, way, and works may be constantly maintained in a full and efficient condition. They also require from the head of each department a certificate to this effect; as well as to the correctness of the stores under his charge and as a further check upon the latter very important item, an officer wholly unconnected with any of these departments is constantly employed in a careful stock-taking, with a view to ascertaining that the amounts as shown in the half-yearly balance sheet, are fully represented by stores actually in existence upon the Company's premises.

The books of eleven subsidiary Companies, which though more or less merged in the general undertaking, yet retain an organization of their own, receive the personal examination of the Auditors, as also the books of the three lines held jointly with the Lancashire and Yorkshire Company, and of the five concerns in which the Great Western Company are partners. The accounts of two railways held jointly with the Manchester Sheffield etc. Company, and one with the Midland Company, also pass under their review.

The London and North Western ledgers are kept by double entry and upon the cash basis,—that is to say, no entries are made until the monies actually appear in the Bankers' Pass Books; this system, therefore, requires that at the end of each half-year the amount of the Company's revenue liabilities, not discharged, should be accurately ascertained and brought to debit of the account. Large sums are thus charged and also credited, in anticipation of the actual payment or receipt of the monies, the details of which require and receive the careful attention of the Auditors.

The charge for stores, etc. consumed at Crewe, and at other large Depots, as also for rails, etc., used by the permanent way department, is, however, always one month in arrear—the books, so far as materials are concerned, being closed one month earlier than the date of the printed half-yearly accounts. This is rendered needful, in order that the adjustment of these voluminous details may not delay the holding of the General Meeting and declaration of the dividend beyond the period fixed by Act of Parliament, but is amply compensated by the anticipation of the charge for the guaranteed dividends on the Lancaster and Carlisle, South Stafford and Kendal and Windermere Railways —these dividends being paid up to 31st January and 31st July in each year, and being fully provided out of the earnings of the preceding half-years ending 31st December and 30th June. The large amount of interest payable on the Debenture Stock and Bonds, is also provided 15 days in advance, the date to which it accrues being so many days later than the expiry of the half-year in which the full amount is charged. Taken together these make very large figures, and have always been regarded by the Auditors as more than sufficient to cover the above named arrears of Stores, and any other charges which (by accident or otherwise) may not have been fully brought against the half-year's Revenue.

The Capital account is very jealously watched, the disbursements being made almost entirely under the authority of estimates previously submitted to and sanctioned by special resolutions of the Shareholders at the General Meetings; each work as voted, has a separate credit opened, against which the expenditure is charged monthly, until the whole operation is complete—this register is laid before the Auditors, who inquire into any case where the outlay is materially in excess of the estimate sanctioned. Any capital money expended, without having been thus previously voted is distinctly and separately shown in the Half-yearly Report, and nothing is allowed to be charged to Capital which ought to devolve upon Revenue, for when any doubt arises, the debit is always carried to the latter account. As illustrating this principle, the following instances are adduced:

Since the year 1851, the whole charge for Interest and Dividends has been borne by the Revenue Account, without any regard being had to the amount of Capital unproductively employed in new lines, etc., under construction, though this at times has been very considerable. The Auditors have no hesitation in stating that, during this period, at least £500,000 has been thus paid out of Revenue.

No money raised by the sale or issue of shares at a premium is ever allowed to swell the amount of divisible profit, the capital expenditure being always reduced by such amounts, the sum now standing at the credit of Capital under this head, arising out of the issue of the last 5 per Cent. Preference Stock alone, being upwards of £240,000.

The building of its own Engines at Crewe has been very advantageous to the Company, a comparison of the charge for the 256 added to the Stock during the past four years, with the prices which they would have cost if purchased, shows a saving of capital outlay to the extent of at least £150,000; while the system which has so long prevailed of renewing, at the cost of Revenue, the small original engines with more powerful machines, has also largely improved the Company's property. During the last ten years 400 engines have thus been renewed, and an additional tractive power, equal, at the very least, to 100 engines, has been acquired, without any charge to capital. In addition to the foregoing, there are now 128 of the old engines, which, though replaced by Revenue, and standing on the books, at their value as old materials only, are yet mostly at active work on the Line, and represent a large additional value to the Company's assets. The building and renewing of the Carriage and Wagon stock have also been effected upon the same principle.

Additional Tools and Machinery to a considerable extent, also all the Carts, Horses, and other stock at work at the various stations on the line, have for many years past been supplied entirely at the cost of Revenue.

The committee may have observed half-yearly charges against Revenue under the head Maintenance of Way, for additional works which might have been placed to the debit of capital. The original timber bridges have also been extensively renewed, during the last ten years, with structures of iron, and stone or brick, and as no charge has ever been admitted to Capital account in this respect, sums large, but not easy to be estimated, have been defrayed by Revenue. In the case of the Penmaenmawr Tunnel, the old loose timber covering was last year replaced with a permanent roof of iron and stonework, at a cost of £9,438 15s 11d, and the St. Helen's Canal, recently taken over by the Company, has been greatly improved, by means of an outlay of £24,600 6s 5d; in neither instance was any charge made to Capital.

In further elucidation, the Auditors would state, that when largely increased traffic requirements compel the removal of original stations and their replacement with more extensive premises, a liberal estimate is made of the cost of the buildings destroyed, this is defrayed out of revenue by instalments, the first being charged as soon as the working of the line receives the benefit of the improved accommodation. The most recent instance of this kind is that of the large sheds at Camden, towards the construction of which Revenue has contributed £15,000, spread over the last three half-years.

With the exception of the 'Renewal of Road Suspense Account', of very remote origin (regularly shown on the face of the Half-yearly accounts), now reduced to £37,688 9s 10d, and in course of steady liquidation upon a scheme approved by the proprietors, there are no arrears of expenditure which can be brought hereafter against Revenue. The heavy annual charge to Revenue for Parliamentary Expenses is mainly due to the fact that, except when Acts are obtained, authorising the construction of Branch lines, no charge whatever is made to Capital Account; the long standing arrear of these expenses is at last, however, about cleared off, and care will be taken to prevent any future accumulations.

The Insurance and Depreciation Funds present favourable results. The 'Fire Insurance' Account, after allowing for the late losses, will show a balance of upwards of £30,000 to credit. The 'Steam-boat Insurance and Depreciation Funds' amount to £130,000, and have been invested in the purchase of five Steamers, now actively at work, and replacing two of the inferior original boats; while the other two original steamers have also been enlarged and improved at the cost of these funds; but over and above all this there is still in hand a free balance of £16,000.

Many similar instances might be mentioned; but the Auditors believe that they have now stated enough fully to demonstrate the system that prevails, and to convince the committee, that complete reliance may be placed upon the truthfulness of the Company's accounts.

The Auditors avail themselves of this opportunity of announcing the resignation of Mr J. E. Coleman, and the appointment in his stead of Messrs Price, Holyland, & Waterhouse, as the Public Accountants, to assist the Auditors, in accordance with the Resolution of the Proprietors, at the Half-Yearly Meeting, held 21st February, 1851.

Henry Crosfield
R. W. Hand

Auditors

13 Economics of the steamship, 1882

(Royal Commission on agriculture [Richmond Commission], minutes of evidence, *B.P.P.*, 1882, XIV, qq. 65952–65)

Mr F. R. Leyland called in and examined

(*The President*) You are, I believe, the owner of the Leyland line of steamers?—I am.

How many vessels are there in that line?—Thirty large steamers.

Where do those steamers trade to?—To Boston in America; to differents ports in the Mediterranean, the Levant, and the Black Sea; to Spain, Portugal, and Egypt.

Your steamers, I believe, bring corn and cattle from America?—They do.

Do you obtain corn or cattle from any other country than America? —Yes, I carry corn from the Black Sea and from the Danube, and cattle from Oporto.

Can you give the Commission the amount of corn which you have shipped from America during the last six months, or for any period? —On an average, we bring from America about 500,000 quarters a year.

How does that amount compare with the amount brought during a corresponding period of last year?—It is about the same, because these vessels from America, being general ships, take a certain quantity to ballast them; they take about 8,000 quarters each voyage, so that it always comes to about the same.

Whence do you obtain your corn?—In America from Boston; the great bulk of the corn that is shipped on the Atlantic sea-board comes from Chicago.

Could you tell us how far the corn is carried in America before it reaches the port from which it is shipped?—From Chicago to Boston is about 1,500 miles; how far it comes before it gets to Chicago I have no means of ascertaining.

Have you the means of knowing what the cost of carriage is in America?—Yes, I can give you the cost of carriage from Chicago right through to Liverpool, because the custom there is to take the corn at Chicago and deliver it at Liverpool at a through rate, including everything. I can give you the averages for three years. In 1879 the average cost per bushel of 60 lbs weight was $11\frac{1}{2}d$ from Chicago to Liverpool; in 1880 the average cost was $14d$: and in 1881 the average cost was $10d$. I can tell you what proportion the steamer got and what proportion the railway got.

Will you give us, first of all, the cost of carriage in America, and then the freightage from America to Liverpool?—It is generally done by division; that is to say, the railway company work at a through rate, and then they arrange with the steamer to take a certain proportion of it. In 1879 the steamer got, of that $11\frac{1}{2}d$, $5\frac{1}{2}d$ and five per cent., which is equal to about $5\frac{3}{4}d$ per bushel; in 1880 the steamer got $5\frac{1}{2}d$ and five per cent.; and in 1881 the steamer got $3\frac{3}{4}d$ and five per cent.; so that, you see, that would leave, on the average, about $4d$ per bushel to the ship and $6d$ per bushel to the railway.

Through how many hands does the corn pass between the American grower and the English consumer?—I cannot tell you that; I can only tell you that, from the time it is purchased in the market at Chicago, it passes simply from the railway into the ship's hands, and then it is conveyed to Liverpool and delivered to the consignee. The tendency of things, of course, is to employ as few middlemen as possible; and the trade has now got into such a groove that an order is sent out from Liverpool straight to Chicago, and the corn is bought there at a certain price, and is delivered free on board in Liverpool; so that there is really no intervention by a middleman.

Do you think that the supply of American corn will be as abundant in the future as it has been during the last few years?—That is a difficult question to answer. I should think that the tendency is decidedly towards a considerable increase, because the length of railways that is now being constructed in America, in every direction, is so great that it must bring an additional area of corn-growing country into direct competition with this country.

Do you think that the cost of the carriage will increase or diminish? —I think that the strong tendency is to diminish, for this reason, that if you take first the sea transport, all this trade is being now carried on by steamers, so that one may confine one's attention to that. The cost of constructing a steamer is now much less than it was 10 years ago. The steamers that I have now employed in that trade carry about 3,000 tons of dead weight. They were built about nine years ago at a cost which, for that period, was considered low, the market being then depressed. Last year 1 built two vessels to carry 4,800 tons of cargo, that is 1,800 tons or 60 per cent more cargo, for less money than I built the first ship for. In addition to that you have to consider, not only that the cost of the construction of the ships is very much less than it has been, but the cost of working ships has diminished probably 30 or 40 per cent owing to the improved style of engines that have been put into the ships; less coal is burnt; fewer men are required to work them; improved appliances, such as steam winches for loading and discharging cargo, take the place of hand labour; and every one knows what an immense amount of saving that is; improved appliances are used even for the working of the sails; and the result is that steamers are now worked at certainly 30 or 40 per cent less cost than they were 10 years ago; and so far as my experience goes the tendency is distinctly all in that direction. As the trade increases—and the American trade, as we all know, has increased enormously within the last few years—larger vessels are employed. Vessels that used to be employed for the American trade were ships constructed to carry a couple of thousand tons, and now it is a very common thing to have vessels employed that carry 4,000 tons, and we all know that a vessel of that size can be worked at very much less cost. . . .

14 Apologia for textile warehouses, 1890

(Reuben Spencer, *The home trade of Manchester with personal reminiscences and occasional notes* (2nd ed. London, 1890), pp. 52–4)

. . .The centres of textile manufacturing industry, which *must* be visited by the wholesale buyers of our warehouses, exist in various parts of our own country, in Scotland, Ireland, and Wales, and also in Germany, France, Italy, Switzerland, Austria, and Russia. The goods selected from all these centres, together with the article of our own manufacture, are distributed from Manchester over the whole of Britain, our colonies, and to the continents of Europe and America, we might say the world. The great and almost immeasurable diversity of taste and culture, and the corresponding variety of requirements to be found within so wide an area of distribution, will necessitate a large number of distributors, who possess capital and ability—men and women who carefully study the wants of their clients, and who can stock their warehouses with suitable materials for meeting their demands. It is absurd to say, as some do, that the trade will be swallowed by a few houses. Of course, the best-managed houses, and the most capable and industrious men who have capital, will rise above others, and secure a larger trade. But that three or six men or firms in Manchester do or ever can possess all the genius and force necessary to absorb and conduct the whole of the textile branch, and satisfy the tastes of the thirty thousand to fifty thousand merchants and drapers, who deal with Manchester houses, is beyond any intelligent man's belief. Take, for instance, mantles, ready-mades, furniture, print, haberdashery, fancy goods, dress goods, or almost any branch: a moment's reflection will show that there is both room and need for a large number of merchants and general dealers, persons or firms, who devote themselves to one, two, or three special articles, apart altogether from the manufacturers. There are at present forty thousand commercial travellers. Now, suppose that, instead of these travellers working from given distributing centres, the whole of the manufacturers in Leicester, Nottingham, Derby, Northampton, Loughborough, Hinckley, Belper, Luton, London, Bolton, Bury, Wigan, Preston, Colne, Clitheroe, Blackburn, Haslingden, Bacup, Accrington, Heywood, Rochdale, Hebden Bridge, Halifax, Bradford, Dewsbury, Batley, Huddersfield, Leeds, Wakefield, Barnsley, Heckmondwike, Sowerby Bridge, Todmorden, Stockport, Ashton, Denton, and a score of other towns in this country, where textile goods are made, together with towns in Scotland, Ireland, and Wales, as well as in the numerous continental centres of textile manufacture—let us suppose that all these were to send their own travellers to the drapers in the cities and towns of this, not to say of other countries, the loss of time and the annoyance to the retail trader would only be exceeded in foolish impracticability by the wasteful expenditure that would be incurred, and for which the people, the real consumers, must pay.

The reader will thus see the necessity there is for wholesale warehouses and fixed centres for distribution. They must exist; and for economy of expenses the fewer general centres the better.

15 A pioneer of the multiple store, 1895

(*Trewman's Flying Post*, Exeter, 30 November 1895)

LIPTON IN THE BISCUIT TRADE

Lipton, the People's Food Provider, has now commenced manufacturing biscuits on an extensive scale in his own Factories, which have been specially built and fitted up with all the latest and improved machinery and travelling ovens of the most up-to-date type, and is now selling at all his Branches and Agencies throughout the Kingdom

BISCUITS AT PRICES HITHERTO UNKNOWN LIPTON THE LARGEST PROVISION DEALER IN THE WORLD. FANCY CAKE AND BISCUIT BAKER. TEA, COFFEE AND COCOA PLANTER, CEYLON.

Fruit Grower, Cocoa and Chocolate Manufacturer, Maker of Soups, Sauces, Potted Meats, Bottled Fruits, Jams, Jellies and Marmalade.

Local Branches—Exeter; 173 Fore Street.
Plymouth; 24 Bedford Street.

BRANCHES EVERYWHERE AGENCIES THROUGHOUT THE WORLD

16 London's traffic problem, 1903

(Royal Commission on London traffic, *B.P.P.*, 1906, XLI, Appendixes to the evidence, Nos. 37, 52)

17 "E-532 Documents"

(*i*) *From a traffic census, 8 a.m. to 12 p.m.*

	Strand, by the Law Courts	Piccadilly, west of Circus
	15 July 1903	31 July 1903
Description of Traffic		
2 wheeled carts	812	642
Do. carriages and hansoms	2,599	3,495
4-wheeled carts and waggons	2,371	2,574
Do. carriages and cabs	819	1,257
Buses [horse-drawn]	5,443	5,707
Cycles	1,090	867
Motor Cars, private	33	172
Do. commercial	16	33
Total Vehicles	13,183	14,747
Total Foot Passengers	51,610	62,370

(*ii*) *The coming of the motor-bus—evidence of a director of the London General Omnibus Company*

...The omnibus proprietors are using their utmost endeavours to provide a means of mechanical traction for their vehicles, and numerous motors are constantly submitted to one or other of the companies or proprietors.

These motors are given a most careful trial, and at the present time the principal companies have contracts pending for the production of specimen motor omnibuses, and they are advised by electrical and motor experts that it is only a question of time, and that in all probability a short one, before a satisfactory motor omnibus will be produced, in which case they submit with confidence that they will not only be able to hold their own with every other form of traction, but will, by their superior mobility, supersede tramcars, whether horse-drawn or electric.

17 Water transport, 1906

(Royal Commission on the canals and inland navigations of the United Kingdom, minutes of evidence, *B.P.P.*, 1906, XXXII, qq. 10154, 10171–86, 10202–8)

(*Chairman*) You are a Director of Messrs. Rowntree & Company's works at York?—Yes.

What you do is, that you use the Ouse?—Yes.

Do you use it to a considerable extent?—For practically all our heavy traffic except coals.

You mean your heavy goods traffic?—Yes.

Has that traffic to be carted two miles from the wharf on the Ouse? —That is so.

Whereas the railway takes away from you and delivers direct into your warehouses?—Yes.

Are the water facilities, even under those disadvantages, of great use to you?—Of very great use.

Especially in the carriage of your raw material?—Yes, almost entirely for raw materials.

Is it practically no use to you in sending out your finished goods? —No, I do not consider it is any use for that.

If the water facilities were improved they would still be chiefly of use to you for your inwards goods?—That is so.

You doubt whether they could ever be much use for your outward traffic?—It is very unlikely, I think.

Can you explain the reason for that?—The reason is because our outwards traffic is mostly in small lots, about one cwt., and it also goes to so many places. We have customers in something like 4,000 different places, and of course the canals do not touch anything like that number, and then in addition to that the customers always want things delivered by return, and they expect to have their goods delivered within two or three days, and the canals could never do that, of course.

Do your consignments of manufactured products average only about a hundredweight?—About a hundredweight, yes.

Then the distribution is so widely spread that you would have to make use of the railway?—We are obliged to.

Is the time taken in the delivery of raw materials to you not usually of great moment?—No, not of very great moment.

What are the principal waterways you use?—We use the Ouse almost entirely; most of our goods come either to Hull or to Goole. They either come direct from the continent to Hull or to Goole, or else from London coastwise to Hull and Goole, and then up the Humber and Ouse.

What are your principal raw materials?—Coal as regards tonnage is our principal raw material—then there are timber, sugar, cocoa, gum, almonds, glucose, lime-juice, cardboard, and wood-pulp. These are the principal heavy materials. . . .

17*

We have still some old works which are in the centre of the town and which are about equi-distant from the wharf and from the railway, so that the comparison is a fair one; there being cartage both from the wharf and from the railway station. I took five classes of goods of which we use large quanities (it is no use taking a lot of small things of which we do not use any great quantity), sugar, glucose, cocoa, gum, and almonds. These come under classes 1, 2 and 3 under the Railway Classification: and working these five out and averaging them the average rate per mile from Goole to York by water transit is 3·04d.

Is that per ton per mile?—Per ton per mile. The distance from Goole to York being thirty-five miles by rail.

Now will you give us the rate per mile?—The average rate by rail is 4·14d, the average rate by water is 3·04d, the distance by rail being thirty-five miles, and the rates per mile being worked out on the railway mileage in both cases. That is to say, the increased cost of rail over water is 30 per cent, on those five things.

Can you give the same from Hull to York?—I took the rates on the same goods from Hull to York, a distance of forty-five miles by rail. The average railway rate is 3·17d, and the average water rate is 2·4d— that is to say, the increased cost of rail over water is 32 per cent.

(*Lord Farrer*) Does that include delivery or not?—It includes delivery. And collection?—Yes.

(*Chairman*) Can you give the same from London to York?—I took the rates on the same goods from London to York, the average railway rate per ton per mile being 2·11d, the average rate per ton per mile by water being 1·23d, that is to say that the increased cost of rail over water is 71 per cent. and there I took the railway mileage in each case, 189 miles.

18 Heyday of co-operation, 1935

(Co-operative Wholesale Society, *The people's yearbook, 1937* (Manchester, 1937), pp. 61–2)

Co-operative Employment

The total number of co-operative employees in all types of society in the year 1935, was 301,717. Of this number, the retail societies employed 210,953. The total figure represents an increase from 164,383 at the end of 1918.

The number of employees engaged in retail distributive societies for 1918 was 119,629. These employees were divided into two categories—those engaged in production and those engaged in distribution. The average wage to the former amounted to £96·09 and to the latter, £82·75. In 1935, amongst the employees of retail societies, there were 39,595 employees classed as engaged in production and service, and their average wages were £140·63; 133,663 were engaged in distribution and their average wages were £126·05; 37,695 were classed as engaged in transport and their average wages were £138·53.

In the Wholesale societies there were 61,614 employees, of whom 51,719 were described as being engaged in production and service, and 9,895 were described as distributive workers. The average wage in the productive and service class was £125·09, and in the distributive class £175·84.

The total wage bill of the Co-operative Movement in 1935 amounted to £38,606,296.

Trade

The total retail trade of the Co-operative Movement amounted, in 1935, to £220,429,517. Owing to the absence of adequate national statistics of distribution, it is impossible to state with accuracy what proportion of the total retail trade of the country is transacted by the Co-operative Movement. The most reliable estimates vary between 9 per cent and 15 per cent.

In some branches of retail trade, however, the movement handles a considerably higher proportion than is indicated by these figures, while there are some branches in which its proportion of the total trade is much lower. Co-operative societies have been particularly successful in developing their trade in domestic necessities—such as groceries, coal, and, in recent years, milk. In the case of the latter article they handled in 1935 more than one-fifth of the total quantity of liquid milk consumed in Great Britain. In the case of fashion goods, furniture, and those commodities which are not the subject of day to day purchasing, they have not been so successful. It will be observed, from the following table, that a more balanced distribution of trade over the various departments is gradually being obtained. This may probably be attributed largely to the increasing concentration of the membership in societies of a size sufficiently large to enable them to undertake a wider range of services than is possible in the case of smaller societies.

Analysis of Co-operative Retail Trade

Department	Trade £ million		Percentage of Total Trade	
	1925	1932	1925	1932
Grocery	126·0	118·9	68·6	59·1
Butchering	12·4	16·8	6·8	8·6
Dairy	4·7	12·3	2·5	6·1
Coal	5·0	9·1	2·7	4·5
Confectionery	*4·8	2·0	*2·6	1·0
Greengrocery and Fish	1·7	3·1	0·9	1·5
Drapery	11·0	16·0	6·0	8·0
Boot and Shoe	3·6	5·1	2·0	2·5
Tailoring and Outfitting	2·7	5·3	1·5	2·6
Chemist	0·3	1·0	0·2	0·5
Furnishing[1]	—	5·6	—	2·8
Sundry and Unclassified	11·3	6·0	6·2	3·0
Total Retail Trade	183·5	201·2	100·0	100·0

*Includes some bread trade later included in Grocery.
[1]Includes Hardware, Earthenware, and Jewellery.

The combined trade of the English C.W.S. and the Scottish C.W.S. in 1935 amounted to £117 million at wholesale prices. It may be estimated that approximately one-third has to be added to this figure in order to arrive at the total trade of the wholesale societies in retail prices. To the resultant figure of £156 million must also be added the trade of the productive societies, the bulk of which is transacted with co-operative societies, and the value of the productions of retail societies in their own factories, and of the services—such as boot and shoe repairing—which they render to their members, and which are included in their total sales.

When allowance is made for all these factors it would appear that only a comparatively small proportion, probably about 10 per cent of the trade of the retail societies does not pass through some co-operative wholesale agency.

19 Railways in decline, 1939

(*The Economist*, 4 March 1939. Excerpts from the statement of the chairman, Lord Stamp, at the annual meeting of the London, Midland and Scottish Railway Company.)

RESULTS OF THE YEAR

Railway traffic receipts

At our last meeting I pointed out that for the fourth successive year our railway traffic receipts had increased and that for the first eight weeks of 1938 we had improved on 1937 by £317,000, and the general prospect over our system was that trade would be just about the same level in 1938 as in 1937.

By the end of the seventeenth week, however, the increase had been wiped out, and for the remainder of the year there was a serious decline, resulting in a final decrease of railway traffic receipts amounting to £3,005,000—a decline which has continued so far in 1939 at the average rate of £78,000 per week.

The reasons for the decline are well known to you—a sharp recession in trade since March, which was accentuated by the international situation, and a further diversion of certain classes of merchandise traffic to other forms of transport, coupled with reductions of rates for vulnerable traffic retained. In addition, large stocks had been built up by various trading interests in 1937, and there was a consequent fall in traffic conveyed in 1938.

Traffic Reductions

I gave you last year the decreases in the various classes of railway traffic receipts in 1937 as compared with 1929, and repeating these figures in relation to those for 1938 will best indicate where we have suffered.

	Reduction per cent of 1929	
	In 1937	In 1938
Passengers	6·4	5·6
Parcels, etc.	4·9	6·1
Merchandise (Class 7 and upwards)	14·4	21·6
Merchandise (Classes 1–6)	4·6	20·7
Coal	3·8	6·1
Livestock	28·8	36·2
Total railway traffic receipts	8·2	12·4

20 The shops of a county town, 1950

(Board of Trade Statistics Division, *Britain's shops: a statistical summary of shops and service establishments* (H.M.S.O., 1952), p. 40. The town is Carlisle, population 68,290.)

	Number	Number per 10,000 population
TOTAL	1043	152·7
RETAIL SHOPS		
TOTAL	821	120·2
Grocers, dairy shops	242	35·4
Butchers, fishmongers, poulterers	101	14·8
Greengrocers, fruiterers	32	4·7
Bread and flour confectioners	36	5·3
Other food and drinks including general shops	3	0·4
Confectioners, newsagents, tobacconists	77	11·3
Boots and shoes	24	3·5
Men's wear	35	5·1
Women's and children's wear	68	10·0
Hardware	18	2·6
Chemist's wares, photographic goods	31	4·5
Other retail shops	154	22·6
SERVICE TRADE ESTABLISHMENTS		
Restaurants, cafés, etc.	60	8·8
Hairdressing	52	7·6
Repair establishments	57	8·3
Motor trade establishments	53	7·8

21 The first motorway, 1955

(*Parliamentary Debates, Commons*, 536 (2 February 1955), cols 1100–4, 1112–13. The first part of the M1 was opened in 1957.)

The Minister of Transport and Civil Aviation (*Mr John Boyd-Carpenter*) I am now in a position to make a statement on the first instalment of the Government's expanded road programme. There have been few major improvements and very little new construction in our road system since 1939, and the problem which now faces us is, therefore, an immense one.

Her Majesty's Government are working to plans which will take a good many years to complete, but which are intended to provide this country with an up-to date road system. . . .

These plans include both entirely new motor roads of the most modern construction and an ordered series of major and minor works on existing through routes, the individual items of which are planned so as taken together to convert these routes into modern all-purpose highways. The construction of new roads is a process which takes some time to begin, because compliance with the statutory procedure designed for the protection of the interests likely to be affected means that a considerable period must elapse before the line of the road can be finally determined and the necessary land acquired. Consequently, the work done in the early years of the programme must necessarily be mainly in connection with the improvement of existing routes.

Items have been selected for an early place in the programme which, in the judgment of Her Majesty's Government, are of the greatest urgency for relieving congestion of traffic, particularly industrial traffic, and for promoting road safety. In this selection we have necessarily had regard to such essential preliminary considerations as the state of preparedness of schemes and the availability of land.

I propose to circulate in the *Official Report* a list of schemes which I hope to authorise during the financial year 1955–56. For trunk roads, for which I am the highway authority, this list includes schemes costing over £100,000 each; but in view of the need for further consultation with local highway authorities which is proceeding in respect of classified road schemes, I have included only two such schemes, each of which involves an Exchequer contribution of over £500,000. One, the Dartford-Purfleet Tunnel, was in the list announced by my predecessor, and the other has been inserted after consultation with the promoting authority.

The list does not, as the House will see, include the Hyde Park boulevard scheme. That is at present under consideration. I should mention another important London scheme which, I hope, to authorise before the end of this financial year. That is the first instalment of route 11, to which I referred in my statement on traffic congestion on 17th November last. I am also circulating in the *Official Report* a list of schemes on trunk and classified roads likely to cost the Exchequer more than £500,000 each which I hope to authorise during the three years 1956–57, 1957–58 and 1958–59. I have not at present gone below the £500,000 mark in the list for these later years, because the final selection of a large number of the many schemes involved requires further consultation with highway authorities and others. . . .

Over and above schemes in particular localities, there are certain major projects of national importance which should be ready for commitments towards the end of this period. They will be projects of great magnitude and the cost will be formidable; indeed, to enable us to proceed as rapidly as we should like the Government have in mind that tolls should be charged in suitable cases. This will enable the

Exchequer to get back something on the money put up and will, of course, include provision for sharing between the Exchequer and local authorities where the latter had also put up money.

The first of these projects will be the first section of a London-Yorkshire motor road; that is, a road confined to motor traffic with severely restricted access and with fly-over crossings and junctions. The first section will extend from the north end of the proposed St Albans by-pass to a point near Rugby, with a spur for Birmingham traffic connecting with the existing trunk road A 45 to Birmingham, which road is itself to have dual carriageways and by-passes at Dunchurch and Meriden. Following on the first section of the London-Yorkshire motor road the Government would wish to put in hand, not only the remainder of that motor road, but also the North-South motor road through Lancashire from Preston to Birmingham.

The effect of the proposals now put forward is that the work authorised in the financial year 1955–56 will represent an ultimate expenditure from the Exchequer of approximately £27 million and the total to be authorised during the following three years will involve an ultimate Government expenditure of about £120 million. This excludes expenditure on the very large projects of national importance to which I have just referred. There will also be the substantial complementary expenditure of local highway authorities on classified road schemes.

The House will realise that on a programme of this sort it takes some time for payments to reach the level of authorisations, the actual expenditure on most schemes being spread over a number of years like the constructional work itself, and falling for the most part two or three years after authorisation. I do not think it would be appropriate for me to forecast now the precise scale of the programme any farther into the future. The actual amount of work to be authorised must clearly depend on the resources available and on the economic and financial condition of the country at the time. But it is the Government's firm intention to continue with a substantial programme of road construction and improvement—at least on the scale I have indicated—until the roads of this country are adequate for the traffic they have to bear.

*   *   *

C Schemes in England and Wales proposed to be authorised in 1956–57, 1957–58, and 1958–59 (each costing the Exchequer over £500,000).

(*i*) *London–Yorkshire Motor Road*

The programme provides for the construction of the London–Dunchurch (for Birmingham) section of the London–Yorkshire Motor Road. It will be built to modern engineering standards and restricted, under the powers given by the special Roads Act, 1949, to motor traffic only. No frontage access will be permitted and the motor road will be carried over or under all roads by means of fly-overs, connections being provided with the principal main roads, designed so as to keep the different streams of traffic in continuous movement. This section starts at the northern end of the proposed St. Albans by-pass, which itself will be constructed to motor roads standards, and passes to the west of Luton in Bedfordshire, thence through the eastern end of Buckinghamshire, entering Northamptonshire near Hanslope. It crosses the trunk road A 45 about half-way between Weedon and Northampton and reaches trunk road A 5 about $1\frac{1}{2}$ miles north of Ashby St Ledgers.

At this point a spur for the Birmingham traffic runs westwards crossing the Northamptonshire/Warwickshire boundary about $2\frac{1}{2}$ miles south of Rugby and terminating at the eastern end of the Dunchurch by-pass. Beyond this point the Birmingham traffic will use the existing trunk road A 45, which will have dual carriageways together with by-passes at Dunchurch and Meriden.

The programme also includes the provisions of dual carriageways from the Hampstead Borough boundary at Finchley Road to the commencement of the St Albans by-pass, as well as the construction of a by-pass to Doncaster which forms the terminal point of the complete motor road.

22 Marketing of greengrocery, 1956

('Report of the [Runciman] Committee on Horticultural Marketing', *B. P. P*, 1956–57, XIV, pp. 19–20, 39–41)

56 It is estimated that there are some 150,000 retail shops handling fruit and vegetables in the United Kingdom. Of this number, some 40,000 are greengrocery shops, or greengrocery with fish, selling horticultural produce as their main business, but most of the remainder handle fruit and vegetables merely as a subsidiary part of their business. Grocers, for example, often sell tomatoes, apples, and oranges; some butchers sell tomatoes; many of the big department stores have a counter selling selected fruits and occasionally vegetables; and everyone is familiar with village shops in the rural areas which sell the

staple and less perishable fruit and a few vegetables along with a comprehensive range of other goods. In the remoter districts, particularly of Scotland, fruit and vegetables are mostly sold from general grocery shops, and there are few greengrocers selling only fruit and vegetables.

57 There are many different types of greengrocery shop, ranging from the small back-street shop in the heavily populated areas of the big towns to which the housewife can go each day to buy the potatoes and vegetables she wants for dinner, to the specialist shop in the town centre stocking the more expensive types of fruit and flowers. Between these extremes are high-street shops which stock a wide range of produce and aim to catch the eye of the passer-by and the housewife who is doing her weekly shopping, and, quite different in type, the shops in the new suburban areas whose customers are regular but do not shop daily. There are also established retail markets where stallholders sell from regular pitches, travelling shops on regular rounds, and street vendors doing a casual trade. In some places, there are producer-retail markets such as the Women's Institute market stalls where locally grown produce, mostly from private gardens and allotments, is sold direct to the public. In the last few years, supermarkets selling under one roof many different products mainly on the self-service principle have also entered the field.

58 All these different types of shops afford different services. The West End florist's shop has little in common with a small side-street shop selling coarse vegetables. But different consumers have different requirements. Some look for good service, some for high quality, others for cheapness.

59 The diversity of the retail trade not only helps to provide the customer with what she wants, whether it be convenience, quality, or cheapness, and ensures active competition in the trade, but it also affords a necessary range of outlets for the producer. As horticultural crops do not grow to plan, there will always be some produce of higher quality and some of lower from the same field or orchard, and part of the crop may be early and part late. If a grower aims to market graded produce of reasonably high quality, it may be vital to him to find an outlet for his second-best. If he can get high prices in the luxury trade for specially selected produce or out of season produce, his return for his crop taken as a whole may be brought up to an economic level in seasons when the general run of the crop fetches only low prices. Furthermore, certain commodities handled in the luxury or specialist trade often form an integral part of the overall plan of

production of a holding that also grows the more staple commodities, and a market must be found for them if the holding is to pay its way. Flowers, for example, play an essential part in the economy of the glasshouse industry. By using his glasshouses for flowers when the tomatoes have been cleared a grower can spread his overheads, keep his staff employed over the whole year, and increase his earning capacity. If there were not a market for glasshouse blooms in the specialist florist trade, it is more than likely that there would be fewer tomatoes in the average greengrocers' shops and that prices would be higher.

60 Shopping habits and tastes vary from place to place. In the industrial areas, housewives often go out to buy their vegetables every day. In the suburban areas where shops tend to be farther away and where there is more space in the houses for storage, housewives tend to do their shopping less frequently. In all areas, however, Friday and Saturday are the days when most produce is bought. Particular local tastes are often firmly rooted. Consumers in Glasgow, so we are informed, seem to like big apples, whereas in London the medium sized apple is more popular; the Lancashire towns and London are good markets for cos lettuce, but other areas will not buy cos lettuce; the Scottish industrial towns consume large quantities of leeks, swedes, and parsley which are purchased in relatively small amounts in the towns of southern England.

61 A large part of the total sales of fruit and vegetables is made by small single shop businesses, and the volume of business handled by multiple firms is proportionately smaller than in most other trades.

*   *   *

140 On our visits to the [wholesale] markets we have noticed great keenness in buying and selling. Retailers often visit several salesmen to get the feel of prices before they buy. A successful buyer is one who can get an article for a penny or two per crate less than his competitors. It is worthwhile for him to bargain for a few pence. Now that there is no general shortage of most commodities, any salesman whose prices were consistently higher than those of his competitors would soon lose custom. Where there is active competition, the market price has a very real meaning. But although there is such a thing as a market price, this does not mean that there is anything like a fixed price for the day for any commodity. Quality can be so variable that there is room for wide variation of price around the average. A consignment of strawberries carelessly picked, badly packed, and possibly damaged

in transit, although it might have looked good when it left the farm may have deteriorated so much by the time it is sold that it may only obtain a price that is low compared with the prices quoted as the market average. On the whole, the best produce sells first. Later in the day, prices—particularly of highly perishable commodities—may have to be reduced in order to clear, and late senders, or those whose produce is below standard, may have to take lower prices. But even although actual prices may have varied through the day for these reasons, the regular buyer is conscious that he has made his purchases in the light of the market price. If he buys early, possibly at higher prices, he knows that the produce is fresh and can be on sale in his shop quickly. If he prefers to wait until later in the day expecting to pay lower prices he must face the risk that supplies will run short or that only inferior produce will be available. If he is a casual trader, he may be quite content to wait until the very end of the day, buying if prices are cheap enough to suit him, but otherwise content not to buy. According to his type of trade, so the buyer makes his purchases.

141 Because, by their command of a wide range of produce, the primary markets attract buyers serving great numbers of consumers, they are able to move great quantities of produce. They are consequently able to establish a balance between supply and demand in a way that smaller markets cannot. A small local market supplied by grower-wholesalers or local merchants can be glutted and prices be broken by an unexpected consignment of quite a small amount—perhaps as little as a lorry load. Growers with more produce to sell than they have a known local outlet for are naturally reluctant to risk overloading such a market. If, however, they send their surplus to a major market that can draw upon the reserves of buying power of a large population, they know that there is a reasonable prospect that the surplus will be absorbed. In times of abundance and temporary over-supply the primary markets absorb produce for which there is no immediate local outlet, and distribute it to wherever an outlet can be found. At such times, wholesale prices may fluctuate, but not to the same extent as they would if produce were sent casually to smaller markets.

142 Again, because of their size and the amount of business done in them, the big primary markets are the best points for wholesalers operating on subordinate markets or having local businesses to buy commodities of which they happen to be short. If they try to buy these locally they may, by doing so force up the price; but their influence on a major market will be proportionately so small that their purchases will not unduly disturb the price. Wholesalers consequently tend to

look to the major primary markets as the main source of supply for produce which may temporarily be scarce in their own locality.

143 Outside London, the main primary markets are Birmingham, Glasgow, Liverpool, and Manchester, each of which handles well over £10 million of produce annually. Other markets which can reasonably be regarded as major primary markets, although the volume of sales is less than at the markets mentioned above, are Bristol, Cardiff, Edinburgh, Hull, Leeds, Newcastle, Nottingham, Sheffield and Southampton. On all these markets there are available at most times of the year all the horticultural commodities that are in general demand. Buyers attending them know that they can expect to find all the produce which their customers are likely to want. They also know that there are enough wholesalers handling each commodity to ensure effective competition in setting prices. It is natural that retailers should generally prefer to obtain their supplies if they can conveniently do so, from major primary markets than from subordinate markets or from secondary distributors.

144 Covent Garden is the most important of all the primary markets handing some £70 million produce annually. It is recognised in the trade that if a particular commodity is to be found anywhere it will be found at Covent Garden. The market receives imported produce from all over the world. It is a convenient collecting centre for produce from important producing areas such as Kent, Essex, and Lincolnshire. The concentration of railway and road services in London enables produce from Cornwall, Worcester, and even Scotland to be sent more quickly to Covent Garden than to other places that may be nearer. For the same reason, buyers from far afield can get supplies reasonably conveniently through Covent Garden. It was almost inevitable that it should develop into a major market. The amount of produce now sold at Covent Garden is much greater than at any other market and buyers from all over the country, even although they may not always buy at Covent Garden, keep a close eye on business there.

145 If because supplies are abundant, prices at Covent Garden are lower than those in the provinces, a rapid movement of produce can be expected from London to the provinces. Provincial wholesalers therefore know that they cannot for long keep prices out of line with those ruling at Covent Garden, allowance being made for costs of transportation. A similar movement of produce from other primary markets will also occur if there are significant differences in prices. Big buyers normally keep in touch with several markets and will buy

in one rather than in another if there is any significant difference in price. The increased use of the telephone and the development of rapid transport by road have made the markets much more sensitive to price changes elsewhere than they formerly were. The effective network of market intelligence built up between individual traders operating on different markets has resulted in a more balanced distribution of produce and has tended to keep prices more uniform throughout the country.

23 The railways—rescue operations

(i) Time of hope, 1955

(British Transport Commission, *Modernisation and re-equipment of British Railways*, (H.M.S.O. 1955), pp. 6–8)

6 The aim must be to exploit the great natural advantages of railways as bulk transporters of passengers and goods and to revolutionise the character of the services provided for both—not only by the full utilisation of a modern equipment but also by a purposeful concentration on those functions which the railways can be made to perform more efficiently than other forms of transport, whether by road, air or water.

7 There need be no doubt about the main components of the expenditure under the Plan, but while it is possible to estimate in round figures the costs involved, obviously any calculations that look so far ahead must be qualified by some reserve. All the figures and estimates given below are based upon conditions ruling in the autumn of 1954. Subject to this, the heads of the Plan may be summarised as follows: *First*, the track and signalling must be improved to make higher speeds possible over trunk routes, and to provide for better utilisation of the physical assets; there will be an extended use of colour-light signalling, track circuits and automatic train control, the further introduction of power-operated signal boxes, and the installation of centralised traffic control where conditions are suitable; and the extended use of modern telecommunication services .... £210 million
Secondly, steam must be replaced as a form of motive power, electric or diesel traction being rapidly introduced as may be most suitable in the light of the development of the Plan over the years; this will involve the electrification of large mileages of route, and the introduction of several thousand electric or diesel locomotives .... £345 million
Thirdly, much of the existing steam-drawn passenger rolling stock must

be replaced, largely by multiple-unit electric or diesel trains; the remaining passenger rolling stock, which will be drawn by locomotives (whether electric, diesel or steam), must be modernised; the principal passenger stations and parcels depots will also require considerable expenditure .... £285 million

Fourthly, the freight services must be drastically remodelled. Continuous brakes will be fitted to all freight wagons, which will lead to faster and smoother operation of freight traffic; marshalling yards and goods terminal facilities will be re-sited and modernised, and in particular the number of marshalling yards will be greatly reduced. Larger wagons will be introduced, particularly for mineral traffic, and loading and unloading appliances will require extensive modernisation in consequence .... £365 million

Fifthly, expenditure will be required on sundry other items, including improvements at the packet ports, staff welfare, office mechanisms, etc; and a sum of at least £10 million for development and research work will be associated with the Plan, making a total of ...... £35 million

Total	£1,240 million
Say	£1,200 million

8 The result will be a transformation of virtually all the forms of service now offered by British Railways. In particular:

(*i*) as regards passenger services, remodelling of the operations will provide fast, clean, regular and frequent services, electric or diesel, in all the great urban areas; inter-city and main-line trains will be accelerated and made more punctual; services on other routes will be made reasonably economic, or will be transferred to road:

(*ii*) as regards freight services, there will be a complete re-orientation of operations designed to speed up movement, to reduce its cost, and to provide direct transits for main streams of traffic; and to attract to the railway a due proportion of the full-load merchandise traffic which would otherwise pass by road.

9 The economic benefit to be derived will be of a decisive order. In the Commission's view the expenditure will ultimately attract a return amounting to at least £85 million a year.

10 The final answer will of course depend not only on working efficiency, but also on the additional amount of remunerative traffic that can be attracted to the railway system. As regards passenger services, the remarkable growth in the volume of personal travel during the last few decades seems likely to continue, so that the market for passenger travel, urban or long-distance, private or business, should

18 "E-532 Documents"

continue to expand. Despite air transport and the private car, therefore, and notwithstanding the fact that the total volume of travel
includes a great deal of movement for which railways cannot be
competitive, there will remain a large pool in which the railways will
take a larger share, once the quality and cost of the services are
transformed. At the same time as the railways attract further traffic
which they are inherently suited to carry (provided that the most
modern equipment is available), certain other traffics, which are now
carried at disproportionately high costs and are inherently more suited
for road transport, will be gradually transferred to road.

11 As to freight transport, the available forecasts of industrial
development seem to show that, even after allowing for some rational-
isation of production to save transport, the total demand will continue
to grow. The extent to which the railways will be able to share in
this demand will depend on their ability to provide improved services
at lower cost; but the possibilities in this direction are great. The
normal trend of increased production should of itself assist the railways, and some of the ground that has been lost to other forms of
transport over the past thirty years should be recovered. On balance,
these increments of traffic should more than counterbalance any local
falling-off in traffics that may follow changes in the pattern of industrial
development, or the handing over to road of traffics that are better
suited to road transport.

(ii) Retrenchment and a modest future, 1963

(British Railways Board, *The reshaping of British Railways* [*Beeching
Report*] (H.M.S.O., 1963), Part I, pp. 2, 3, 11, 57-9)

It is, of course, the responsibility of the British Railways Board so
to shape and operate the railways as to make them pay, but, if it is
not already apparent from the preceding paragraphs, it must be clearly
stated that the proposals now made are not directed towards achieving
that result by the simple and unsatisfactory method of rejecting all
those parts of the system which do not pay already or which cannot
be made to pay easily. On the contrary, the changes proposed are
intended to shape the railways to meet present day requirements by
enabling them to provide as much of the total transport of the country
as they can provide well. To this end, proposals are directed towards
developing to the full those parts of the system and those services
which can be made to meet traffic requirements more efficiently and
satisfactorily than any available alternative form of transport, and

towards eliminating only those services which, by their very nature, railways are ill-suited to provide.

The point at issue here is so important that it is worthwhile to emphasise it by expressing the underlying thought in a different way.

The profitability or otherwise of a railway system is dependent on a number of external influences which may change markedly from time to time, important among them being decisions affecting the freedom of use, cost of use, and availability of roads. For this and other reasons, it is impossible to plan the maximum use of railways consistent with profitability, for years ahead, without some risk that it will prove, in the event, that services have been over-provided and that overall profitability is not achieved. On the other hand, to retain only those parts of the existing system which are virtually certain to be self-supporting under any reasonably probable future conditions would lead to a grave risk of destroying assets which, in the event, might have proved to be valuable.

Confronted with this dilemma, arising from the impossibility of assessing future conditions and future profitability very reliably, the Railways Board have put forward proposals for reshaping the system which are conservative with regard to closures and restrainedly speculative with regard to new developments, but which are all directed towards shaping the system to provide rail transport for only that part of the total national traffic pattern which costing and commonsense consideration show to have characteristics favourable to rail transport. ...

The railways emerged from the war at a fairly high level of activity, but in a poor physical state. They were able to pay their way, because road transport facilities were still limited, and they continued to do so until 1952. From then onwards, however, the surplus on operating account declined progressively. After 1953 it became too small to meet capital charges, after 1955 it disappeared, and by 1960 the annual loss on operating account had risen to £67·7 m. This rose further to £86·9 m. in 1961.

In 1955, a modernisation plan was embarked upon. It was a plan to modernise equipment, but it did not envisage any basic changes in the scope of railway services or in the general mode of operation of the railway system. It was expected that the substitution of electric and diesel haulage for steam, concentration of marshalling yards, reduction in number and increased mechanisation of goods depots, re-signalling, and the introduction of other modern equipment, would make the railways pay by reducing costs and attracting more traffic.

18*

By 1960, however, it had become apparent that the effects of modern-isation were neither so rapid nor so pronounced as had been forecast, that the downward trend in some railway traffics would persist, and the operating losses were likely to go on increasing unless radical changes were made.

*   *   *

In April 1961, British Railways had about 7,000 stations open to traffic, equivalent to one for every 2½ miles of route. ...

An analysis of the passenger receipts arising at passenger stations, excluding some very little used ones and unstaffed halts, was made from very complete records kept in 1960. ... One third of the stations produced less than 1 per cent of the total passenger receipts and one half of the stations produced only 2 per cent. At the other end of the scale 34 stations, or less than 1 per cent of the total, produced 26 per cent of the receipts.

Of the freight stations, one third produced less than 1 per cent of the station freight receipts and one half of the stations produced less than 3 per cent.

The total revenue derived from the least used half of the total number of stations and the cost of running them is set out below.

	£m per annum
Receipts from:	
Originating passenger, parcels and other coaching train traffics at the least used 50 per cent of all passenger stations	4.8
Freight traffic forwarded from the least used 50 per cent of all freight stations	1.7
Estimated cost of least used 50 per cent of all stations ..	9.0

*   *   *

The thought underlying the whole Report is that the railways should be used to meet that part of the total transport requirement of the country for which they offer the best available means, and that they should cease to do things for which they are ill suited. To this end, studies were made to determine the extent to which the present pattern of the railways' services is consistent with the characteristics which distinguish railways as a mode of transport, namely: the high cost of their specialised and exclusive route system, and their low cost per unit moved if traffic is carried in dense flows of well-loaded through trains.

As a result, it is concluded that, in many respects, they are being used in ways which emphasise their disadvantages and fail to exploit their advantages.

The proposals for reshaping the railways are all directed towards giving them a route system, a pattern of traffics, and a mode of operation, such as to make the field which they cover one in which their merits predominate and in which they can be competitive.

To this end, it is proposed to build up traffic on the well-loaded routes, to foster those traffics which lend themselves to movement in well-loaded through trains, and to develop the new services necessary for that purpose. At the same time, it is proposed to close down routes which are so lightly loaded as to have no chance of paying their way, and to discontinue services which cannot be provided economically by rail. These proposals are, however, not so sweeping as to attempt to bring the railways to a final pattern in one stage with the associated risks of abandoning too much or, alternatively, of spending wastefully.

Although railways can only be economic if routes carry dense traffic, density is so low over much of the system that revenue derived from the movement of passengers and freight over more than half the route miles of British Railways is insufficient to cover the cost of the route alone. In other words, revenue does not pay for the maintenance of the track and the maintenance and operation of the signalling system, quite apart from the cost of running trains, depots, yards and stations. Also, it is found that the cost of more than half of the stations is greater than the receipts from traffic which they originate.

Amongst traffics, stopping passenger services are exceptionally poor. As a group, they are very lightly loaded and do not cover their own movements costs They account for most of the train miles on much of the lightly loaded route mileage, but also account for a considerable train mileage on more heavily loaded routes, and are one of the main causes for the continued existence of many of the small and uneconomic stations.

Fast and semi-fast, inter-city passenger trains are potentially profitable and need to be developed selectively, along with other forms of traffic on trunk routes. High peak traffics at holiday periods are, however, very unremunerative. They are dying away and provision for them will be reduced.

Suburban services feeding London come close to covering their full expenses, but give no margin to provide for costly increases in capacity, even though they are overloaded and demand goes on increasing.

Suburban services feeding other centres of population are serious loss makers, and it will not be possible to continue them satisfactorily without treating them as a part of a concerted system of transport for the cities which they serve.

Freight traffic, like passenger traffic, includes good flows, but also includes much which is unsuitable, or which is unsuitably handled by the railways at present. The greater part of all freight traffic is handled by the staging forward of individual wagons from yard to yard, instead of by through-train movement. This is costly, and causes transit times to be slow and variable. It also leads to low utilisation of wagons and necessitates the provision of a very large and costly wagon fleet.

Coal traffic as a whole just about pays its way, but, in spite of its suitability for through train movement, about two-thirds of the total coal handled on rail still moves by the wagon-load. This is very largely due to the absence of facilities for train loading at the pits, and to the multiplicity of small receiving terminals to which coal is consigned. Block train movement is increasing, but substantial savings will result from acceleration of the change. This depends, in turn, upon provision of bunkers for train loading at the pits, bunkers for ship loading at the ports, and of coal concentration depots to which coal can be moved by rail for final road distribution to small industrial and domestic consumers.

Wagon-load freight traffic, other than coal, is a bad loss maker when taken as a group, but over half of it is siding-to-siding traffic, much of which moves in trainload quantities, and this makes a good contribution to system cost. One third of the remainder moves between sidings and docks, and this falls just short of covering its direct costs. The remaining 30 per cent of the whole passes through stations, at one or both ends of its transit, and causes a loss relative to direct expenses which is so large that it submerges the credit margin on all the rest.

Freight sundries traffic is also a bad loss maker. It is handled at present between over 900 stations and depots, which causes very poor wagon loading and a high level of costly transhipment of the freight while in transit. Railways handle only about 45 per cent of this traffic in the country, and do not select the flows which are most suitable for rail movement. If they are to stay in the business, British Railways must concentrate more upon the inter-city flows and reduce the number of depots handling this form of traffic to not more than a hundred.

Study of traffic not on rail shows that there is a considerable tonnage which is potentially good rail traffic. This includes about 8 m. tons which could be carried in train-load quantities, and a further 30 m. tons which is favourable to rail by virtue of the consignment sizes, lengths of haul, and terminal conditions. In addition, there is a further 16 m. tons which is potentially good traffic for a new kind of service —a Liner Train service—for the combined road and rail movement of containerised merchandise.

Preliminary studies of a system of liner train services, which might carry at least the 16 m. tons of new traffic referred to above and a similar quantity drawn from traffic which is now carried unremuneratively on rail, show such services to be very promising and likely to contribute substantially to support of the main railway network, if developed.

5

Finance

By the standards of continental Europe, of Latin America or of the United States of America, England has enjoyed a remarkable degree of financial stability over the past two or even three centuries. Since the foundation of the Bank of England in 1694, the government has always honoured its obligations to pay interest on and to repay at due time its internal debts. A family that had enjoyed an unearned income of £2,000 a year in 1760 would still, with the same money income, have been in an enviable position in 1914, though there has been a sharp fall in the value of money since then. In the nineteenth century it was not uncommon for banks to fail (*A6*), but wholesale collapse such as happened in the United States as recently as 1933, has been avoided. Property owners have found this state of affairs much to their liking, and financial stability is undoubtedly a sign of economic strength. Whether it is also a cause of prosperity as bankers would argue is more debatable. In the nineteenth century the question would scarcely have seemed worth asking. Nowadays when unorthodox finance seems as likely to induce economic growth as traditional policies the matter is less clear.

Within the general pattern of stability there have been substantial changes. Unit banks and family businesses have given way to branch banks and joint-stock companies (*A5*, *B2*, *3*, *4*). In the eighteenth century a substantial slice of the national debt was held by the Dutch; by the early twentieth century foreign investment by the British had reached enormous proportions (*B1*, *5*). Since 1914 Britain has borrowed as well as lent, and with the relative weakening of her international economic position, devaluation and some repudiation of external debt has occurred. Hire-purchase companies and other financial institutions have competed with the banks for customers and business (*A10*, *11*).

A. BANKING AND CURRENCY

1 The suspension of cash payments, 1797

(Reports of Committees on the Bank of England, 1797 and 1826, *B.P.P.*, 1826, III, pp. 142, 255–6)

The alarm of Invasion [in 1796–1797] which, when an immediate attack was first apprehended in Ireland, had occasioned some extraordinary demand for cash on the Bank of England, in the months of December and January last, began in February to produce similar results in the north of England. Your Committee find, that in consequence of this apprehension, the farmers suddenly brought the produce of their lands to sale, and carried the notes of the Country Banks, which they had collected by those and other means, into those banks for payment; that this unusual and sudden demand for cash reduced the several banks at Newcastle to the necessity of suspending their payments in specie, and of availing themselves of all the means in their power of procuring a speedy supply of cash from the metropolis; that the effects of this demand on the Newcastle banks and their suspension of payments in cash, soon spread over various parts of the country, from whence similar applications were consequently made to the metropolis for cash; that the alarm thus diffused not only occasioned an increased demand for cash in the country, but probably a disposition in many to hoard what was thus obtained; that this call on the metropolis, through whatever channels, directly affected the Bank of England, as the great repository of cash, and was in the course of still further operation upon it, when stopped by the Minute of Council of the 26th of February.[1]

＊　　　＊　　　＊

Your Committee find, that the Court of Directors of the Bank did, on the 26th October 1797, come to a Resolution, a copy of which is subjoined to this Report.

Your Committee, having further examined the Governor and Deputy Governor, as to what may be meant by the political circumstances mentioned in that resolution, find, that they understand by them the state of hostility in which the nation is still involved, and particularly such apprehensions as may be entertained of invasion, either in

---

[1] The Minute of February 26, 1797 suspended the obligation of the Bank of England to pay coin for its notes.

Ireland or in this country, together with the possibility there may be of advances being to be made from this country to Ireland; and that from these circumstances so explained, and from the nature of the war, and the avowed purpose of the enemy to attack this country by means of its public credit, and to distress it in its financial operations, they are led to think that it will be expedient to continue the restriction now subsisting, with the reserve for partial issues of cash, at the discretion of the Bank, of the nature of that contained in the present Acts; and that it may be so continued, without injury to the credit of the Bank, and to the advantage of the nation.

'*Resolved*, that it is the opinion of this Court,[2] that the Governor and Company of the Bank of England are enabled to issue Specie, in any manner that may be deemed necessary for the accommodation of the public; and the Court have no hesitation to declare that the affairs of the Bank are in such a state, that it can with safety resume its accustomed functions, if the political circumstances of the country do not render it inexpedient: but the Directors deeming it foreign to their province to judge of these points, wish to submit to the wisdom of Parliament, whether, as it has been once judged proper to lay a restriction on the payment of the bank in cash, it may, or may not, be prudent to continue the same?'[3]

2 ... And its consequences, 1810

(Report from the select committee on the high price of gold bullion, *B.P.P.*, 1810, III, pp. 22–3, 30–1)

Another Director of the Bank, *Mr Harman*, being asked, If he thought that the sum total of discounts applied for, even though the accommodation afforded should be on the security of good bills to safe persons, might be such as to produce some excess in the quantity of the Bank issues if fully complied with? he answered. 'I think if we discount only for solid persons, and such paper as is for real *bona fide* transactions, we cannot materially err'. And he afterwards states, that what he should consider as the test of a superabundance would be, 'money being more plentiful in the market'.

It is material to observe, that both *Mr Whitmore* and *Mr Pearse* state that 'the Bank does not comply with the whole demand upon them for discounts, and that they are never induced, by a view to their

---

[2] Copy of a Resolution of the Court of Directors of the Bank of England at a meeting on Thursday, 26 October 1797.

[3] The Bank of England resumed cash payments in 1819.

own profit, to push their issues beyond what they deem consistent with the public interest'.

Another very important part of the Evidence of these Gentlemen upon this point, is contained in the following Extract:

'Is it your opinion that the same security would exist against any excess in the issues of the Bank, if the rate of the discount were reduced from £5 to £4 per cent?'. Answer—'The security of an excess of issue would be, I conceive, precisely the same. '*Mr Pearse*—'I concur in that Answer'.

'If it were reduced to £3 per cent?'.—Mr Whitmore, 'I conceive there would be no difference if our practice remained the same as now, of not forcing a note into circulation'. *Mr Pearse*—'I concur in that Answer'.

Your Committee cannot help again calling the attention of the House to the view which this Evidence presents, of the consequences which have resulted from the peculiar situation in which the Bank of England was placed by the suspension of Cash payments. So long as the paper of the Bank was convertible into specie at the will of the holder, it was enough, both for the safety of the Bank and for the public interest in what regarded its circulating medium, that the Directors attended only to the character and quality of the Bills discounted, as real ones and payable at fixed and short periods. They could not much exceed the proper bounds in respect of the quantity and amount of Bills discounted, so as thereby to produce an excess of their paper in circulation, without quickly finding that the surplus returned upon themselves in demand for specie. The private interest of the Bank to guard themselves against a continued demand of that nature, was a sufficient protection for the public against any such excess of Bank paper, as would occasion a material fall in the relative value of the circulating medium. The restriction of cash payments, as has already been shewn, having rendered the same preventive policy no longer necessary to the Bank, has removed that check upon its issues which was the public security against an excess. When the Bank Directors were no longer exposed to the inconvenience of a drain upon them for Gold, they naturally felt that they had no such inconvenience to guard against by a more restrained system of discounts and advances; and it was very natural for them to pursue as before (but without that sort of guard and limitation which was now become unnecessary to their own security) the same liberal and prudent system of commercial advances from which the prosperity of their own establishment had resulted, as well as in a great degree the commercial prosperity of the whole Country. It was natural for the Bank Directors to believe, that nothing but benefit could accrue to the public at large, while they saw the growth of Bank profits go hand in hand with the accommodations granted to the Merchants. It was hardly to be expected of the Directors of the Bank, that they should be fully aware of the consequences that might result from their pursuing, after the suspension of cash payments, the same system which they had found a safe one before. To watch the operation of so new a law, and to provide against the injury which might result from it to the public interests, was the province, not so much of the Bank as of the Legislature: And, in the opinion of Your Committee, there is room to regret that this House has not taken earlier notice of all the consequences of that law.

By far the most important of these consequences is, that while the convertibility into specie no longer exists as a check to an over issue of paper, the Bank Directors have not perceived that the removal of that check rendered it possible that such an excess might be issued by the discount of perfectly good bills. So far from perceiving this, Your Committee have shewn that they maintain the contrary doctrine with the utmost confidence, however it may be qualified occasionally by some of their expressions. That this doctrine is a very fallacious one, Your Committee cannot entertain a doubt. The fallacy, upon which it is founded, lies in not distinguishing between an advance of capital to Merchants, and an addition of supply of currency to the general mass of circulating medium. If the advance of capital only is considered, as made to those who are ready to employ it in judicious and productive undertakings, it is evident there need be no other limit to the total amount of advances than what the means of the lender, and his prudence in the selection of borrowers, may impose. But, in the present situation of the Bank, entrusted as it is with the function of supplying the public with that paper currency which forms the basis of our circulation, and at the same time not subjected to the liability of converting the paper into specie, every advance which it makes of capital to the Merchants in the shape of discount, becomes an addition also to the mass of circulating medium. In the first instance, when the advance is made by notes paid in discount of a bill, it is undoubtedly so much capital, so much power of making purchases, placed in the hands of the Merchant who receives the notes; and if those hands are safe, the operation is so far, and in this its first step, useful and productive to the public. But as soon as the portion of circulating medium, in which the advance was thus made, performs in the hands of him to whom it was advanced this its first operation as capital, as soon as the notes are exchanged by him for some other article which is capital, they fall into the channel of circulation as so much circulating medium, and form an addition to the mass of currency. The necessary effect of every such addition to the mass, is to diminish the relative value of any given portion of that mass in exchange for commodities. If the addition were made by notes convertible into specie, this diminution of the relative value of any given portion of the whole mass would speedily bring back upon the Bank, which issued the notes, as much as was excessive. But if by law they are not so convertible, of course this excess will not be brought back, but will remain in the channel of circulation, until paid in again to the Bank itself in discharge of the bills which were originally discounted. During the whole time they remain out, they perform all the functions of circulating medium; and before they come to be paid in discharge of those bills, they have already been followed by a new issue of notes in a similar operation of discounting. Each successive advance repeats the same process. If the whole sum of discounts continues outstanding at a given amount, there will remain permanently out in circulation a corresponding amount of paper; and if the amount of discounts is progressively increasing, the amount of paper, which remains out in circulation over and above what is otherwise wanted for the occasions of the Public will progressively increase also, and the money prices of commodities will progressively rise. This progress may be as indefinite, as the range of speculation and adventure in a great commercial country.

It is necessary to observe, that the law, which in this Country limits the rate of interest, and of course the rate at which the Bank can legally discount, exposes the Bank to still more extensive demands for commercial discounts. While the rate of commercial profit is very considerably higher than five per cent as it has lately been in many branches of our Foreign trade, there is in fact no limit to the demands which Merchants of perfectly good capital, and of the most prudent spirit of enterprise, may be tempted to make upon the Bank for accommodation and facilities by discount. Nor can any argument or illustration place in a more striking point of view the extent to which such of the Bank Directors, as were examined before the Committee, seem to have in theory embraced that doctrine upon which Your Committee have made these observations, as well as the practical consequences to which that doctrine may lead in periods of a high spirit of commercial adventure, than the opinion which *Mr Whitmore* and *Mr Pearse* have delivered; that the same complete security to the public against any excess in the issues of the Bank would exist if the rate of discount were reduced from five to four or even to three per cent.

*　　*　　*

Upon a review of all the facts and reasonings which have been submitted to the consideration of Your Committee in the course of their Enquiry, they have formed an Opinion, which they submit to the House:—That there is at present an excess in the paper circulation of this Country, of which the most unequivocal symptom is the very high price of Bullion, and next to that, the low state of the Continental Exchanges; that this excess is to be ascribed to the want of a sufficient check and control in the issues of paper from the Bank of England; and originally, to the suspension of cash payments, which removed the natural and true control. For upon a general view of the subject, Your Committee are of opinion, that no safe, certain, and constantly adequate provision against an excess of paper currency, either occasional or permanent, can be found, except in the convertibility of all such paper into specie. Your Committee cannot, therefore, but see reason to regret, that the suspension of cash payments, which, in the most favourable light in which it can be viewed, was only a temporary measure, has been continued so long; and particularly, that by the manner in which the present continuing Act is framed, the character should have been given to it of a permanent war measure.

Your Committee conceive that it would be superfluous to point out, in detail, the disadvantages which must result to the Country, from any such general excess of currency as lowers its relative value. The effect of such an augmentation of prices upon all money transactions for time; the unavoidable injury suffered by annuitants, and by creditors of every description, both private and public; the unintended advantage gained by Government and all other debtors; are consequences too obvious to require proof, and too repugnant to justice to be left without remedy. By far the most important portion of this effect appears to Your Committee to be that which is communicated to the wages of common country labour, the rate of which, it is well known, adapts itself more slowly to the changes which happen in the value of money, than the price of any other species of labour or commodity. And it is enough for Your Committee to allude to some classes of the public servants, whose pay, if once raised in consequence of a depreciation of money, cannot so conveniently be reduced again to its former rate, even after money shall have recovered its value. The future progress of these inconveniencies and evils, if not checked, must at no great distance of time, work a practical conviction upon the minds of all those who may still doubt their existence; but even if their progressive increase were less probable than it appears to Your Committee, they cannot help expressing an opinion, that the integrity and honour of Parliament are concerned, not to authorize, longer than is required by imperious necessity, the continuance in this great commercial Country of a system of circulation, in which that natural check or control is absent which maintains the value of money, and, by the permanency of that common standard of value, secures the substantial justice and faith of monied contracts and obligations between man and man.

Your Committee moreover beg leave to advert to the temptation to resort to a depreciation even of the value of the Gold coin by an alteration of the standard, to which Parliament itself might be subjected by a great and long continued excess of paper. This has been the resource of many Governments under such circumstances, and is the obvious and most easy remedy to the evil in question. But it is unnecessary to dwell on the breach of public faith and dereliction of a primary duty of Government, which would manifestly be implied in preferring the reduction of the coin down to the standard of the paper, to the restoration of the paper to the legal standard of the coin.

3 Bill brokers and country banks, 1810

(Report from the select committee on the high price of gold bullion, *B.P.P.*, 1810, III, pp. 122–4, 138–9)

(*i*) *Discount of bills*

Thomas Richardson, Esq called in, and Examined

I believe you are a bill broker?—Yes.

You are also an agent for country banks?—Yes.

Have country banks increased in number since the restriction on the Bank of England?—Very considerably.

Can you tell in what proportion?—No, I never made any calculation.

Do you know how many country banks there are?—No, I do not, it might be easily ascertained from the printed Lists of Country Bankers.

Are you aware that the notes of the country bankers in circulation are much increased?—I have no doubt of it; very considerably.

Are those notes which are made payable in London increased? —Yes, I should think very much.

Do you mean the notes of country banks generally are increased? —Yes, both descriptions; those made payable in London, and those which are not.

What means have you of knowing they are increased?—General observation.

What is the nature of the agency for country banks?—It is two-fold; in the first place to procure money for country bankers on bills when they have occasion to borrow on discount, which is not often the case; and in the next place, to lend the money for the country bankers on bills on discount. The sums of money which I lend for country bankers on discount are fifty times more than the sums borrowed for country bankers.

Do you send London bills into the country for discount?—Yes.

Do you receive bills from the country upon London in return, at a date, to be discounted?—Yes, to a very considerable amount, from particular parts of the country.

Are not both sets of bills by this means under discount?—No, the bills received from one part of the country are sent down to another part for discount.

And they are not discounted in London?—No. In some parts of the country there is but little circulation of bills drawn upon London, as in Norfolk, Suffolk, Essex, Sussex, etc. but there is there a considerable circulation of country bank notes, principally optional notes. In Lancashire there is little or no circulation of country bank notes; but there is a great circulation of bills drawn upon London at two or three months date. I receive bills to a considerable amount from Lancashire in particular, and remit them to Norfolk, Suffolk, etc. where the bankers have large lodgements, and much surplus money to advance on bills for discount.

Do you not send bills drawn in London by one merchant upon another, to be discounted in the country?—Yes, to a considerable amount.

Are not bills of that description called notes, in London?—Generally so.

How do you get your remittances for those bills that you so send to be discounted?—In bills that have three or four days to run, or by orders for cash on bankers in London.

What part of the country are they sent into?—Norfolk, Suffolk, etc. and small sums into some parts of Yorkshire.

Are not the returns sometimes made in bills at two months, or other dates? —It is very seldom the case, unless it be in exchange for a bill of a much longer date.

Do not transactions of this nature take place to the amount of several hundred thousand pounds a year?—I have never had any transaction of the sort last described. In the modes of discounting previously mentioned, many millions go through my hands in a year.

How many millions pass through your hands in the course of the year? —I should certainly speak within bounds if I say seven or eight millions.

Do the country bankers in general keep agents in London, exclusive of the bankers on whom they draw?—No, not of the description of which I am.

Are not the agents principally employed for the purpose of lending the money of the country bankers on discount on bills accepted in London?—We are employed both by those who have money to lend, and those who want to borrow money.

You have stated, that seven millions of money pass through your hands annually; what proportion of that may you have lent for country bankers on discount?—A million and a half. I speak of the sum outstanding upon discount at one time,on account of country bankers, which, multiplied about four or five times in the year, owing to the bills being from two to three months, will amount to the aggregate sum which I have mentioned.

Then it follows that the seven millions which have passed through your hands, have been lent for country bankers on discount?—Yes, I have no transactions whatever but which relate to discount.

Do you know, in point of fact, whether such transactions as you have now described, were in practice previous to the suspension of the cash payments of the Bank?—Yes, they were.

Do you know whether they were practised to a similar extent?—No, they were not.

In what proportion, compared with the present time?—I cannot form any exact criterion.

Can you state to the Committee the cause of such difference?—I believe it to be on account of the increase of country paper, and also Bank of England paper.

Are the bills so discounted on behalf of the country banks, such as the Bank of England would refuse to discount?—At least two-thirds of them, on account of their having more than 65 days to run.

Are there any other reasons for which you think the Bank would refuse discounting such bills?—Yes.

State them.—Some houses have more occasion for discount than others; the Bank only take a limited amount. The business of some houses arises principally at one period of the year when they make their sales; they then want larger accommodations than the Bank would afford them, and many of the bills being indirect, by which I mean not discountable at the Bank without two London indorsements.

Do you ever discount bills for London bankers through the medium of your country correspondents?—I do not believe that it is a general practice for the London bankers to apply for any such discounts.

Will you state what sum of money belonging to country bankers has been employed by you in the last year in the purchase of Exchequer bills, and other Government securities?—In Exchequer bills I do not think £1,000.

In what other securities?—Occasionally we buy stock for country bankers, but only to a very limited amount.

Do you guarantee the bills you discount, and what is your charge per cent? —No, we do not guarantee them; our charge is one-eighth per cent brokerage upon the bill discounted; but we make no charge to the lender of the money.

Do you consider that brokerage as a compensation for the skill which you exercise in selecting the bills, which you thus get discounted?—Yes, for selecting of the bills, writing letters, and other trouble... If a manufacturer has sold his goods at six months, and learns that money is plentiful in London, and that he can have his bill discounted, he will send it to be discounted.

Does not that accommodation tend to increase the business of the country manufacturer?—Yes, no doubt of it; he goes to market again with his ready money.

Can you state what it may cost to raise money by discount in the manner you have described?—It will cost six and a half per-cent per annum to the merchant, supposing the transaction to take place four times in the year, the banker five and a half per-cent per annum.

Will you explain that difference?—The merchant pays from one-eighth to one-fourth per cent for obtaining the bill on the banker in London; the country banker, unless he draws upon his London banker, pays no commission, as he pays away the bill he receives, and indorses it.

Have there been many losses incurred upon bills thus discounted? —No, there have not by us, except to a small amount indeed.

Were there any losses incurred upon such bills, before the restriction upon the cash payments in the Bank?—Yes, many more in the same proportion.

Were not many losses incurred in the year 1793?—To a very large amount.

How do you account for the greater proportion of losses before than since the restriction of the Bank?—I think that many of the country bankers have many losses by taking bills themselves; but those who do their business in London by means of a broker, who understands it, have but few losses.

(ii) Notes and guineas

Vincent Stuckey, Esq called in, and Examined

In what branch of trade are you concerned?—I am concerned in three country banks, viz. Bristol, Bridgewater, and Langport, all in the County of Somerset.

Do all those banks issue notes?—Yes.

State to the Committee the nature of their circulation.—Their circulation of course is chiefly confined to the neighbourhood from whence they are issued; but we conceive they have a more extensive circulation than many other banks, because every note, of whatever value, is made payable in London as well as at the place from whence it is issued.

Has the amount of the circulation of those banks much increased in late years?—We have only opened the Bristol Bank about three years; from that period, till within these six months, the circulation has been increasing; now it is almost stationary. The Langport Bank has been opened nearly forty years, the circulation of that has considerably increased within these last seven years. The Bridgewater Bank has been opened about seven years, and the circulation of that continued increasing for the first six years.

It therefore appears that the circulation of those banks has considerably increased of late years; has it been within your own observation that other banks in the same district have increased in their circulation in the same proportion?—We know but little of the increase of circulation of other banks, and we conceive ours in a considerable degree to have arisen from an increased credit and the liberality with which we have treated our customers.

Do you think that the increased circulation of your notes has tended to diminish the circulation of other paper in their vicinity, or do you

19*

not think that other banks have also added to the amount of theirs?
—It is very probable that other banks have added to the amount of
their circulation; but we conceive ours to have arisen, and to continue,
for the reasons I have before stated.

Have you the means of knowing whether there has been any material
increase in the number of banks in the West of England, and the
amount of the circulation of the paper of country banks in that district,
during these few years last past?—There is no doubt but a very considerable increase has taken place in the number of banks, I cannot
speak so positively as to their circulation; but although many banks
have been opened in our immediate neighbourhood, we have not
found our own circulation decreased.

Is it the practice of the banks in your district to issue notes upon
real security upon mortgage?—We are not fond of lending upon
mortgage, and seldom do it; we generally issue our notes by discounting good bills, or by lending cash for a short period to agriculturists
upon their own security, or the best that under all circumstances we
think proper.

Do country banks find it necessary to keep a deposit of Bank of
England notes in proportion to the issues of their own paper, and to
the probable demands which may be made upon them for the payment of that paper?—We have hitherto kept but a small quantity of
Bank of England notes, but a large proportion of guineas.

Have you lately found any material increased demand for guineas?
—At Bristol we have found an increased demand, but very little increased at Bridgewater or Langport.

Do guineas to any great extent circulate in the West of England?—I
should imagine not to any considerable extent.

Do you know whether Bank of England notes circulated in the
country have increased or diminished since 1797?—I have no means
of ascertaining that fact; but the circulation of Bank of England notes
is very small, the people in the country generally preferring the notes
of country bankers, whom they conceive to be men of responsibility
in the country.

Is it not your interest as a banker, to check the circulation of Bank
of England notes, and with that view do you not remit to London
such Bank of England notes as you receive beyond the amount which
you may think it prudent to keep as a deposit in your coffers?—Unquestionably.

You have stated, that you have a considerable deposit of guineas;
would you give guineas in exchange for your own notes to any stran-

ger who might require them?—We should not give them guineas for the whole of the notes, but we certainly should give them some, and at this present time.

Do you at present receive in the currency of your trade many payments in guineas?—At Bristol very few, at Bridgewater and Langport we frequently receive them.

What is the principle by which you regulate the issue of your notes? —We always keep assets enough in London, consisting of Stock, Exchequer bills and other convertible property, sufficient to pay the whole of our notes in circulation.

4 Bank Charter Act, 1844

(7 and 8 Vic., c. 32)

An Act to regulate the Issue of Bank Notes, and for giving to the Governor and Company of the Bank of England certain Privileges for a limited Period.

Be it enacted that from and after the thirty-first day of August, one thousand eight hundred and forty-four, the issue of Promissory Notes of the Governor and Company of the Bank of England, payable on demand, shall be separated and thenceforth kept wholly distinct from the general Banking business of the said Governor and Company; and the business of and relating to such issue shall be thenceforth conducted and carried on by the said Governor and Company in a separate department, to be called 'The Issue Department of the Bank of England', subject to the rules and regulations hereinafter contained; and it shall be lawful for the Court of Directors of the said Governor and Company, if they shall think fit, to appoint a committee or committees of directors for the conduct and management of such Issue Department of the Bank of England, and from time to time remove the members, and define, alter, and regulate the constitution and powers of such committee, as they shall think fit, subject to any bye-laws, rules or regulations which may be made for that purpose: provided nevertheless, that the said Issue Department shall always be kept separate and distinct from the Banking Department of the said Governor and Company.

II And be it enacted, that upon the thirty-first day of August, one thousand eight hundred and forty-four, there shall be transferred, appropriated, and set apart by the said Governor and Company to the Issue Department of the Bank of England securities to the value of fourteen million pounds, whereof the debt due by the public to the said Governor and Company shall be and be deemed a part; and there shall also at the same time be transferred, appropriated, and set apart by the said Governor and Company to the said Issue Department so much of the gold coin and gold and silver bullion then held by the Bank of England as shall not be required by the Banking Department thereof; and thereupon there shall be delivered out of the said Issue Department into the said Banking Department of the Bank of England such an amount of Bank of England notes as, together with the Bank of England notes then in circulation, shall be equal to the aggregate amount of the securities, coin and bullion so transferred to the said Issue Department of the Bank of England; and the whole amount of Bank of England notes then in circulation, including those delivered to the Banking Department of the Bank of England as aforesaid, shall be deemed to be issued on the credit of such securities, coin, and bullion so appropriated and set apart to the said Issue Department; and from thenceforth it shall not be lawful for the said Governor and Company to increase the amount of securities for the time being in the said Issue Department, save as hereinafter is mentioned, but it shall be lawful for the said Governor and Company to diminish the amount of such securities, and again to increase the same to any sum not exceeding in the whole the sum of fourteen million pounds, and so from time to time as they shall see occasion; and from and after such transfer and appropriation to the said Issue Department as aforesaid it shall not be lawful for the said Governor and Company to issue Bank of England notes, either into the Banking Department of the Bank of England, or to any persons or person whatsoever, save in exchange for other Bank of England notes, or for gold coin or for gold or silver bullion received or purchased for the said Issue Department under the provisions of this Act, or in exchange for securities acquired and taken in the said Issue Department under the provisions herein contained: provided always, that it shall be lawful for the said Governor and Company in their Banking Department to issue all such Bank of England notes as they shall at any time receive from the said Issue Department or otherwise, in the same manner in all respects as such issue would be lawful to any other person or persons.

✳   ✳   ✳

IV And be it enacted, that from and after the thirty-first day of August, one thousand eight hundred and forty-four, all persons shall be entitled to demand from the Issue Department of the Bank of England, Bank of England notes in exchange for gold bullion, at the rate of three pounds, seventeen shillings and nine-pence per ounce of standard gold: Provided always, that the said Governor and Company shall in all cases be entitled to require such gold bullion to be melted and assayed by persons approved by the said Governor and Company at the expense of the parties tendering such gold bullion.

V Provided always, and be it enacted, that if any banker who on the sixth day of May one thousand eight hundred and forty-four was issuing his own bank notes, shall cease to issue his own bank notes, it shall be lawful for Her Majesty in Council at any time after the cessation of such issue, upon the application of the said Governor and Company, to authorize and empower the said Governor and Company to increase the amount of securities in the said Issue Department beyond the total sum or value of fourteen million pounds and thereupon to issue additional Bank of England notes to an amount not exceeding such increased amount of securities specified in such Order in Council, and so from time to time: provided always that such increased amount of securities specified in such Order in Council shall in no case exceed the proportion of two thirds the amount of bank notes which the banker so ceasing to issue may have been authorized to issue under the provisions of this Act; and every such order in Council shall be published in the next succeeding *London Gazette*.

*   *   *

XII And be it enacted, that if any banker in any part of the United Kingdom who after the passing of this act shall be entitled to issue bank notes shall become bankrupt, or shall cease to carry on the business of a banker, or shall discontinue the issue of bank notes, either by agreement with the Governor and Company of the Bank of England or otherwise, it shall not be lawful for such Banker at any time thereafter to issue any such notes.

*   *   *

XIV Provided always, and be it enacted, That if it shall be made to appear to the Commissioners of stamps and taxes that any two or more banks have, by written contract or agreement (which contract or agreement shall be produced to the said Commissioners), become united within the twelve weeks next preceding such twenty-seventh day of April as aforesaid, it shall be lawful for the said Commissioners to ascertain the average amount of the notes of each such bank in the manner hereinbefore directed, and to certify the average amount of the notes of the two or more banks so united as the amount which the united Bank shall thereafter be authorized to issue, subject to the regulations of this Act.

5 Joint-stock banking, 1849

(James W. Gilbart, *A practical treatise on banking containing an account of the London and country banks* ... *a view of joint-stock banks and the branches of the Bank of England* (5th ed. 1849), I, 165–7)

When the law existed in England that no bank should have more than six partners, the branch system scarcely existed. In some cases, a bank had a branch or two a few miles distant, but no instance occurred of a bank extending itself throughout a county or a district. But with joint-stock banking arose the branch system—the head office was placed in the county town, and branches were opened in the principal towns and villages around. The credit of the bank being firmly established, its notes circulated freely throughout the whole district. The chief advantages of this system are the following:

There is greater security to the public. The security of the whole bank is attached to the transactions of every branch; hence there is greater safety to the public than could be afforded by a number of separate private banks, or even so many independent joint-stock banks. These banks could have but a small number of partners—the paid-up capital and the private property of the partners must be comparatively small; hence the holder of a note issued by one of the independent joint-stock banks could have a claim only on that bank: but if that bank, instead of being independent, were a branch of a large establishment, the holder of a note would have the security of that large establishment; hence the branch system unites together a greater number of persons, and affords a more ample guarantee.

The branch system provides greater facilities for the transmission of money. The sending of money from one town to another is greatly facilitated, if a branch of the same bank be established in each of those towns, for all the branches grant letters of credit upon each other. Otherwise you have to ask the banker in the town from which the money is sent, to give you a bill upon London, which is transmitted by post; or you request him to advise his London agent to pay the money to the London agent of the banker who resides in the town to which the money is remitted. This takes up more time, and is attended with more expense. A facility of transmitting money between two places usually facilitates the trade between those places.

The branch system extends the benefits of banking to small places where independent banks could not be supported. An independent bank must have an independent board of directors who in most cases will be better paid—the manager must have a higher salary, because he has a heavier responsibility, and a large amount of cash must be kept unemployed in the till, because there is no neighbouring resource in case of a run. There must be a paid-up capital, upon which good dividends are expected: a large proportion of the funds must be invested in exchequer bills, or other Government securities, at a low interest, in order that the bank may be prepared to meet sudden calls; and the charge for agencies will also be more. On the other hand, a branch has seldom need of a board of directors, one or two being quite sufficient—the manager is not so well paid: there is no necessity for a large sum in the till, because in case of necessity the branch has recourse to the head office, or to the neighbouring branches; nor is a large portion of its fund invested in Government securities that yield but little interest, as the head office takes charge of this, and can manage it at a less proportional expense. Besides, at some branches, the manager attends only on market days, or once or twice a week. The business done on those days would not bear the expense of an independent establishment.

The branch system provides the means of a due distribution of capital. Some banks raise more capital than they can employ, that is, their notes and deposits amount to more than their loans and discounts. Others employ more capital than they raise, that is, their loans and discounts amount to more than their notes and deposits. Banks that have a surplus capital usually send it to London to be employed by the billbrokers. The banks that want capital must either restrict their business, or send their bills to London to be re-discounted. Now, if two banks, one having too much, and the other too little capital, be situated in the same county, they will have no direct intercourse, and will consequently be of no assistance to each other; but if a district bank be established, and these two banks become branches, then the surplus capital of one branch will be sent to be employed at the other—thus the whole wealth of the district is employed within the district, and the practice of re-discounting bills in London will be proportionably diminished.

6 Perils of industrial banking, 1858

(Select committee on the Bank Acts, *B.P.P.*, 1857–8, V, qq. 3456–65. The witness was K. D. Hodgson, a director of the Bank of England.)

... I went down to Newcastle on Monday night, the 23d of November, and I arrived there on Tuesday morning. I went to the bank at about 11 o'clock, and saw there Mr Jonathan Richardson, Mr Ogden, Mr Hawkes, Mr Sellick, Mr Matthew Bigg, and Mr George Ridley, the Member for Newcastle, who was the chairman, but who, though the chairman, had never been an acting director; he was the chairman of the company, but being in London almost continually, he took no part in the management of the bank. I stated to Mr Richardson and Mr Ogden, who were the principal persons with whom I had conversation there, that I had been sent down by the Bank of England to examine into their books, and to see whether it was possible to render them such assistance as would enable them to go on; but that the first condition of the Bank of England doing anything was, that they should prove themselves solvent; and I therefore requested that they would show me their books. I found that they only balanced once in every three months, and as this was towards the end of November, I was obliged to content myself with the balance of the 30th of September: and the only way in which I could arrive at the actual state of the accounts was by taking this balance of the 30th of September, and on every important account inquiring and endeavouring to ascertain whether any change had taken place in its condition since that period. The result was, that I found the estimated liabilities, as then stated, amounted to 2,600,000*l*, of which there were 1,350,000 *l* of deposits, 1,150,000*l* accounts current, and they had re-discounted 1,500,000*l*, of which they expected that 100,000*l* would come back upon them, and for which they would ultimately be liable, making altogether 2,600,000*l*. Their assets were of a very peculiar nature, and I soon discovered that they were of a kind the early realisation of which would be almost impossible. They held in securities about 1,000,000*l* of different kinds, of which I will give the Committee a note afterwards. They held in trade bills, that is to say, small bills on shopkeepers of Newcastle, about 250,000 *l*, bills which were probably good in themselves, but which were not available anywhere out of Newcastle; they were not bills which could have been discounted in any other part of the money-market. They had in overdrawn accounts, 1,664,000*l* without any specific securities attached to them. Of these 1,664,000*l*, there were 400,000*l* which it was very candidly confessed must be considered as totally bad, and which ought to have been written off long before, but which still remained in the account as good debts. The capital of the concern was 656,000*l* nominally, but in reality it was considerably less than that; because in 1847 they had been in trouble, and in order to get out of that trouble they had made a call, I think, of 5*l* or 10*l* a share, which was not paid upon some of the shares, which shares were forfeited and taken by them into the stock of their bank, to be reissued should occasion warrant their doing so. The consequence was, that the actual capital of the bank was something like 600,000*l*. This statement was, of course, one which at once showed me that any attempt to help them, short of taking up the whole concern and liquidating it for them, would be perfectly useless, and I stated so to them. It was evident that the whole capital was gone, and, looking at the character of the securities. I came to the conclusion, not only that the capital was gone, but that the bank was totally insolvent. Being very much struck with the extraordinary loss which had taken place in the bank, which, when a private bank, I knew to have been a very flourishing one, I inquired whether there was not some old sore of which nothing had as yet been said. I was told that there was one; there was rather a disinclination to mention what it was, but I felt it my duty to press it, and they told me they had a very large debt with the Derwent Iron Company. It is known by the name of the Derwent Iron Company, but in that country such companies have several names; it is sometimes called the Consitt Iron Works, which is the place where the works themselves are situated, but its general term is the Derwent Iron Company. I inquired the amount of this debt and found, much to my astonishment, that it amounted to 750,000*l*, the capital of the bank being 600,000 *l*.

Was there any security for that debt?—There was a kind of security which consisted of 250,000 *l*, of what were called Derwent Iron Company's debentures, which were, however, in reality nothing but the promissory notes of the directors, there being very few persons in this Derwent Iron Company. It is the custom in Northumberland to take the style of a joint-stock company when there are only a few persons in it. The same thing is done whether it is a private partnership or a joint-stock company, therefore these so-called debentures were merely the promissory notes of the partners in the concern. The Northumberland and Durham Bank had also 100,000*l*, mortgage on the plant, and the remaining 400,000 *l* was totally unsecured. In addition to this original debt which was there mentioned to me, of 750,000*l*, there is now another charge upon it of 197,000 *l*, resulting from bills which have not been paid, and which, in order that the Derwent Iron Company might get them discounted, the Northumberland and Durham Bank had endorsed or otherwise guaranteed. These have now come back, so that the total liability for which the Derwent Iron Company is indebted to the Northumberland and Durham Bank is about 947,000*l*; very nearly 1,000,000*l* ....

(*Sir G. C. Lewis.*) I think you stated that there was some peculiarity about the securities; that there was 1,000,000*l* of securities:—Yes; 1,000,000*l* of securities were taken of the most extraordinary nature for any bank to hold that I ever saw, as I think the Committee will agree in thinking, when I mention that that 1,000,000*l* of securities, which was the only tangible asset, with the exception of the local bills, which they had against the 2,600,000*l* of liabilities, consisted of 350,000*l* of the Derwent Iron Company's obligations, 250,000*l* being debentures, and 100,000*l* mortgage on the plant. They had besides these, 100,000*l* on a building speculation at Elswick, near Newcastle, which however was not a primary mortgage, there being a mortgage of 20,000*l* on that land belonging to Mr Hodgson Hinde. They had also another 100,000*l* on other building land and houses in the neighbourhood of Newcastle. They had about 350,000*l* in securities of collieries, works and manufactures of different sorts, and they had about 50,000*l* in Derwent navigation bonds guaranteed by the railway, but which railway was the only security to which they could look in any given time to realise any sum of money; that made about 1,000,000*l* altogether.

(*Chairman*) By 'a given time' you mean that railway bonds had some defined period, but that as regarded the other securities there was no period which could be mentioned within which they could be certain to be realised—Exactly.

Do you remember the time for which the railway debentures were still current?—I think they were dependent on an Act of Parliament, which was not opposed, and which was therefore expected to come into force within six months from that time.

At that time would they be realisable?—They would.

(*Mr G. C. Glyn*) What railway company was it?—I think it was the North-Eastern.

(*Mr Cayley*) You do not mean that the other securities were scarcely marketable at all, do you?—They were absolutely unmarketable.

7 Outbreak of war, 1914

(Charles Wright and Charles E. Fayle, *A history of Lloyd's from the founding of Lloyd's Coffee House to the present day* (Macmillan, 1928), pp. 400–2. The document consists of excerpts from an insurance broker's diary.)

On Wednesday, 29th July, 1914, took about £17,000 in short bills to Bank for discounting. After some demur the bills were discounted at Bank Rate (3%) subject to our paying half the rise should the rate be raised next day.

30th July. Took another batch and found the Bank Director in a state of great agitation, unwilling to look at any bills, stating that crisis was developing badly and that in 24 hours we should hear of serious developments.

Friday, July 31st. It was evident from the course of negotiations that Germany and Russia were shaping for War but no definite announcement was made. This being so, I asked the Bank Director what was the serious development he promised. He replied that a moratorium would be enacted and £1 notes issued. He said that numbers of people had been playing the fool by making a run on the banks for gold. One man had drawn £1,500 and announced he was putting it into the National Safe Deposit! Customers at suburban branches had drawn out gold and then handed it back in a sealed packet for safe custody! I then took the bills to another great bank at which we had an account, but there everyone appeared to be so agitated that I left the place without stating my business.

August 1st. (*Saturday before Bank Holiday*). Lloyd's and Stock Exchange closed to-day. Negotiations evidently taking a highly warlike direction.

August 2nd. (*Sunday*). War in being between Germany and Austria, France and Russia.

August 3rd. (*Bank Holiday*). Sir Edward Grey's speech. Evident British intervention almost certain.

August 4th. Business resumed at Lloyd's. A day of agitation. War declared at midnight. Stock Exchange closed.

August 7th. Banks which had been closed re-opened. Another interview with Bank Director who was unable to forecast developments. He said Banking system of the whole Country was insolvent. The Treasury was supporting the Banks and the whole structure now rested on Government credit. He said that all the big financial houses had returned their bills and that the foreign bills returned were so numerous that with all the notaries working at them it was physically impossible to 'note' them. He attributed the collapse to the entire cessation of foreign remittances. He said that he could not guarantee we should be allowed to draw against our own balance.

August 10th. Meeting of Brokers. Financial position discussed. I read my letter to our bankers reciting my interview shewing that Banks were empowered to plead moratorium against their own customers and that customers' balances were not necessarily available for drawing. I argued this fatal to credit of Insurance Market because the default of Underwriters would create widespread alarm. There was, however, a feeling that it would be impossible for some brokers to go on unless they could plead moratorium. I advised calling a representative meeting to review the position and to consider whether an appeal should be made to the Treasury. I was deputed to see the Chairman of Lloyd's. I saw him and he said the Committee of Lloyd's would meet and consider the advisability of calling the proposed Conference. In the afternoon another customer of our bank called and said that the Chairman of the Bank had used more reassuring language than the Director whom I had seen. This customer and I then saw the Chairman who confirmed what the other Director had said but ended by intimating that our balance would in fact be available and further, that without pledging himself he hoped to advance against securities. He said that many of the Banks would have stopped but for the moratorium.

August 11th. The Committee of Lloyd's met last evening and decided to do nothing. On learning this, I made up my mind to take my own course and not to promote further meetings but to attend any that might be called. We paid Lloyd's Underwriters the premiums for the June quarter and this action was followed by others, which had a salutary effect. Since the moratorium was established we have received many payments including £10,000 from one Shipping Company and £5,000 from another. There seems a somewhat better feeling today. The State War Insurance Department is doing a large business.

August 12th. A further improved tone to-day. More accounts are being paid at Lloyd's. The Banking prospects are also improving and negotiations seem to be proceeding for the re-opening of the Stock Exchange. A great deal of War insurance is being effected.

August 13th. The Government has announced a great Scheme under which the Bank of England will discount good bills, the State guaranteeing the Bank against loss. This will have an enormous effect and will go far to re-establish the finance of the Country. There is a further improved tone. We had notice of two large consignments of gold from Paris to Constantinople going July 24th and 27th not having arrived. Our names interested to the extent of £16,000.

August 14th. £140,000 out of £180,000 of the consignments mentioned above arrived. Payments in the City being made more freely. All people doing a large foreign business still badly hampered. No captures by Germans being announced a very considerable war insurance business is developing at Lloyd's. It is interesting to note the public confidence in the Institution notwithstanding the commercial ruin that surrounds it.

8 Back to gold, 1925

(Gold Standard Act, 1925: 15 and 16 Geo. V, c. 29)

Be it enacted by the King's most Excellent Majesty, by and with the advice and consent of the Lords Spiritual and Temporal, and Commons, in this present Parliament assembled, and by the authority of the same, as follows:

1—(1) Unless and until His Majesty by Proclamation otherwise directs—

(*a*) The Bank of England, notwithstanding anything in any Act, shall not be bound to pay any note of the Bank (in this Act referred to as a 'bank note') in legal coin within the meaning of section six of the Bank of England Act, 1833, and bank notes shall not cease to be legal tender by reason that the Bank do not continue to pay bank notes in such legal coin:

(2) So long as the preceding subsection remains in force, the Bank of England shall be bound to sell to any person who makes a demand in that behalf at the head office of the Bank during the office hours of the Bank, and pays the purchase price in any legal tender, gold bullion at the price of three pounds, seventeen shillings and tenpence halfpenny per ounce troy of gold of the standard of fineness prescribed for gold coin by the Coinage Act, 1870, but only in the form of bars containing approximately four hundred ounces troy of fine gold.

9 Collapse of prices, 1931

([Macmillan] committee on finance and industry, Report, *B.P.P.*, 1930–31, XIII, pp. 114–15)

... During the first quarter of this year [1931] the quantity of our exports fell off by more than 30 per cent, whereas the reduction in the quantity of our imports was only 6 per cent. Nevertheless, as a result of the catastrophic fall of raw material prices, the visible balance of trade has been actually less adverse to us than in recent years, the net position in terms of money moving £5 millions in our favour, so that less of our surplus under other heads (i.e., from foreign interest, shipping, etc.), is being required to-day to finance our imports than in 1930 or in 1929.

265 The same point can be strikingly illustrated by what has happened in the case of the single commodity wheat. At the price prevailing in December, 1930, the annual cost of our wheat imports would be about £30 millions less than it was in 1929, and £60 millions less than in 1925. It is obvious what a large contribution this single item represents to the national cost of supporting the present volume of unemployment. It is a great misfortune both for us and for the raw material countries that we should have a great volume of unemployment through their inability to purchase from us as a result of the fall in the price of their produce. But merely from the point of view of our balance of trade it is not to be overlooked that the latter fact not only balances the former but may even outweigh it. We conclude that the underlying financial facts are more favourable than had been supposed, and that Great Britain's position as a creditor country remains immensely strong.

10 The growth of hire purchase, 1958

([Radcliffe] committee on the working of the monetary system, Report, *B.P.P.*, 1958–9, XVII, pp. 72–3).

202 A large amount of short-term and medium-term credit becomes available to private consumers, and to a much smaller extent to business concerns, through hire purchase or credit sale. This normally involves the contraction of a debt to a retailer, dealer or finance house on the purchase of some durable commodity or item of equipment: the debt covers a high proportion of the purchase price and is repaid in regular and fixed instalments over an agreed period. Under hire purchase arrangements the ownership of the goods does not pass to the consumer until the debt has been discharged; under credit sale arrangements the consumer becomes the owner when he makes the first down-payment, and the balance remains as an unsecured debt payable in instalments. There is no other difference of practical importance between hire purchase and credit sale, and in what follows we do not distinguish between them, including them both under the generic term 'hire purchase'. The retailer may finance the transaction from his own resources or he may, under a block discount arrangement, lodge his hire purchase agreements with a finance house and obtain an advance against those documents proportionate to their nominal value. The customer then continues to pay his instalments

to the retailer, who acts as agent for the finance house and pays off the advance at a rate and over a period based on the instalments due to him.

203 The total amount of credit provided in this way has tended to rise with the gradual extension of the practice of hire purchase, and at the end of 1958 was estimated to have reached a total of over £600 mn. This was heavily concentrated on a relatively limited range of goods. Over half had been advanced for the purchase of motor vehicles of all kinds, including commercial vehicles and private cars bought by business firms, a further 25 per cent for furniture, and most of the remainder for radio and television sets, electric and gas cookers, and refrigerators. Hire purchase credit for the purchase of industrial and farm equipment was relatively small; and the total amount extended for business purposes, including the purchase of motor vehicles, was probably not more than about £100 mn.

204 This is not large in relation to the credit available to industry from other sources. Bank advances alone are now over £2,600 mn. and trade credit is at least on the same scale. In relation to other medium-term credit for the financing of durable equipment, however, hire purchase assumes a much more important position. Its use for this purpose has been restricted until recently by control over the minimum down-payment and the period of repayment to be required, and it appears to have been used largely for the purchase of a few of the more expensive and larger types of equipment with a limited life such as commercial vehicles, tractors, and earth-moving equipment, although it was originally introduced for the finance of more durable items such as railway carriages and wagons. It is also of much more importance to small and rapidly growing firms than to others. Of a sample of small firms applying to the Board of Trade for loans from the Revolving Fund for Industry 37 per cent used hire purchase for acquisitions of plant and machinery (excluding motor vehicles) in 1955, and 50 per cent in 1956.

205 As a source of consumer credit hire purchase occupies a much more prominent position. There are, of course, other sources such as credit accounts, check trading, and the ordinary unpaid bill as well as bank overdrafts and other forms of borrowing. But where the credit is needed for periods of at least six months and not more than two or three years, hire purchase is the chief way in which it is provided: and it is hardly too much to say that, just as a mortgage is the normal method of borrowing for the purchase of a house, so hire purchase is becoming the main source of credit for equipping it.

11 Competition for the banks, 1965

(National Board for Prices and Incomes, *Report No 34, Bank Charges*, pp. 9–10, Cmnd. 3292, 1967)

9 Neither in providing a means for making payments, nor in acting as intermediaries between borrowers and lenders are the banks named in this Reference unique. Payment in notes and coin is the main alternative to writing cheques on bank deposits. Instruments such as postal and money orders may be used to transfer money, and financial institutions such as the offices of overseas banks, merchant banks, trustee savings banks and the C.W.S. Bank provide facilities for the transmission of payments competitive with those of the commercial banks.

10 The proposed Post Office Giro offers the prospect of sharpened competition for the banks in providing a mechanism for transmitting money. If the public finds the Giro facilities attractive, it is increasingly likely to prefer to hold cash or balances in the Post Office Savings Bank rather than to hold bank deposits. But these possibilities can of course be offset by suitable competitive responses on the part of the banks themselves.

11 Despite the existence of alternative means of making payment, the banks covered by this Reference, as far as the transmission of money is concerned, dominate the field of large transactions and overshadow their institutional competitors.

12 In their role as financial intermediaries the banks increasingly face competition from other institutions. Although the deposit liabilities of banks count as money and the liabilities of other institutions— i.e. the obligations incurred by them—are not usually so regarded, a range of financial institutions issue liabilities which are highly liquid in that they arise from funds placed with them that can be recalled at relatively short notice. These institutions offer liabilities that from the point of view of the investor are in varying degree almost as good to hold as bank deposits. In addition funds placed with these institutions have other attractions—e.g. a comparatively high interest rate. This means that there is available a range of financial instruments which are significantly competitive in the eyes of individual investors with bank deposits.

B SAVINGS AND INVESTMENT

1 Borrowing from the Dutch, 1774

(Charles Wilson, *Anglo-Dutch commerce and finance in the eighteenth century* (Cambridge University Press, 1941), p. 188, quoting a Dutch writer of 1774)

England has discovered the art not only of keeping her money at home and thereby raising up as many branches of commerce and handicraft as she wants, but also the art of ever finding fresh money, not amongst her own People because it cannot be obtained there, but amongst Strangers—amongst us, who are so philanthropic that we would rather do good to our neighbours than to ourselves, and, both in home and foreign affairs, have become such lovers of foreigners that we overlook the welfare of our own House, protest against it as one will; England possesses a three-fold art: it uses our money to make itself great, and greases our palm continually with paper money: we content ourselves with the Revenues, giving no thought to the Ruin of our Children and Posterity. The English State sinks away under its debt, but the substance of that debt is spread amongst the nation, which becomes rich and prospers, although the State is poor. Our State, on the other hand, is very rich and has, so to say, no Debt, and always finds amongst its own inhabitants as much money, and more, than it wants—more, I say, for those very inhabitants might well wish that it would contract some new Loans, but it does not find them necessary: for what could it use them for? And in this question, they wish that there could be devised in this country some funds which would be good and durable, beneficial to the Country and the People, in which Capital could be invested: but these must, we think, be sought and found in the country itself, because the one fundamental rule of our State and Union is *not to expand*.

2 Beginnings of modern company law, 1844

(The Companies Act, 1844—7 and 8 Vic. c. 110)

IV And be it enacted, That before proceeding to make public, whether by way of Prospectus, Handbill, or Advertisement, any Intention or Proposal to form any Company for any Purpose within the Meaning of this Act, whether for executing any such Work as aforesaid under the Authority of Parliament, or for any other Purpose, it shall be the Duty of the Promoters of such Company and they or

20*

some of them are hereby required to make to the Office hereby provided for the Registration of Joint Stock Companies (and hereinafter called the Registry Office) Returns of the following Particulars according to the Schedule (C) hereunto annexed; that is to say,

1 The proposed Name of the intended Company; and also,

2 The Business or Purpose of the Company; and also,

3 The Names of its Promoters, together with their respective Occupations, Places of Business (if any), and Places of Residence;

*   *   *

XIX And be it enacted, That it shall be lawful for the Committee of Privy Council for Trade and they are hereby empowered to appoint a Person to be and to be called the Registrar of Joint Stock Companies, and, if the said Committee see fit, an Assistant Registrar, Clerks, and other necessary Officers and Servants; and that every such Registrar and Assistant Registrar, Clerks, and Officers shall be entitled to hold their Offices during the Pleasure only of the said Committee. ...

XX And be it enacted, That from the Hour of Ten of the Clock in the Morning until Five of the Clock in the Afternoon, and at such other Times as the said Committee of Privy Council for Trade shall appoint, such Registrar, or in the unavoidable, or, as aforesaid, permitted Absence of the Registrar, then such Assistant Registrar, shall give his Attendance at the said Office every Day throughout the Year, except *Sundays, Good Friday, Christmas Day*, and any other general Holiday or Fast Day appointed by Her Majesty in Council.

*   *   *

XXIV And be it enacted, That if before a Certificate of provisional Registration shall be obtained the Promoters or any of them, or any Person employed by or under them, take any Monies in consideration of the Allotment either of Shares or of any Interest in the Concern, or by way of Deposit for Shares to be granted or allotted; or issue, in the Name or on behalf of the Company, any Note or Scrip, or Letter of Allotment, or other Instrument or Writing to denote a Right or Claim, or Preference or Promise, absolute or conditional, to any Shares; or advertise the Existence or proposed Formation of the Company; or make any Contract whatsoever for or in the Name or on behalf of such intended Company; then every such Person shall be liable to forfeit for every such Offence a Sum not exceeding Twenty-five Pounds; and that it shall be lawful for any Person to sue for and recover the same by Action of Debt.

XXV And be it enacted, That on the complete Registration of any Company being certified by the Registrar of Joint Stock Companies such Company and the then Shareholders therein, and all the succeeding Shareholders, whilst Shareholders, shall be and are hereby incorporated as from the Date of such Certificate by the Name of the Company as set forth in the Deed of Settlement, and for the Purpose of carrying on the Trade or Business for which the Company was formed, but only according to the Provisions of this Act, and of such Deed as aforesaid, and for the Purpose of suing and being sued, and of taking and enjoying the Property and Effects of the said Company; and thereupon any Covenants or Engagements entered into by any of the Shareholders or other Persons with any Trustee on the Behalf of the Company, at any Time before the complete Registration thereof, may be proceeded on by the said Company and enforced in all respects as if they had been made or entered into with the said Company after the Incorporation thereof; and such Company shall continue so incorporated until it shall be dissolved, and all its Affairs wound up; but so as not in anywise to restrict the Liability of any of the Shareholders of the Company, under any Judgment, Decree, or Order for the Payment of Money which shall be obtained against such Company, or any of the Members thereof, in any Action or Suit prosecuted by or against such Company in any Court of Law or Equity; but every such Shareholder shall, in respect of such Monies, subject as after mentioned, be and continue liable as he would have been if the said Company had not been incorporated; and thereupon it shall be lawful for the said Company, and they are hereby empowered as follows; that is to say.

1 To use the registered Name of the Company, adding thereto 'Registered'; and also,

2 To have a Common Seal (with Power to break, alter, and change the same from Time to Time), but on which must be inscribed the Name of the Company; and also,

3 To sue and be sued by their registered Name in respect of any Claim by or upon the Company upon or by any Person, whether a Member of the Company or not, so long as any such Claim may remain unsatisfied; and also,

4 To enter into Contracts for the Execution of the Works, and for the Supply of the Stores, or for any other necessary Purpose of the Company; and also,

5 To purchase and hold Lands, Tenements, and Hereditaments in the Name of the said Company, or of the Trustees or Trustee

thereof, for the Purpose of occupying the same as a Place or Places of Business of the said Company, and also (but nevertheless with a Licence, general or special, for that Purpose, to be granted by the Committee of the Privy Council for Trade, first had and obtained,) such other Lands, Tenements, and Hereditaments as the Nature of the Business of the Company may require; and also,

6 To issue Certificates of Shares; and also,

7 To receive Instalments from Subscribers in respect of the Amount of any Shares not paid up; and also,

8 To borrow or raise Money within the Limitations prescribed by any special Authority; and also,

9 To declare Dividends out of the Profits of the Concern; and also,

10 To hold General Meetings periodically, and extraordinary Meetings upon being duly summoned for that Purpose; and also,

11 To make from Time to Time, at some General Meeting of Shareholders specially summoned for the Purpose, Bye Laws for the Regulation of the Shareholders, Members, Directors, and Officers of the Company, such Bye Laws not being repugnant to or inconsistent with the Provisions of this Act or of the Deed of Settlement of the Company; and also,

12 To perform all other Acts necessary for carrying into effect the Purposes of such Company, and in all respects as other Partnerships are entitled to do:

* * *

XXXVIII And be it enacted, That every Joint Stock Company completely registered under this Act shall annually at a General Meeting appoint One or more Auditors of the Accounts of the Company (One of whom at least shall be appointed by the Shareholders present at the Meeting in Person or by Proxy), and shall return the Names of such Auditors to the Registrar of Joint Stock Companies; and that if an Auditor be not appointed on behalf of the Shareholders, or if he shall die, or become incapable of acting, or shall decline to act at the prescribed Period, or if such Return be not made, then on the Application of any Shareholder of the Company it shall be the Duty of the Committee of Privy Council for Trade and they are hereby authorized to appoint an Auditor on behalf of the Shareholders; and that such Auditor shall continue to act till the next General Meeting; and the due Appointment of such Auditor shall be returned to the

Registrar of Joint Stock Companies, and that thereupon it shall be
his Duty to register the same; and that it shall be lawful for the Commissioners of the Treasury and they are hereby empowered to appoint
that the Company shall pay to such Auditor such Salary or Remuneration as to the said Commissioners shall appear suitable, having
regard to the Duties of his Office, and that thereupon such Auditor
shall be entitled to recover such Salary from the Company as and
when it shall become due, according to the Terms of the Appointment thereof.

* * *

XL And be it enacted, That throughout the Year and at all reasonable Times of the Day it shall be lawful for the Auditors and they
are hereby authorized to inspect the Books of Account and Books of
Registry of such Company; and that the Auditors may demand and
have the Assistance of such Officers and Servants of the Company
and such Documents as they shall require for the full Performance
of their Duty in auditing the Accounts.

* * *

XLIII And be it enacted, That within Fourteen Days after such
Meeting it shall be the Duty of such Directors and they are hereby
required to return to the said Registry Office a Copy of the Balance
Sheet, and of the Report of the Auditors thereon; and that thereupon
it shall be the Duty of the Registrar of Joint Stock Companies and he is
hereby required to register or file the same with the other Documents
relating to such Company.

3 The spread of limited liability

(i) *Gullible investors, 1877*

(W. P. Frith, *My autobiography and reminiscences* (7th ed. 1889), pp.
356–8. Frith is here describing his series of pictures, *The Race for
Wealth*, painted in 1877.)

Encouraged by the success of the 'Road to Ruin', I immediately
embarked in a new venture: a series of five pictures representing the
career of a fraudulent financier, or promoter of bubble companies;
a character not uncommon in 1877, or, perhaps, even at the present

time. I wished to illustrate also the common passion for speculation, and the destruction that so often attends the indulgence of it, to the lives and fortunes of the financier's dupes. I planned my first scene in the office of the financier—eventually called the spider—the principal flies being an innocent-looking clergyman, who with his wife and daughters are examining samples of ore supposed to be the product of a mine—a map of which is conspicuous on the wall—containing untold wealth. The office is filled with other believers: a pretty widow with her little son, a rough country gentleman in overcoat and riding-boots, a foreigner who bows obsequiously to the great projector as he enters from an inner office—in which clerks are seen writing—whilst a picture-dealer attends with 'a gem', which he hopes to sell to the great man, whose taste for art is not incompatible with his love of other people's money. Other flies buzz round the web.

The second picture represents the spider at home. He is here disco-vered in a handsome drawing-room, receiving guests who have been invited to an entertainment. He stands—in evening dress   extolling the merits of a large picture to a group of his guests, one of whom, a pretty girl, shows by her smothered laugh that she appreciates the vulgar ignorance of the connoisseur, whose art terms are evidently ludicrously misapplied. The double drawing-room contains many figures, some of whom may be recognised as the clients in the first scene at the office; others are of 'the upper ten', whose admiration of success, combined with the hope of sharing in it, so often betrays them into strange company.

If 'misfortune makes a man acquainted with strange bedfellows', the converse is no less true; for who has not been startled by the ap-pearance of an uncouth and vulgar figure in what is called 'high society', who, on inquiry, has proved to have had but one cause for his admis-sion, namely, the possession of great wealth, and the reputation of having acquired it by successful speculation—the secret of which his hosts hope to ascertain and practise?

After this moral reflection, for which I must ask pardon, I proceed with my description. My host's wife, of a vulgar type, receives more guests announced by the butler, the open door allowing evidence of the approaching banquet to be seen. Hungry guests examine their watches, other guests arrive, and the company goes to a dinner which must be left to my reader's imagination.

In the third of the series the crash has come. The foolish clergyman sits at his breakfast-table, with his head bent to the blow. His wife, with terrified face, reads the confirmation of her worst fears in the newspaper, which a retreating footman has brought. Two daughters have risen terror-stricken from their chairs, and a little midshipman son looks at the scene with a puzzled expression, in which fear predominates. The catastrophe is complete: the little fortune has been invested in the mine, and the whole of it lost. But my hero has been over-bold; he has produced ore which his impending trial proves to be the product of a mine, but not of the one in which his unhappy victims took shares. He is arrested, and takes his place in the dock at the Old Bailey, where we must now follow him, and also arrive at the fourth of the scenes in 'The Race for Wealth'. See the financier there standing, with blanched face listening to the evidence given by the clergyman, which, if proved, will consign him to penal servitude. His victims—recognisable as those in his office in the opening of my story—stand ready to add their testimony. The widow, the foreigner, the country gentleman, are there; and so also are some of his aristocratic guests, one of whom studies his miserable face by the aid of an opera-glass. The counsel and the jury examine the real and the spurious specimens of ore. The evidence is overwhelming, the verdict is pronounced; and that it is 'Guilty' is proved by the final scene, where in prison-garb the luckless adventurer takes his dismal exercise with his fellow-convicts in the great quadrangle of Millbank Gaol.

(ii) The Duke of Plaza-Toro Limited, 1889

(W. S. Gilbert, *The Gondoliers*, Act I)

(*Duke*) ... Although I am unhappily in straitened circumstances at present, my social influence is something enormous; and a Company, to be called the Duke of Plaza-Toro, Limited, is in course of formation to work me. An influential directorate has been secured, and I shall myself join the Board after allotment.

(*Casilda*) Am I to understand that the Queen of Barataria may be called upon at any time to witness her honoured sire in process of liquidation?

(*Duchess*) The speculation is not exempt from that drawback. If your father should stop, it will, of course, be necessary to wind him up.

(*Casilda*) But it's so undignified—it's so degrading! A Grandee of Spain turned into a public company: Such a thing was never heard of!

(*Duke*) My child, the Duke of Plaza-Toro does not follow fashions —he leads them.

4 The London stock exchange, 1878

(*i*) *Its objects and practical operation*

(Report of commissioners appointed to enquire into the London stock exchange, *B.P.P.*, 1878, XIX, pp. 1, 10.

We find that what is known as the London Stock Exchange is a voluntary association of those who deal in the various securities which pass by the common name of 'stocks and shares'.

This association, as at present constituted, has been in existence about 75 years. It has been the result of a natural growth resulting in great measure from the enormous increase in number and variety of foreign stocks and of the stocks, shares, and debentures, etc., connected with industrial undertakings in modern times; and whereas its members in the year 1864 did not number more than about 1,100, they now number more than 2,000.

The main objects with which this large body of persons have associated themselves together appear to have been the easy and expeditious transaction of business, and the enforcement among themselves of fair dealing.

To these ends a building has been provided for their exclusive use, and a set of rules formed for the admission and expulsion of members, and the control of their conduct both between individual members and towards the public.

We will presently call attention to the constitution of the Stock Exchange in detail, but we desire here to give our opinion that in the main the existence of such an association and the coercive action of the rules which it enforces upon the transaction of business, and upon the conduct of its members has been salutary to the interests of the public, while in the administration of its laws the Committee for General Puposes (which is the governing body of the association) has so far as we have been able to discover from the evidence acted uprightly, honestly, and with a desire to do justice.

* * * * * *

But it is in relation to the 'floating' of new companies and foreign loans that the action of the Stock Exchange has deservedly attracted more attention, and been the subject of more unfavourable remark than in any other particular.

We have thought it right to devote considerable time and attention to this part of the subject of our inquiries, and it will be seen that we have taken much evidence on the subject. For it is undoubted that within the last 20 or 30 years enormous sums of money, representing the savings and accumulation of the individual industry of this country have been dissipated and lost in the attraction of new but unsound investments.

The leading causes of this result are no doubt to be found in the craving for high rates of interest and unreasonable profits on the investment of capital on the one hand, and the dishonest contrivances of promoters to take advantage of these cravings on the other; but the means by which investors have been induced to become the victims of these contrivances are alleged to be intimately connected with, if not chiefly to reside in, operations conducted on the Stock Exchange.

(ii) The markets for securities

(Report of the commissioners appointed to enquire into the London stock exchange, evidence of J. H. Daniell, the government broker, *B.P.P.*, 1878, XIX, qq. 494–500.)

With regard to the dealings on the Exchange, are there not different markets for the different classes of securities?—There are.

Into how many markets is it divided?—I should say into about nine or ten. There is the English market, which would be for consols, Exchequer bills, and India bonds, and all Indian Government securities.

That would be one market?—Yes, the second would be in Indian Guaranteed railway securities, they are dealt in by different people. Then there would be the foreign market, for foreign bonds of all descriptions. There are then the joint stock bank market, the colonial bond market, the foreign obligation market, dealing in foreign obligations, and foreign railway shares, the two together; there is now what we call the heavy railway market, such as the London and North-western and the Lancashire and Yorkshire stocks, and these large stocks—they are called the heavy railway stocks—they are the stocks of the large companies, namely the two which I have mentioned, and the Midland, the London and South-western, and the Great Western; those five particular companies are in one market.

And the Great Northern?—That is another market.

And the North-eastern?—That is another market. The North-eastern is one market, embracing the Great Eastern and the North British. Another market is the Great Northern, the Caledonian and the Sheffield. Then there is the mining market. Then there is the Guaranteed Preference railway security market. There is also the American market, which is a very large one; and there is another market, namely the Tramway market.

Then there must be, I suppose, a miscellaneous market?—The miscellaneous markets go to a great extent among all the others. There is no particular market known as the miscellaneous market. I think that there are no particular dealers who lay themselves out solely for miscellaneous things. The tramway market includes several miscellaneous securities.

There are no men who are prepared to make a price for miscellaneous securities only, so as to form a separate market?—I think not.

5 Foreign investment

(George Paish, 'Great Britain's investments in other lands', *Journal of the Royal Statistical Society*, LXXII (1909), pp. 468–74, 480, 490–1)

Further, the Commissioners [of Inland Revenue] compare this total of nearly 80,000,000*l* with the income received from similar sources in previous years since 1886–87.

Income of British investors from Indian, colonial, and foreign government bonds and stocks, municipal stocks, and railway securities

Year	India Government stocks, loans, and guaranteed railways	Colonial and foreign Government securities	Colonial and foreign securities (other than Government) and possessions 'coupons' and railways out of the United Kingdom, other than those included in column 2	Total
	£	£	£	£
1886–87	7,793,097	16,243,321	20,471,581	44,508,002
'87–88	7,972,606	16,757,736	22,218,029	46,978,371
'88–89	8,026,310	17,388,562	24,581,936	49,999,808
'89–90	7,811,310	17,528,582	26,970,320	52,310,212

	£	£	£	£
'90–91	8,028,524	16,608,700	30,851,608	55,488,832
'91–92	7,784,370	14,919,017	31,995,383	54,728,770
'92–93	7,790,642	15,333,817	32,046,013	55,170,502
'93–94	7,856,721	15,950,233	31,311,075	55,118,029
'94–95	8,021,797	15,927,769	29,556,692	53,506,258
'95–96	8,019,720	16,419,933	30,461,426	54,901,079
'96–97	8,065,866	16,790,472	31,462,629	56,318,967
'97–98	8,168,258	17,205,934	31,265,474	56,639,666
'98–99	8,238,820	18,233,429	33,217,651	59,709,903
'99–1900	8,281,704	18,394,390	33,590,792	60,266,886
1900–01	8,567,639	18,685,410	33,078,476	60,331,525
'01–02	8,880,908	19,245,888	34,432,683	62,559,479
'02–03	9,048,777	19,935,643	34,844,295	63,828,715
'03–04	8,695,929	20,263,072	36,906,305	65,865,306
'04–05	8,760,185	20,880,837	36,421,087	66,052,109
'05–06	8,862,807	22,069,260	42,967,198	73,899,265
'06–07	8,768,237	22,270,846	48,521,033	79,560,116

The Commissioners are careful to point out that this income from capital invested abroad is by no means the total received. They say:

'Beyond this earmarked figure there exists a large amount of income from abroad which in many cases cannot (in the absence of details which the taxpayer alone could furnish) be identified as such in the assessments and which is therefore included in the sum of 373,057,495*l* appearing under the head of 'Businesses, professions, etc., not otherwise detailed'. The fact that this unidentified income from foreign countries and British colonies and possessions is of some magnitude will be appreciated when it is considered that it includes the profits derived from the following sources:

Concerns (other than railways) situate abroad but having their seat of direction and management in this country, e.g., mines, gas-works, waterworks, tramways, breweries, tea and coffee plantations, nitrate grounds, oil fields, land and financial companies, etc.

Concerns jointly worked abroad and in this country, such as electric telegraph cables, and shipping.

Foreign and colonial branches of banks, insurance companies, and mercantile houses in the United Kingdom.

Mortgages of property and other loans and deposits abroad belonging to banks, insurance companies, land, mortgage and financial companies, etc., in this country.

Profits of all kinds arising from business done abroad by manufacturers, merchants, and commission agents resident in the United Kingdom'.

In brief, the income which the Commissioners earmark as coming from abroad is that received from Indian, colonial and foreign government stocks, municipal securities and railways, and the great additional income the country derives from its investments in a vast number of miscellaneous undertakings of all kinds is excluded.

Before proceeding farther, may I digress for one moment in order to direct your attention to the expansion in the income from capital invested abroad that is earmarked by the Commissioners? I have already indicated the governments to which we have lent large sums in recent years. The income contained in column 4 of the foregoing tabular statement is derived partly from municipal stocks and partly from railways, chiefly the latter, and the immense expansion has arisen mainly from the very large amounts of capital we invested in the railway undertakings of the United States, of Argentina, of Canada, of Mexico and of other countries in the eighties and early nineties. A great deal of the capital we then placed abroad was for several years relatively unproductive in consequence of the severe trade depression in the new countries during the nineties. The expansion in trade of recent years, especially in the agricultural countries, has brought with it large profits for the railways and a greatly increased income for the investors of this country. You will recollect that a few years ago many of the railways of the United States, of Canada and of Argentina and of other countries paid no dividends whatever on their share capitals and that in recent years many of these railways have paid dividends of 7 per cent—hence the great expansion in income we have derived. In a small degree the expansion in the income shown to have occurred in 1905 and 1906 was the result of fresh investments of capital, a new period of active investment of British capital in other lands having commenced in the summer of 1904.

To return—It is impossible for anyone to ascertain the income from the capital our merchants, our manufacturers, and others have privately placed abroad, but it is possible to collect a great deal of information in addition to that compiled by the Commissioners.

There are several thousand public companies operating abroad that have raised capital publicly in this country, and these companies issue statements of their capital and of their profits.

To amplify the statement of the Commissioners, I have obtained the reports of as many of these companies as it is possible for a private person to secure. I have classified them, and have ascertained the amount of their capital and of their profits in the aggregate and by groups.

Further, I have sought to analyse the income of 79,000,000*l* which the Commissioners have earmarked, in order to obtain a comprehensive picture both of the sources of the income we receive and of the capital publicly subscribed in this country for investment in other lands.

I will now set out the results of my investigations as regards income, and will leave the question of the amount of capital we have embarked abroad to a later stage, in order not to create confusion.

The income derived from Indian government loans issued and held in this country is 5,017,000*l*, from colonial government loans 13,933,000*l*, from foreign government loans 8,338,000*l*.

Beyond lending to governments, we have lent large sums to municipalities. Our income from this source is 2,500,000*l*, of which 1,500,000*l* is derived from colonial and 1,000,000*l* from foreign cities. The latter include a number of South American municipalities growing rapidly in population, notably Rio de Janeiro, Buenos Ayres, Montevideo, Rosario and Santa Fé. We have also lent money to several cities in the United States, to the city of Mexico and to several Japanese towns.

In providing capital for railway construction this country has performed a great work. Most of the loans to colonial governments have been for railway construction, the major part of India's indebtedness to us is for railways, and a portion of the loans we have made to foreign governments has been used for a similar purpose. But beyond the money for railway construction which we have supplied to governments, we have formed a great many companies to construct and work railways in other lands.

From the capital we have supplied to railway companies working in the colonies, notably in Canada, we receive an income of 7,600,000*l* a year, from those working in India we derive nearly 4,800,000*l* per annum. The railways of Argentina, Brazil, Uruguay, Mexico, Chile and other foreign countries yield over 13,000,000*l* a year to us in the aggregate, and from the railways of the United States our investors receive no less than 27,000,000*l* a year.

The aggregate of these totals which I have compiled from the companies' reports and as far as possible from independent investigations amounts to 82,777,000*l*, and slightly exceeds the sum of 79,560,000*l* returned by the Commissioners of Inland Revenue.

This brings me to the great income we derive from the vast number of miscellaneous companies for which we have provided capital and which largely contribute to the production of the world's supplies of natural wealth. I have taken out the income of 2,172 of the miscellaneous companies the capital of which has been supplied by our investors and I have ascertained the great income that comes to this country for interest and profit upon this capital. The income we derive from capital invested in colonial and foreign banking companies is over 7,000,000*l.* The major portion of this comes from banks doing business in India and the colonies, but a substantial income comes from banks operating in Egypt, China, South America and Mexico.

Breweries and distilleries yield an annual income of only 732,000*l* a year—an insignificant sum having regard to the amount of capital invested. Our investors appear to have burnt their fingers rather badly in the prices they paid for foreign breweries.

In consequence mainly of the Treasury investment in the shares of the Suez Canal Company, the income from canals and docks situated abroad reaches 1,174,000*l* a year.

Commercial and industrial companies yield an income of nearly 5,000,000*l* a year. It must not be supposed that this income is derived from manufacturing firms competing with British traders. One of the most noteworthy characteristics of the British investor is his objection to place capital in any enterprise, or in any country for the matter of that, the development of which appears to be against the interests of the Motherland.

From electric lighting and power companies we derive an income of 314,000*l*.

The income received from financial, land and investment companies amounts to over 6,000,000*l* a year, and will probably show great expansion in future. A large sum of money has been expended upon the acquisition and development of land in various parts of the world which at present gives no return, and the income from this capital has yet to be received. The income, from this source is derived as to two-thirds from the colonies, and as to one-third from foreign countries.

Gas companies bring to us an income of nearly 1,200,000*l*.

Colliery and iron ore undertakings yield 500,000*l* a year. Having regard to the new capital placed in these undertakings in South Africa, in Australia, in India and elsewhere, this source of income should largely expand.

Gold, copper, diamond, silver, lead, tin and other mining enterprises yield an income of nearly 26,000,000*l* a year. Of this large figure gold mining comes easily first with a return of nearly 15,000,000*l* a year, copper second with over 5,000,000*l* a year, diamonds and other precious stones third with 4,400,000*l* a year, while silver, lead, zinc, tin, etc., produce an income of 1,100,000*l*. In consequence of the new capital embarked in mining companies and which has not yet become productive a large expansion in our income from mining may be looked for.

Nitrate companies yield an income of 1,600,000*l*.

Oil companies produce profits of 642,000*l* per annum notwithstanding the misfortunes of the Russian oil companies in which a relatively large sum is placed.

Rubber companies give 446,000*l* a year, a return which will largely increase as the new plantations mature.

Tea and coffee plantation companies, mainly tea, situated in India and Ceylon, yield a profit of 1,794,000*l* per annum.

Transmarine cable companies and colonial and foreign telephones yield upwards of 2,000,000*l* a year.

Tramways give an income of 1,809,000*l*, and waterworks over 400,000*l* a year

In the aggregate, British investors receive an income of nearly 58,000,000*l* from the miscellaneous companies the accounts of which I have analysed. This income is referred to, but is not disclosed by the report of the Commissioners of Inland Revenue.

Thus, if the Commissioners were to bring together all the income of companies trading abroad and distributing interest and profits in this country, the total would be about 140,000,000*l*.

This great sum does not include the interest upon money deposited in Indian, colonial and foreign banks by persons residing in this country, or the large amount of income derived from capital privately placed abroad, amounting to several hundred millions of pounds.

On the other hand, it is essential to recollect that foreign investors draw a good deal of income from capital placed in British companies of one kind and another, and in balancing the account it is essential not to omit this important offset.

I have taken no account of the profit on the capital invested in our mercantile marine. The country's income from shipping is an inquiry of great magnitude and of great importance and one which could not be advantageously dealt with in this paper.

The capital invested

This brings me to the second stage of my inquiry. How much capital have the British people placed in other lands to yield them their great income of 140,000,000*l* a year? Two considerations have to be heeded in this matter. The first is that capital, other than that placed in loans, rarely becomes fully productive until several years after it is expended. Indeed, only in recent years has the capital invested in the eighties and nineties yielded its full harvest of interest and profit. The income received in 1907–08 did not correspond to the capital expended at that time. Probably the corresponding income will not be realised until some time during the decade ending with 1920. By that time the new lines of railway will become self-supporting, new mines will be fully opened up, new rubber plantations will be yielding their harvest, and so on.

The second is that the capital we have sent abroad does not necessarily correspond to the capital possessed by this country in other lands. On the one hand, we sometimes make bad investments, but on the other we more often find very good ones, and on balance the securities purchased ultimately possess a greater value than the price paid for them. A noteworthy instance of this is the nation's purchase of Suez Canal shares. The sum paid for these shares was 4,076,565*l*, and their current value is 33,730,000*l*. Doubtless many other instances will occur to you. There can be no doubt that in spite of our losses the value of the capital in other lands which the country now possesses is much greater than the capital actually remitted by our investors.

To obtain an income from abroad of nearly 140,000,000*l* per annum we have invested a sum of about 2,700,000,000*l*, and this capital is yielding an all round return of 5·2 per cent.

I have no doubt that as soon as the large amount of capital sent out in recent years bears more fruit the annual income from our foreign investments will reach a much greater figure. Indeed, having regard to the investment of capital in Canada, in Argentina, in India, in the United States, and in other lands in the last few years and the additional sums likely to be sent out in the next few years, I look forward to a very rapid growth in our income from interest and profits on capital invested abroad.

Of this total of about 2,700,000,000*l* nearly 1,700,000,000*l* has been expended upon railway construction either by companies or by governments which have raised loans in this country for that purpose.

The capital has been supplied in about equal portions to the countries beyond the seas within the British Empire and to foreign lands. The totals are 1,312,000,000*l* to India and the colonies, and 1,381,000,000*l* to foreign states, mainly to the United States, to Central and South America, to Japan and to China. . . .

The capital and income contained in the following statements are compiled from reports covering the year 1907, and in some cases the early portion of 1908. Before the close of the current year (1909) the total will probably reach 3,000,000,000*l*.

In the twelve months to June, 1908, the new capital subscribed in this country for India, for the colonies and for foreign countries amounted to 110,000,000*l* and in the period from 1st July, 1908, to 30th June, 1909, the additional capital subscribed has reached no less than 175,600,000*l*. This makes a total of no less than 286,000,000*l* subscribed in the last two years.

For a period of enterprise similar to that through which we are now passing, we have to go back to the later eighties when we were investing vast sums in the Australasian Colonies, in the United States, in Argentina, in India, in Canada, and in other lands. At that time the country was not nearly as wealthy, nor were its annual savings available for investment in securities nearly as great. Nevertheless, in the seven years from 1884 to 1890, inclusive, we placed abroad over 400,000,000*l* of capital. In the seven years from 1898 to 1904—a period of extravagance and war expenditures in this country—our investments in other lands were not much over 100,000,000*l*. Since 1904 we have been endeavouring to make up for lost time, and in four and a half years we have placed no less than 400,000,000*l* of new savings in profitable Indian, colonial and foreign securities. There are good prospects that in the seven years from 1905 to 1911 inclusive we shall place new capital in other lands to the extent of at least 700,000,000*l*, or an average of about 100,000,000*l* a year.

*   *   *

In conclusion, I would direct your attention for one moment to the immense advantages to this country of its investments in other lands. The investment in the last sixty years of about 2,500,000,000*l* of British capital has occurred simultaneously with a vast growth of British trade and prosperity, and in my opinion this growth of our trade and prosperity is largely the result of our investment of capital in other countries. By building railways for the world, and especially for the young countries, we have enabled the world to increase its production of wealth at a rate never previously witnessed and to produce those things which this country is specially desirous of purchasing—foodstuffs and raw materials. Moreover, by assisting other countries to increase their output of the commodities they were specially fitted to produce, our investors have helped those countries

21*

to secure the means of purchasing the goods that Great Britain manufactures. Thus, by the investment of capital in other lands we have, first, provided the borrowing countries with the credit which gave them the power to purchase the goods needed for their development, and secondly, enabled them to increase their own productions so largely that they have been able to pay us the interest and profits upon our capital and also to purchase greatly increased quantities of British goods.

The large sums of capital which Great Britain is now supplying to other lands will ensure greatly increased incomes to her own people of all ranks and classes, will widen the Indian, colonial and foreign markets for the goods she manufactures and will greatly assist in providing her dense and constantly growing populations with plentiful supplies of foodstuffs and of raw materials. [In the discussion that followed, the speaker cleared up various points including the problem of foreigners who subscribed to overseas issues raised in London.] ... He made inquiry as to the proportion of the income from foreign investments assessed to income-tax in England received by foreigners, and he thought that on the average about 10 per cent of it belonged to foreigners. That would mean that 270,000,000 *l* of capital belonged to foreigners out of the total sum calculated of 2,700,000,000*l*. But against this sum had to be placed the capital privately invested abroad by this country, and he thought the amount of income derived from that source greatly exceeded the amount of income paid over to foreign investors. He had come to the conclusion that they derived an income from something like 500,000,000*l* of capital privately invested in other countries. He thought the figure of 2,700,000,000*l* of British capital invested abroad was, if anything, an under-statement, but he only wished to place before the members a figure for which he had documentary evidence.

6 Finance for industry, 1931

([Macmillan] committee on finance and industry, Report, *B.P.P.*, 1930–31, XIII pp. 165–7, 173–4)

384 Industry is yearly becoming more internationalised, and British industry, if it is not to be stranded in a back-water, will find institutions closely in touch with international finance invaluable in many ways as intermediaries. British companies in the iron and steel, electrical, and other industries must meet in the gate their great American and German competitors who are generally financially

powerful and closely supported by banking and financial groups, with whom they have continuous relationships. British industry, without similar support, will undoubtedly be at a disadvantage. But such effective support cannot be obtained merely for a particular occasion. It can only be the result of intimate co-operation over years during which the financial interests get an insight into the problems and requirements of the industry in question and the industrial interests learn the value of the support which financial interests can give. The future development of British industry, particularly in the establishment of enterprises abroad, is greatly dependent on co-operation of this character, since a knowledge of foreign conditions, farsighted planning and large supplies of capital will be or, may be, all required. And in the realm of foreign investment it is primarily towards British-owned enterprises abroad that we should wish to see our energies and capital turned rather than merely towards subscribing to foreign Government and municipal loans, which absorb our available foreign balance while doing little for our industry and commerce. So far as heavy industry is concerned, we may find ourselves cut more and more out of the world if our competitors, advancing with combined and powerful industrial and financial resources, develop abroad one 'tied' enterprise after another, or in the alternative purchase from us enterprises previously 'tied' to us.

385 In the last few exceedingly difficult years it would have been of high value if the leaders, for instance, of the steel or shipbuilding or other industries had been working in the closest cooperation with powerful financial and banking institutions in the City with a view to their reconstitution on a profitable basis. The tasks still confronting us require great financial as well as industrial experience. It has been represented to us strongly in evidence that a great deal remains to be done in more than one important industry in overcoming sectional and individual opposition to desirable amalgamations and reconstructions designed to eliminate waste and cheapen costs. It was stated to us that very important economies and much greater efficiency are possible if there are concerted movements to that end. We believe this to be the case and we believe also that these results can often only be obtained if there is some independent authority (such as a financial institution) able to review an industry as a whole from outside, to suggest plans of re-organisation, and to assist in the provision of finance. But industrialists on their side must be willing to co-operate. There are grounds for supposing that progress has been often very slow from the fact that the industrialists interested

have not been ready themselves to take any wide view or recognise the enormous changes now taking place in the world and in methods of production. The greatest possible reduction of costs is particularly essential in our basic industries and in industries which are in competition with foreign countries since only by this means can their future be secured. We are not in a position to go into detail in this matter, but we desire to express a strong opinion that sectional interest should not be allowed to stand in the way of re-organisations which are in the national interest.

386 In a second direction we see advantages in a closer co-operation of industry and finance. It is all-important to the community that its savings should be invested in the most fruitful and generally useful enterprises offering at home. Yet, in general the individual investor can hardly be supposed to have himself knowledge of much value either as to the profitable character of the security of what is offered to him. How easily he can be misled in times of speculative fever by glittering—even tawdry—appearances is proved by the experience of 1928, as the following striking figures will show.

In that year the total amount subscribed for capital issues whether of shares or debentures of 284 companies was £117,000,000. At 31st May, 1931, the total market value of these issues as far as ascertainable was £66,000,000 showing a loss of over £50,000,000 or about 47 per cent. In fact the public's loss has been greater since many of these shares were no doubt sold by the promoters at a high premium. Still more striking perhaps, 70 of the above companies have already been wound up and the capital of 36 others has no ascertainable value. The issues of these 106 companies during that year amounted to nearly £20,000,000.)

387 That you cannot prevent a fool from his folly is no reason why you should not give a prudent man guidance. We believe that our financ al machinery is definitely weak in that it fails to give clear guidance to the investor when appeals are made to him on behalf of home industry. When he is investing abroad he has the assistance of long-established issuing houses, whose reputation is world-wide. When subscriptions to a foreign issue are invited by means of a public prospectus, it is almost certain that that issue will be vouched for by one of these issuing houses whose name will be evidence that it has been thoroughly examined and the interests of the investors protected as far as possible. For the issuing house's issuing credit, which can easily be affected, is involved, and it is very highly to its own interest to make sure that the issue is sound. If, as must from time to time happen,

something goes wrong with the loan or the borrower the issuing house regards it as its duty to do everything it can to put matters straight, and, indeed, to watch continuously the actions of the borrower to see that the security remains unimpaired. These duties are sometimes very onerous and involve a great deal of labour and expense, as well as judgment, skill and experience.

388 Contrast this with nearly all home industrial issues. There are it is true, one or two first-class houses in the City which perform for certain first-class companies the same functions as the older issuing houses perform for foreign borrowers. In addition these latter are to a limited extent entering the domestic field, though, for the reasons we give later, their direct interest must probably remain limited. Again, the advice of stockbrokers, when asked for, may be a safeguard but it is scarcely sufficient to take the place of the responsibility of a first-class issuing house. With these exceptions the public is usually not guided by any institution whose name and reputation it knows. For in the main all issues are made in the same way. Each of them has in the most prominent position and in much the heaviest type on the prospectus the name of a joint stock bank—in the great majority of cases one of the 'Big Five'—as receiving subscriptions to the issue. It is perfectly natural that an inexpert investor, seeing the name of a well-known bank on the prospectus, should believe that the bank in question vouches in some way for the issue. But he is mistaken. None of the Big Five regard themselves, except in very rare cases, as fathering the issue or in any way responsible for it beyond seeing that the prospectus complies generally with the law and that the issue is on the face of it respectable. None of them would wish the public to assume that they vouched for the issue and the figures of profits and assets, etc., given in the prospectus, in the same way as the issuing houses vouch for their issues. They frequently know the company through having had banking relations with it over a series of years. On the other hand, they may know practically nothing of it. The *real* issuer may be the company itself, if it is a strong and good one, or a finance company or syndicate—few large, many small, some good, some indifferent, some bad—and sometimes it is a company or a syndicate got together for the sole purpose of making a particular issue. The investor may not realise that the all-important point is not the name of the bank receiving subscriptions but that of the company or syndicate really responsible for the issue.

✳   ✳   ✳

404 It has been represented to us that great difficulty is experienced by the smaller and medium-sized businesses in raising the capital which they may from time to time require, even when the security offered is perfectly sound. To provide adequate machinery for raising long-dated capital in amounts not sufficiently large for a public issue, i.e., amounts ranging from small sums up to say £200,000 or more, always presents difficulties. The expense of a public issue is too great in proportion to the capital raised, and therefore it is difficult to interest the ordinary investor by the usual method; the Investment Trust Companies do not look with any great favour on small issues which would have no free market and would require closely watching; nor can any issuing house tie up its funds in long-dated capital issues of which it cannot dispose. In general, therefore, these smaller capital issues are made through brokers or through some private channel among investors in the locality where the business is situated. This may often be the most satisfactory method. As we do not think that they could be handled as a general rule by a large concern of the character we have outlined above, the only other alternative would be to form a company to devote itself particularly to these smaller industrial and commercial issues. In addition to its ordinary capital, such a company might issue preference share capital or debentures secured on the underlying debentures or shares of the companies which it financed. The risks would in this manner be spread, and the debentures of the financing company should, moreover, have a free market. We see no reason why with proper management, and provided British industry in general is profitable, such a concern should not succeed. We believe that it would be worth while for detailed inquiries to be made into the methods by which other countries attempt to solve this particular problem.

6

Foreign trade

Between 1760 and the 1870s England came to depend increasingly on foreign trade—for raw materials, for foodstuffs and for markets for manufactured goods. Since the 1870s there has been relatively little further change in the country's dependence on foreign trade. Imports have tended to grow—but no faster than the national income; exports have grown fast at some periods—1900–13 and since 1945; slowly at others—1874–1900; and exports actually fell between the wars. On the whole the ratio of merchandise trade to national income has changed little in the past hundred years. Recent concern about the balance of payments should not blind us to the fact that for most of the period since 1760 England has not needed to worry unduly about her international accounts. It is true that there has been a deficit on the balance of visible trade at least since reliable statistics were first published in 1854, but the deficit has until recently been fully made up by invisible earnings. In the more comfortable climate of the nineteenth century it was possible to welcome a growth of imports, as a sign of prosperity (7). 'Fear of imports', to use Heckscher's phrase, is a product of an insecure trading position. Britain's share of world trade in manufactures has remorselessly declined from its peak in the 1870s. In part the decline may have been due to inefficiency in the export trades, but it was largely inevitable. A similar deterioration may be detected in the pattern of imports. As the nineteenth century wore on and in the twentieth century particularly, imports of manufactures have increased rapidly. So too, and almost as ominously, have imports of raw materials as native supplies of metallic ores have been exhausted, and as coal has given way to imported oil. On the other hand agriculture has come to the rescue of the balance of payments with a production drive that has reduced the dependence on imported food compared with the 1930s.

The rapid growth of nineteenth-century exports owed much to the enterprise of a distinct class of businessman, the merchants (6 and *11*). The rise of metal exports and the fall of textiles has led to a reduction in the role of the merchant and his place has been largely taken by salesmen directly employed by manufacturing firms.

1 The slave trade, 1766

(Gomer Williams, *History of the Liverpool privateers and letters of marque with an account of the Liverpool slave trade* (Heinemann, 1897), p. 531)

Shipped by the grace of God, in good order and well condition'd by James [surname illegible], in and upon the good Ship call'd the *Mary Borough*, whereof is Master, under God, for this present voyage, Captain David Morton, and now riding at Anchor at the Barr of Senegal, and by God's grace bound for Georgey, in South Carolina, to say, twenty-four prime Slaves, six prime women Slaves, being mark'd and number'd as in the margin,† and are to be deliver'd, in the like good order and well condition'd, at the aforesaid Port of Georgia, South Carolina (the danger of the Seas and Mortality only excepted), unto Messrs Broughton and Smith, or to their Assigns; he or they paying Freight for the said Slaves at the rate of Five pounds sterling per head at delivery, with Primage and Avrage accustom'd. In Witness whereof, the Master or Purser of the said Ship hath affirm'd to three Bills of Lading, all of this tenor and date; the one of which three bills being accomplish'd, the other two to stand void; and so God send the good ship to her desir'd port in safety, Amen.

Dated in Senegal, Ist February, 1766,

David Morton

2 A West African agency, 1767

(T. S. Ashton, ed. *Letters of a West African trader, Edward Grace, 1767–70* (Council for the Preservation of Business Archives, 1950), pp. 2–5)

Amable Doct.   23rd May 1767

The same Motives which last year induced us to engage with you in a Commerce to Galam do still subsist, and the good Opinion we entertained of your Integrity and ability to conduct this Business to our mutual advantage is increased since your return from Africa: we have therefore with the greater pleasure purchased a proper Vessel and Cargo of your own choosing, in order to carry this Scheme of Trade more fully into Execution.

By the Contract made with you last year, you would have been entitled to one Sixth part of the Profit which might have arisen on

---

† Marked on the Right Buttock

the Adventure, but as you now go out sole Manager and Conductor of the Business, we readily agree that you shall share with us one fifth part of all the Benefit arising therefrom, for the Space of four Years; at the Expiration of which Term, you will be at Liberty, either to engage with us anew, or to return to England, or to what other place you please.

Mr George Hicks has engaged to go out with you and to remain at Galam in our Service subject to all your lawful Commands for the Space of four years; from the good disposition appearing in this young Man, we hope he will prove not only a useful Servant, but an agreeable Companion and we pray God to preserve you both in health and to give you Success; we are to pay him Twenty Pounds the first year, Twenty five the second, thirty the third and forty Pounds for the fourth years Service, either in Goods at Cost, or in Gold, which Goods he has Liberty to dispose of to the best advantage for his own Account, besides which we have engaged to send him out Goods to the amount of Fifty Pounds at Prime Cost, at the Expiration of the Second year, if he demands them, which he is likewise at Liberty to trade with for his own particular Emolument.

You are not unacquainted with our dislike to the Slave trade, of all others it is the last we would engage in; the Principal Objects we have in view, are Gold, Cotton Yarn, Cotton and Ivory, besides which Tortoiseshell, Leopard, Lion and other Skins, if to be procured on low Terms, may prove advantageous.

The vegitable Silk (call'd Faveton by the Natives) we are afraid cannot be employed here to any advantage therefore we would have you send only a small Quantity thereof by way of Sample, but if it is of a Texture sufficiently strong and the Natives can spin it into a kind of yarn fit for manufacturing Stocking or other Things we recommend you to send us as much thereof so spun as possible.

You will recollect the conversation we have had concerning the Earth, and the impregnated Stones from the Gold Mines: on your arrival at Galam, you will do well to send some Person to the Mines, in order to bring you a quantity of that Earth and those Stones, which we hope you will be able to send in small Casks by the Vessel you go out in; Neither will you forget to send us as large a quantity of Red and yellow Wood as you can. Spun Cotton or Cotton Yarn the more you send the more agreeable. Bees Wax may turn to good account, if a large quantity can be procured, and Bamboo Canes we think may prove a good Object of Trade; besides which you will do well to send us Samples of the different produce of Stones Minerals and other produce of the Country; in a Word we should be glad to be furnished with every kind of thing to be procured in order to enable to direct you what to send the ensuing year.

The Vessel wherein you embark, will we suppose be detained at Galam, about two months, in which case we entreat you to send us an account of your Health and of the situation of things by a Marabou Messenger, and if Mr Berville should go with you to Galam, we are persuaded on his return you will advise us of everything necessary for our Information.

As some of the Goods you take with you, such as Brandy and Sugar, and a small assortment of other Articles, with a quantity of Wine you may take in at Sta. Cruz Theneriff, may be disposed of to good advantage at Senegal we leave you at liberty to dispose of part thereof at Senegal, if it's consistent with our Interest, also to leave with Captn Williamson such Articles you may think proper for him to sell there for our Account, taking a Receipt of him for the same, and sending us a duplicate thereof.

You will not fail writing to us, of everything concerning our affairs, as frequently as possible, always remembering to send Duplicates and Triplicates of your Letters.

If from the process we have furnished you with of making Indigo, anything can be done, you will not fail (at your Leisure) attending thereto.

Great Injustice has been done with respect to the last Cargoe under Mr Brackstones direction, it was indeed idle in him to undertake a Business he knew nothing of: had he survived, notwithstanding his Ignorance, things would have turn'd out better, for we have great reason to suspect that the Black Sailors after his decease plunder'd the Effects, you will therefore make strict enquiry as to this matter, not that we expect to have any Redress but it may be of Service to know who were the Agressors and to avoid employing them.

Tho' we have no Inclination to follow the Slave trade, yet as we cannot expect large Returns in any other Article this year, and as opportunity most probably will offer to purchase fifty to one hundred Slaves on advantageous Terms before the Vessel returns from Galam we do not forbid you to invest part of our Effects for this purpose, provided you can Contract with responsible People at Senegal to receive them at about £15 to £16 per Head on the return of the Vessel, payable in Dollars at five shillings each, or in gold at four Pounds per Ounce. This or whatever return you can make will be very acceptable and encourage us to prosecute the undertaking with Spirit and Vigour.

Pray make our Compliments to Mr Berville and thank him in our Names for the good advice given you with regard to your Naturalization, and the precaution he judged necessary for us to take before your departure.

As we have found from your Conversation that the River at Galam produces large quantities of Mussels we do imagine that Pearls are to be found therein, you will therefore employ some people of the Country to bring to you a quantity, which if upon opening should afford Pearls they will be very valuable, but if not, the Expence will be trifling and this Speculation ended.

As you are perfectly acquainted with the Country and its Trade, your own good Sense and discretion will point out to you the readiest Means to promote our joint Interest, wherefore commiting our Effects to your Trust and good Management, with an intire confidence in your Integrity, we have not to add, but our best Wishes for your Health and prosperity being

Very sincerely

[Edward Grace]

3 Social status of the merchant, c. 1788

(Edward Gibbon, *Memoirs of My Life*, ed. Georges A. Bonnard (Nelson, 1966), Appendix I, pp. 199–200)

...The good-sense of the English has embraced a system more conducive to national prosperity; the character of a merchant is not esteemed incompatible with that of a Gentleman; and the first names of the peerage are enrolled in the books of our trading Corporations. The descent of landed property to the eldest son is secured by the common law: and though Kent, under the name of *Gavelkind*, retains a more equal partition, this provincial custom is defeated by the practise of settlements and entails. The pride and indolence of younger brothers might frequently acquiesce in the life of a William Wimble so incomparably described by the Spectator: but a more rational pride must often prevail over their indolence and urge them to seek in the World the comforts of independence. Since the auspicious reign of Elizabeth the commerce of England had opened a thousand channels of industry and wealth, and the more splendid resources which now divert a Gentleman's younger sons from the mercantile profession were much less frequent and beneficial. After the reformation the Church assumed a graver and less attractive form, and though many might be content to sleep in the possession of a patrimonial living, the bench of bishops was long filled by indigent scholars; before the gentry, or at least the nobility became fully sensible of the value of a calling which bestows riches and honours without requiring either genius or application. In every age the youth of England has been distinguished by a martial spirit; and the subjects of Elizabeth and her successors sought every occasion of danger and glory by sea and land. But these occasions were rare and voluntary: nor could they afford such an ample and permanent provision as is now supplied by an hundred regiments, and an hundred ships of the line. Our civil establishment has gradually swelled to its present magnitude; nor did India unfold her capacious bosom to the merit or fortune of every needy adventurer. The common alternative was the bar and the counter; but the success of a lawyer unless he be endowed with superior talents is difficult and doubtful: the various occupations of commerce are adapted to the meanest capacity, and a modest competency is the sure reward of frugality and labour, since those humble virtues have so often sufficed for the acquisition of riches.

4 Foreign trade in the early nineteenth century

(Committee on Orders in Council, *B.P.P.*, 1812, III, pp. 38, 40, 41, 132–3, 522–3.)

[*Evidence of Joseph Shaw, Chairman of Birmingham Chamber of Foreign Commerce and exporter of hardwares*]

Have you had occasion to make any estimate, founded upon your own inquiries, of the number of workmen employed in the Birmingham manufactory[1]—and the neighbouring towns? I never particularly estimated for the whole of them, but in the year 1808 I took an estimate of the people employed in the American trade. ... Those that could be ascertained to be (as nearly as could be) exclusively employed in the American trade were 50,000, exclusive of the nail trade, which employed from twenty to thirty thousand [of whom two-thirds were engaged in the American trade.]

*   *   *

Can you state to the Commitee, from your observation, what proportion the foreign trade generally bears to the trade for home con-

---

[1] Brassfounding, hardware, plated ware, jewellery, etc.

sumption? . . . I should think it was considerably more than one half, including the United States.

Do you think it would amount to two-thirds? I should think not far from it. . . . Do you think the foreign trade is equal to two-thirds of the whole manufacture?—When the foreign trade is the same as in the year 1810, not in its present state; it is now very different.

*   *   *

To what cause do you ascribe the diminution of your trade to the Continent?—The risk of sending goods into many ports of the Continent is too great.

*   *   *

Then it is the French, Berlin, and other decrees that have produced this diminution of your trade to the Continent?—To my own particular trade. I cannot say how it is as to others.

[*Evidence of John Bailey, exporter and home factor of Sheffield goods*]

What are the principal articles manufactured at Sheffield?—They are very numerous, I can present a list of them to the House; the principal articles are cutlery, files, edged tools, saws, and a great variety of other heavy articles.

Can you speak to the population of Sheffield, and such parts of the neighbouring parishes as are concerned in the Sheffield manufacture?—The population of the parish of Sheffield, as returned by the overseers in the year 1811, was 53,000 odd; but including those parts of parishes in which Sheffield goods are manufactured, the population amounts to 60,000 at least.

Can you tell what proportion of hands are employed in manufacturing for the American market?—For the American market, about 4,000 male adults, and 2,000 women and children, making a total of 6,000.

How many do you estimate are employed in manufacturing for the home trade?—Six thousand male adults, and one thousand women and children.

How many do you calculate are employed in the remaining parts of the Sheffield trade, namely, manufactures for the foreign market, exclusive of the American?—Two thousand male adults, and one thousand women and children.

This last market includes Spain and Portugal?—Spain, Portugal, the West Indies, South America, and Canada, with some few other parts.

What proportion does the American market bear to the home market, as far as regards the Sheffield goods?—The American exports amount, as nearly as I have been able to ascertain, to one-third of the whole manufactures of Sheffield; the home trade to, I think, three-sixths.

[He adds that the American trade had been affected by the Orders in Council and the Non-importation Act of the United States. The home trade with towns in the American trade had been injured also. Goods to the value of £400,000 were waiting in Sheffield and Liverpool warehouses.]

[*Evidence of Robert MacKerrell, London merchant, dealing in cottons and muslins, and manufacturer of Paisley*]

Can you inform the Committee what the state of the trade was in the years 1808, 1809, 1810, and 1811?—In 1807 we felt the whole effect of the Berlin decree, we were entirely excluded from the Continent; I speak with regard to my own transactions and those of a vast number of my friends. We had in 1807, and previous to that, trades to the South of Europe, particularly in Portugal, which were uninterrupted, but which were likewise put an end to by the French invasion in November of that year. In 1808 the trade revived considerably; a great quantity of our goods, and of English merchandise, was introduced into the Continent through Heligoland; considerable exports were made to the Baltic; the trade in the Mediterranean increased very considerably; a very great trade was opened to this country in consequence of the Royal Family of Portugal removing to the Brazils, which likewise made an opening to Spanish South America. In 1809 the trade through Heligoland was most extensive; Bonaparte had his hands full with the Emperor of Germany and with the Spaniards, and had no time to attend to the coast; the trade to the Mediterranean increased very much; the quantity of goods taken out that year greatly exceeded any previous year, for reasons that at that time we could not account for. The trade to the Brazils was equally extensive with the year before, vast exportations took place to South America, and in general, trade in the line in which I am engaged was reckoned a fair trade; the markets were never heavy.

[The Orders in Council increased the English export trade to the South of Europe, and Africa and the Levant were supplied with English substitutes for Continental cottons and linens.]

What has been the state of your trade for the last eighteen months, and, as far as you have been informed, of the country in general? —The state of the trade during the last eighteen months has been depressed; for the last twelve months it has been recovering, but for the six months previous it was very much depressed indeed.

To what do you attribute that depression?—We attribute the depression of trade which took place to the effect of the Berlin and Milan decrees. [Northern Europe, the Baltic, etc., were shut against English trade and English ships were sequestered even in Swedish ports.]

5 South American markets after independence, 1826

(R. A. Humphreys, ed. *British consular reports on the trade and politics of Latin America, 1824–1826* (Royal Historical Society, Camden, third series, LXIII, 1940), pp. 195–7. The writer is referring to trade with Peru.)

All the manufacturies of Manchester have been sent to these markets and particularly plain white cottons, or shirtings and calicos, and printed calicos, commonly called plates; but the supply of these articles, as well as of the calicos of Blackburn for the use of printers has been so excessive that they have not yielded the advantage which was reasonably expected from the low cost at which they are produced by the newly invented machinery. Velveteens, satteens, nankeens, quiltings, etc., which are produced by manual labor and of limited supply have had a ready and profitable sale.

The cambric muslins of Preston, Bolton, etc, have come in large quantities, but those which are not still on hand have been sold at ruinous prices from the preference given to Scotch white and colored dresses.

The Rochdale baizes and coatings are used throughout the interior of this continent by the peasantry and partly by the better classes, but as they were supposed to interfere with the native manufactures the Government has lately imposed a prohibitory duty on them. Stuffs have never given a profit; Cassimere shawls at first paid well but the market is now overstocked; woollen clothes also used to sell profitably until the Franch and German manufactures were introduced, and the stocks on hand are considerable.

Sheffield and Birmingham goods have arrived in comparatively limited supplies, and when well assorted have left a fair profit.

The Nottingham and Leicester manufactures of hosiery and lace have so far exceeded the demand that a heavy loss has been the result of most shipments; and though English silk stockings are greatly preferred to French, yet the price at which the latter are offered is so low, viz. from 8 to 10 dollars the dozen, that I cannot contemplate a favorable return in this commodity.

The Norwich bombazeens and other articles are consumed by the better classes, but are not in much demand.

English linens, principally from Barnsley, and consisting of ducks, drills, etc. have been sent to this country, but the cheapness of cotton manufactures is a bar to their extensive use.

Earthenware and glass have also arrived in large quantities and when selected with attention have in most instances sold advantageously. The demand still continues, and although the lower price at which German glass can be afforded has secured it a temporary sale, the superior quality of the English manufactures will probably soon command a preference at proportionate prices.

Iron has been sent with various success, but the demand is much more limited than formerly owing to the decline in building. The late importations have been almost wholly made by the mining companies for their own use.

Immense supplies have been exported from Scotland to these countries of nearly every article of its manufacture, and the cotton goods mixed with silk are particularly adapted to the consumption of this continent.

Shawls of muslin, printed shawls, and Paisley worsted shawls have all been sold in large quantities; the prices are now extremely low, not only from the excess of importation, but because the first have met with much competition in the Cassimere shawls of Yorkshire, and French woollen shawls. The last, however, have been injudiciously selected, as the size is too small to envelop the head.

Muslin dresses also have been superabundantly supplied, 40,000 having arrived in one vessel and to one House; but though owing to this circumstance a severe loss must have been entailed upon the shippers, the consumption is so great from the article being suited to the market and this branch of trade not interfered with by other nations, that when the shipments are in proportion to the demand, they cannot fail to ensure a fair profit.

Dresses, printed, which have been sent in almost equal abundance, have been in most cases of a more expensive quality than is required, and have not met a ready sale. In this branch the French and Germans have endeavoured to compete with us, but hitherto unsuccessfully.

Lappets, at the opening of the trade, sold largely and well, but the shipments which have been made would have sufficed a population of ten times the amount of this Republic. Forced sales have been the natural consequence, but as the article has thus found a consumption in remote parts of the country, the demand will probably be permanent. No foreign nation has interfered with this branch of commerce.

Book muslins, Japan sprigs, and other descriptions of muslins have, when the quality has been sufficiently good, answered well; but British exporters have yet to learn the extent of expence to which the females of this country who consume this class of manufactures will go for their dresses. It is a singular fact that the greater part of the French dresses trimmed with gold or silver, which sell at 15 to 80 dollars each, are made of Scotch book or mull muslin.

Printed handkerchiefs have been largely introduced, but the patterns have been commonly ill selected. They are by no means so saleable in Peru as in Chile, where every gaucho wears one or two of them round his head; but the demand is likely to increase as they are preferred to the Pullicat, Masuliputam, and Madras handkerchiefs which have been hitherto in use. The French make a good article of this class, which, however, has not met a succesful sale.

Ginghams are not at present in estimation among the Peruvians; and in Lima the *saya* and *manto* are for the present peculiarly unfavorable to the general use of dresses of this description. But as they are extensively consumed by the Buenos Airesans and little less so by the Chilenians, it seems probable that at no very distant period Peru will also adopt them. The Germans have introduced into the two former countries an article of this class greatly superior both in quality and in colors, but the price will prevent their superseding the British manufacture.

Shirtings, madapolams, etc, or white cottons, have commonly been unproductive like those of Manchester, from the superabundance of supply.

Shoes have been sent from Scotland in extensive shipments to Buenos Aires and Chile, but owing to the existing high duties in Peru the few that have arrived have not paid.

Scotch linens, though imported in moderate quantities, have usually sold low because better goods of German manufacture have been offered at an inferior price.

The cotton manufactures of Ireland may be considered to be in their infancy; and although from this cause as well as from the want of capital the supplies to these countries have been limited, it is to be feared that the large supplies from Manchester and North America must have prevented their realizing a profitable sale. The calico printers of Ireland are, however, improving rapidly, and some of the most favorite patterns in Buenos Aires were of their production.

22*

Irish linens were imported into these countries on the opening of the trade in rather large quantities, and much loss has resulted from them, partly from the heavy expence of transport into the interior, but principally from the interference of those of Germany, which have been introduced and sold at rates only equal to the prices of cotton goods. It is difficult to suppose that the Germans or French can have profited by this trade, and as the lowness of the price has very materially increased the consumption of linen goods, the Irish linen trade, if the bounty be not withdrawn, may be expected shortly to give a more favorable result. Fine linens for shirts, etc. are saleable in small lots, and are preferred to the German and French.

Low linens for bagging may probably soon be sent to advantage, as the daily increasing value of the hides of this country will render it an object for the farmer to apply this article to many purposes for which the hide is now used.

6 Expected profits of foreign trade, 1833

(B. W. Clapp, *John Owens, Manchester merchant* (Manchester University Press, 1965), p. 40. This extract is from a letter written by Owens to Messrs Hodgson and Robinson, his agents in Buenos Aires.)

The carriage, shipping charges and freight will average 4 per cent. Insurance 2 per cent, Interest (at least) 4 per cent—making for charges 10 per cent—now the least profit we would think equivalent for the risque and trouble attending shipments to your place would be 7½per cent, making on the whole 17½per cent advance on the invoice.... Of course we are aware that owing to fluctuations in the supply or demand and circumstances which can be neither foreseen nor prevented this profit cannot in all cases be obtained, but we mean in a general way and on the ordinary run of business.

7 A welcome growth of imports, 1872

(Leone Levi, *History of British commerce and of the economic progress of the British nation, 1763—1870* (London, 1872), pp. 476-7)

The increase of commerce in the United Kingdom during the last one hundred years is something wonderful. In 1763 the population probably was 10,000,000. In 1870 it was 31,000,000, showing an increase of 210 per cent. But if the population has increased three times, the imports increased thirty times, from 10,000,000*l* to 303,000,000*l*; the exports nearly twenty times, from 13,000,000*l* to 244,000,000*l*;

the navigation of ports fifteen times, from 1,500,000 tons to 36,000,000 tons; and the shipping belonging to the kingdom fourteen times, namely, from 550,000 tons to 7,100,000 tons. The whole trade of the kingdom actually doubled itself during the last fifteen years, from 260,000,000*l* in 1855 to 547,000,000 *l* in 1870. This is the rate at which British commerce has been increasing; but large figures give an imperfect idea of their meaning. A trade amounting to about 550,000,000*l* a year in a population of 31,000,000,means immense activity, large increase of comforts, and great accumulation of wealth.

The fifteen millions of tons of shipping which entered at ports in the United Kingdom laden with precious produce from all parts of the world, estimated in value at 303,000,000*l*, brought large quantities of raw materials for our manufactures, of articles of food for the masses of the people, and of foreign merchandise to satisfy the increasing wants of the community. Of raw materials our manufacturers stand in absolute need. Whatever shortens the supply of such articles as cotton, silk, and even wool, limits the power of production. A bad crop of cotton in the far distant regions beyond the ocean, a disease in the cocoon, or any other calamity which increases the price of these articles, is so much actual loss to whole communities in Lancashire and Yorkshire, and through them to the whole kingdom. In 1840 there were entered for home consumption in the United Kingdom 4,545,000 cwt cotton, 48,421,000 lbs wool, and 1,896,000 lbs flax and hemp. In 1870 the consumption was 9,836,000 cwt cotton, 171,000,000 lbs wool, and 5,300,000 lbs flax, hemp, and jute. The world, we are thankful to say, has ample stores of produce to supply us with food, and, thanks to free trade, our people can get it whenever wanted. A large portion, indeed, of our population now depends on foreign corn,[1] and we could not well do without the oxen and bulls, sheep and lambs, bacon and beef, butter and cheese, sugar and coffee, fish and eggs, which come in so great quantities. Ever since 1840 the

---

[1] In an able paper on the home produce, imports, and consumption of wheat, by J. B. Lawes, F R S and Dr Gilbert, it was shown that from 1852-3 to 1868-9 the average area under crop in wheat was 3,922,586 acres, the average yield $28\frac{1}{4}$ bushels per acre, and the total produce 13,810,013 quarters, from which deducting $2\frac{1}{4}$ bushels per acre for seed, left available consumption 12,706,785 quarters. To these there was added an average annual importation of 6,375,272 quarters, making in all 19,082,057 quarters available for consumption. The average annual population of the United Kingdom having been 28,816,816, there were available for consumption per head 3·5 bushels from home produce 1·8 bushels from imports; total, 5·3 bushels per head.

increase in the consumption of foreign articles of food has been very large. The consumption of butter has increased from 1·05 lbs to 4·15 lbs per head; of cheese, from 0·92 lbs to 3·67 lbs per head; of corn, from 42·47 lbs to 124·30 lbs per head; of tea, from 1·22 lbs to 3·81 lbs per head; of sugar, from 15·20 lbs to 41·93 lbs per head. What folly, what crime, was it by law to hinder the people from getting what will sustain life. And our people are well pleased to use foreign clocks and watches, foreign gloves and silks, and other articles of finery, which our neighbours near or far can produce cheaper or better than we can. The interest of the largest number should always be the first consideration in any sound legislation. Of the 303,000,000*l* of imports nearly 140,000,000*l* consisted of raw materials, 100,000,000*l* of articles of food 30,000,000*l* of manufactured articles, and the remainder of other products and merchandise.

8 Prices and exports, 1872-86

(Sir Thomas H. Farrer, *Free trade versus fair trade* (3rd ed. Cobden Club, 1886), pp. 186–91)

We have already seen, in Chapter XXVIII, what an impetus our trade received in the period between 1840 and 1860. We know also how much the trade of France, as well as of England, grew after the treaty of 1860; and we may fairly ask our opponents, who are calling for a reversal of the policy which produced those benefits, to show not only that we have since that time been deprived of them, but that we should not have suffered that loss if we had not been Free Traders. We have a right to call upon them to define the specific evil of which they complain, and then to prove that it is due to Free Trade. I need not say that no such definition, no such proof, is forthcoming, and we are left with nothing but a vague shadow to fight with.

Let us, however, take such facts as we can lay hold of, and see how far they bear out the notion that we are losing our markets in the world.

Let us admit that our exports, as measured in nominal values, considerably diminished since those roaring years of prosperity, 1872 and 1873. They were 256 and 255 millions in those years, and 191 and 223 millions in 1879 and 1880. In 1882 they were 241 millions, and in 1884 233 millions, further decreasing to 213 millions in 1885. Let us admit, too, that this decrease of exports has been the sign and result of a real depression, and that both profits and wages have decreased since those so-called prosperous years. This in itself has nothing to do with the question at issue, unless it can be shown to arise from a permanent loss of market for our manufactures. Nothing whatever of the kind has been shown, or can be shown. But it can be shown that the prosperity of the earlier years of the decade is exaggerated; that the depression is exaggerated also; and that there are ample causes to account both for one and the other without assuming any falling off in the general demand for, or supply of, English goods.

The prosperity of 1872 and 1873 has been immensely exaggerated. All persons engaged in producing coal and iron made, no doubt, enormous profits, but they were led by those profits into an extravagant expenditure, partly on personal expenses and luxuries, but still more on plant and machinery for increasing the output, which has flooded the market with excessive supply, and from which no adequate return has yet been received. This expenditure of capital in fixed and, at first unremunerative investments, is one cause of subsequent depression. But whilst coal and iron masters made fortunes in those years, manufacturers and others who had to use coal and iron had to bear heavy outgoings, and their profits were reduced accordingly. Prices being high all round, people with fixed incomes suffered accordingly. Even the high wages of the time went less far than lower wages do when prices are lower. A great deal of the prosperity was apparent rather than real.

The statistics made the exports appear larger than they really were, because prices were so high. The quantities of goods exported, and the labour necessary to produce them, were as large in the subsequent years of depression as they had been in the years of inflation, but appear to be less because prices are so much lower. The exports of British produce were 255 millions in 1873, and 223 millions in 1880. If the exports of 1880 were valued at the prices of 1873 they would be 311 millions, or larger than those of any previous year.

Imported raw material, e.g., cotton and wool, was much dearer in the period of inflation than in the subsequent period of depression, and consequently that portion of the exports which is due to British labour and capital differed in the two periods much less than appears at first sight by the figures of the total exports. For instance, the raw cotton imported in 1873 was about the same in quantity as the raw cotton imported in 1879. But the raw cotton used in our manufactures exported cost us 14 millions more in 1873 than the same quantity cost us in 1879.

The prices and exports of the inflated years were due to causes which were temporary and accidental, and brought with them a necessary reaction. Amongst other causes may be mentioned.—

Expenditure of capital in this country on plant and machinery, not even yet fully reproductive.

Investments of English capital abroad, some of which were wholly unproductive—e.g., the bad foreign loans; and some of which were not immediately productive—e.g., American railways, but which are now in various ways bringing us a large return of imports.

Advances made to assist France in paying the German indemnity, which caused a large export from France and England to Germany at the time, and large exports from France to England and to Germany at a later time. ...

All these causes have little to do with the permanent demand for goods; all of them largely increased our exports at the time; some of them proved in the end losses, whilst others have helped that increase in our subsequent imports which Fair Traders seem to dread even more than losses. The inflation, as well as the depression, is therefore fully accounted for without any reference to closed markets or decrease in permanent demand.

It is a complete mistake to suppose that extraordinarily large exports, very high prices, and a great demand for labour are necessarily signs of great and permanent prosperity; they are only signs of great activity. They may be caused by a continuous demand, and by good and reproductive investments of capital, in which case they are elements of permanent prosperity. But they may be caused by bad investments, by payment of debt, or by unproductive expenditure on war, or by other causes which may lead to absolute loss. If I employ a thousand men to dig a hole and fill it up again, I shall cause high wages, high prices, and great prosperity in my neighbourhood for a time; but my capital will be lost, and when the work is at an end there will be a sad reaction and relapse. These are very elementary truths, but they seem to be forgotten by many popular expounders of statistics.

In addition, there is another cause for a chronic and permanent diminution in the values of both exports and imports, to which I can only advert very shortly. Good statisticians are of opinion that the value of gold, as compared with commodities, is steadily on the rise, and that it has been so since the effect of the gold discoveries was exhausted. The question is too long and difficult to be discussed here. But if the statisticians are right, as they probably are, the rise in the value of gold will account not only for some diminution in the figures of value by which we estimate our trade, but for a general fall of prices, and also of wages, which are too frequently and hastily attributed to commercial depression and to a decline in the producing and consuming powers of mankind.

The effect of the appreciation of gold must be slow and gradual, and is often concealed by the greater immediate effect of fluctuations due to other causes. But its result in producing a feeling of depression is probably out of proportion to its real effect. Upon the real wealth of the country as a whole it has not necessarily any effect at all. If, as Hume has said, every one had to-morrow half as many sovereigns in his pocket as he has to-day, he would be neither richer nor poorer. He would buy or sell for half-a-sovereign what he has to buy or sell for a sovereign to-day. But a time of falling prices is notoriously a period of depression. It is a bad time for the larger merchants who carry on the great trades of the country. When they have to borrow money to complete their purchases, and it rises in value before they have to repay it, whilst the commodities which they buy fall in money value at the same time, they suffer actual loss; and this loss probably operates on their expectations and makes them less daring in business. Other classes gain what the merchant loses. The retail dealer, who can generally postpone a proportionate reduction of his retail prices, and the consumer who ultimately gets the benefit of the fall in price, share the gain between them. But it is the wholesale trader who carries on the large speculative business of the country, and who is listened to as its representative, and he is out of pocket and out of heart at a time of falling prices.

Since my second edition was published, the subject of the precious metals has assumed greater importance. The opinion that gold has risen in value, and that prices, and even wages are much affected by the rise, continues to gain ground, and remedies of different kinds are proposed. It is out of the question here to approach the question of the currency, which is only remotely connected with the more immediate subject of this book; but it may not be impertinent to urge that, although it is quite true that a greater or smaller quantity of coins does not add to or take from the wealth of a country, yet an increase of the precious metals and a consequent rise of prices causes a rise in spirits, a sanguine feeling, and a tendency to speculation, which results in an actual increase of business; whilst a diminution in the precious metals causes a lowering of spirit, a feeling of depression, and an indisposition to speculate, which results in an actual decrease of business. If this is the fact, it would be well worth while that the much-disputed facts concerning the present relations of gold and silver to each other and to commodities should be made the subject of a special authoritative inquiry. If such an inquiry should not result in suggesting any effectual remedy, it would still be useful in clearing men's minds as to the facts, and in removing such impediments to a resumption of business as arise from vain imaginations. So far as depression is a 'mental attitude', such an inquiry might operate as a cordial and a cure.

There is another cause of chronic depression, or rather for a feeling of chronic depression, to which it is worth while to advert. We are told that a great change is taking place in the mode in which foreign trade is conducted. Before the times of steam and telegraph there was a long interval of time and space between the commencement and the end of a transaction, between the original purchase and the final sale. This afforded great scope for the merchant or middleman; and his profits depended upon the judgment and skill with which he could forecast distant and future markets. At the present time the state of markets is known everywhere at once by telegraph, and the period of transport is abridged by steam. Stocks in hand need no longer be as large as formerly. The original vendor and final purchaser are brought nearer together, and the opportunities for the skill and judgment of the middleman are curtailed. The world gains on the whole by the change, but the old-fashioned merchants, who have done so much to make England what she is, suffer or are extinguished, and from that powerful and important class we have the natural cry that trade is bad, although at the same time the bulk of transactions is greater than ever.

Both these causes, viz., an appreciation of gold, and a change in the methods of trade, effect the same class, a class which is naturally influential in specially expressing its feelings of depression. That such a class should suffer is to be lamented. But the losses of this class are the gains of other classes, and it would be wrong to suppose that they constitute any real diminution in the wealth or prosperity of the country as a whole.

9 Character of Britain's foreign trade, c. 1900

(A. L. Bowley, *A short account of England's foreign trade in the nineteenth century, its economic and social results* (rev. ed. Sonnenschein & Co. 1905), pp. 133–5)

Our foreign trade is to a great extent an elaborate machinery for supplying us with food. Of the £470 millions'-worth of goods which come to our shore, at least £200 millions'-worth is food; while £115 millions'-worth is of raw materials, which can be definitely reckoned, and of the remaining £155 millions, if it was possible to distinguish

materials for manufacture from materials for sale, part would be found
to be materials for our factories, rather than goods for consumption.

Let us consider how this enormous annual bill for food is paid,
regarding the whole nation as a single family. Tabulating imports and
exports thus:

Imports for Consumption or Manufacture, Roughly Grouped

*Annual Average
1898–1902*

Food	£190 million
Wine and Tobacco	10 million
Raw Textiles	55 million
Other Raw Materials and Articles, mainly unmanufactured	75 million
Articles mainly Manufactured	110 million
	£440 million
Interest on Capital (approximate estimate)	85 million
	£355 million,

to be paid by exports, and earnings of shipping.

Exports of Home Produce

*Annual Average
1898–1902*

Textile Products	£100 million
Iron and Steel and their Products	50 million
Coal	30 million
Other Merchandise	90 million
	£280 million
Earnings of Shipping (approximate estimate)	85 million
	£355 million

In the manufacture of textiles, so great is the increase of value, that
after the home market has been supplied, the exports of yarn and cloth
are worth £45,000,000 more than the raw materials. The cotton
workers thus pay for the raw cotton we use at home, and have this
surplus towards our food supply.

Workers in mines, iron foundries, steel, machinery, etc., add to the
value of the ores extracted and the ores and metals imported to such
an extent that we are supplied, and a surplus of £50,000,000 is ready to
pay for imports.

These sums added to the earnings of our ships go far to make up the £190,000,000 sufficient to cover our bill for provisions.

This done, we may reckon that half the miscellaneous goods imported are in return for miscellaneous exports, while the other half are the interest due on our foreign investments.

We do not mean to imply that the destination of our textiles is the country which supplies our corn, with similar conditions for the other case; but only to supply data for comparing the relative importance of different trades. As a matter of fact, our exports do not always go to the countries from which we obtain imports. The United States are paid for their corn partly by manufactures, partly by goods from China and other countries, who thus pay their debts to us. Again, food imports are often the payment of interest on capital invested; and in many cases it is not easy to trace the equilibrium of trade. The broad result is a balancing of all the individual transactions, accomplished by bills of exchange on individual dealers, and involving an infinitude of operations.

10 Exports in decline, 1931

(Alexander Loveday, *Britain and world trade: quo vadimus and other economic essays* (Longmans, 1931), pp. 148–50)

A considerable proportion of the literature which has appeared in recent years concerning British industry and trade is somewhat puzzling to the outsider. The analysis of the situation has changed from year to year. A certain doubt and anxiety has been steadily growing since 1923 when the authors of *Is Unemployment Inevitable?* reached the considered opinion that they did not believe that 'the abnormal unemployment of the last few years will become chronic or inevitable'.

But at any rate until quite recent months writers have tended to attribute the difficulties of the United Kingdom largely to the alleged impoverishment of Europe or to the competitive advantages which certain countries have reaped from inflation. The chaos which existed in Europe during the first few years after the Armistice no doubt hampered the exports of this country and of all others to which Europe constituted an important market. But while others developed their trade outside Europe we lost a large part of our trans-oceanic markets and during the last two or three years European demand has probably been up to the pre-war standard.

I propose to endeavour to show in the first place that our failure to secure trade has been greatest—outside the British Empire—in regions which have been least affected by the war; that our markets have not been mainly captured by countries enjoying the alleged benefits of inflation; that these favoured nations have not increased their trade by more or indeed by as much as have others which have pursued a monetary policy similar to our own. This last fact constitutes of course no proof and implies no probability that our monetary policy has not adversely affected our foreign trade and our industrial production. It implies merely that there must be some deeper lying cause which has affected us and not our neighbours—which has magnified the ill effects or diminished the good effects of deflation. That ultimate cause probably most students of the questions would agree is lack of suppleness in the mechanism of production, and that ultimate cause has put a check, not only on our export trade, but on our whole production of wealth. We are to-day standing still or but slowly crawling forward in a progressive world. We are standing or crawling with an awkward rigidity in a world in which suppleness is becoming a constantly more imperative quality, because demand is becoming constantly more mobile.

If the alleged reduced purchasing power of Europe was the true cause of our difficulties then we might expect to find that while our trade with other parts of the world was fairly maintained our exports to Europe had suffered a substantial contraction. But this is the reverse of the truth. It is practically certain that the imports of Europe to-day are greater than they were in 1913. Europe is obtaining from the United Kingdom a slightly smaller proportion of her total purchases, and after allowing for price changes possibly a slightly smaller quantum of goods. But during the last two or three years such reduction as there may have been in the aggregate of her purchases from us was wholly due to our enfeebled competitive power and not to any general reduction of imports by Europe. Moreover the falling off in our exports, if any, must have been relatively small. It can scarcely have amounted to 5 per cent in 1927 or 1928. On the other hand our share in the trade of the other continents of the world has been very greatly reduced as may be seen at a glance from the following figures:

Proportion of total imports obtained from U. K.

	%	%	%	%
	1913	*1925*	*1927*	*1928*
By:				
Europe	8·4	9·2	8·1	7·7

	1913	1925	1927	1928
North America	16·8	11·2	10·3	10·1
South America	27·9	21·2	18·7	18·8
Africa	36·7	34·6	32·9	31·3
Asia	31·5	19·9	18·6	18·9
Oceania	53·4	46	43·2	44·4
World	14·5	13·3	12·1	11·7

Therefore our share in European purchases in 1928 was about 8 per cent smaller than in 1913, our share in those of Asia, which still account for over one-fifth of our exports and of North America, was 40 per cent lower, of South America over 30 per cent and of Oceania and Africa some 15 per cent lower. We have held our position better in Europe than elsewhere and better in the British Empire than in other parts of the trans-oceanic continents. We contribute about one-fifth less than we did to the total world trade viewed in this way. To attribute this loss to the weakness of European demand is definitely misleading. The figures render it abundantly clear that our loss of status has been greatest in the American continent and Asia where the effects of the war were least.

11 Manchester merchants, 1946

(Bd of Trade, *Working party reports: cotton* (H.M.S.O., 1946), p. 93)

In the case of cotton textiles the distributive function is in a very special sense a distinct function of great importance, since, as is clear from the facts recorded in Chapter IV, the notable feature in the industry's structure is that, for the main part, the determination of the nature and quantity of the goods to be produced is in the hands of the 'converters', who belong essentially to the distributive side and represent a separate interest. It is a fair description of the position to say that the producers of cotton goods, with the exception of a relatively few vertically integrated firms, represent an industry whose sales department is detached from it. Other industries—even highly organised industries—are, as is pointed out in a later paragraph, very much affected by wholesale and retail selling charges which are quite outside the manufacturer's control; but in many cases the manufacturer can, through his own sales department, both study and influence consumer demand and thus have a basis on which to plan and operate a production policy for the utilisation of his plant to the best advantage. But the cotton textile producers as a whole (i.e., save for the exceptions noted above) have hitherto relied upon others to fulfil this vitally important function.

12 Outlook for exports, 1952

(*The Observer*, 14 September 1952, reprinted in *Re-thinking our future: the outline of a new arduous but hopeful policy for Britain* (The *Observer*, 1952), pp. 30–2)

The last half-century has been a time of almost feverish economic development all over the world. Other industrial countries have copied from us and in some lines of production they have gone ahead. The developing countries, so-called, cannot yet imitate many of the more complex industrial processes. But they have discovered that with many of the simpler processes of manufacture and assembly (particularly of consumer or 'shop' goods), they can compensate with cheaper labour for their lack of technical experience.

In country after country, too, the natural desire of business men to make profits and of workers to live better has been spurred on by nationalist feeling. To be a colony in the nineteenth century meant to be a producer of primary products—food and raw materials. Industries producing shop goods for their home markets are to-day encouraged and protected everywhere by newly independent Governments, as a matter more of national pride than of strict supply-and-demand economy.

All this has been going on a long time. But the events of the last twenty years have helped us to shut our eyes to it. In the thirties, by the accident of the Depression, we were able to buy our food and raw materials cheap; we needed to sell exceptionally few of our exports to pay our way. After the last war, our exporters were spoiled by a combination of favourable trends. First, by post-war shortages in the buying countries. Second, by the temporary absence of competition from Germany and Japan. And, third, by the sterling balances—the vast sums piled up by Commonwealth and other countries during the war, which encouraged them to buy extravagantly from us.

Outlook for Textiles

Now, all three trends have come, or are rapidly coming, to an end. It seems very likely that 1950 and 1951 will be viewed as the 'golden years' for British exports, and that those easy markets will never come again. Similarly, the subsequent slump that we have been experiencing will look less and less like a temporary post-Korean accident, and more and more like the resumption of a long-term trend.

If this is so, we shall never again be able to boost our export earnings merely by cutting off supplies from the home market. With more and more products, the only result would be a glut of 'frustrated exports' cluttering up our shops at home. In this situation, and with tough competition to overcome, our only course is to modernise our export trade to fit a world bent on large-scale industrialisation.

The biggest single group in our exports last year was still the textile group. Exports of manufactured textiles—mainly cotton and woollen goods—made up about a fifth of our total exports. Yet these are manufactures in which the cost of the imported raw material is high, compared with that of many other industries. The skills required—except in the production of the most elaborate cloths—are fairly simple. They can be quickly acquired by Japanese, Indian, Egyptian or Brazilian workers.

The market for textile products, too, is highly vulnerable—as Lancashire well knows—to sudden changes. A country, such as Australia, which abruptly decides to cut down its imports, will pick first on finished textiles as something that consumers can temporarily go short of.

The trend is inexorable. World trade in textile manufacture has been steadily declining since 1899. Of that shrinking trade, other countries, Japan especially, have been able to wrest an ever larger share from Britain. They have been able to do so because they have been able to profit by our experience and because they had the great advantage of cheap labour.

If we look back a hundred years or so, it should not strike us as odd that textiles in the 'slump' conditions of mid-1952 accounted for only about a sixth of our exports. After 1820, when they made up two-thirds of British exports, they fell steadily to a half in the 1860s: and to a little over a quarter before the last war. Within the lifetime of many of us, it is not impossible that we shall, in fact, be importing more cheap textiles than we export.

To say that we should stop leaning so heavily on textile exports does not mean that the whole export side of the industry must, in time, be closed down. There are sections of the woollen and worsted trade, and even a few in cotton, where manufacturers have read the writing on the wall and have hastened to develop mixtures of traditional materials with the new synthetic fibres, which will keep them ahead of their competitors at home and in some overseas markets. But, by and large, we, as a nation, must not try to blunt the economic pincers which will eventually make mass cotton exports a thing of the past.

New Technical Fields

The passing of old industries is always a sad business. It is always hard to deny the call for protection and 'temporary' subsidy or special aid of one kind and another. Particularly so when, as with textiles, the appeal is backed by powerful forces in industry, by strong trade unions and a sizeable group of MPs. Yet, for all that, we cannot in our present straits afford to close our eyes to harsh facts.

Textiles are the biggest and most obvious problem in modernising our exports. But other industries are faced with similar economic forces. Some of them are managing to survive and develop in spite of them. The radio industry, for example, enjoyed after the war a tremendous demand for its products. Radio sets were exported in millions to hungry overseas markets. Now the pinch is beginning and many young industrial countries are finding that the assembly of radios by mass-production methods is not beyond the ability of even an untrained labour force. As new factories spring up they are protected by solicitous Governments with quotas and tariffs. Our own radio industry, therefore, has to concentrate more and more on the production of the more delicate and difficult components, on transmitting rather than receiving equipment, and on the development of altogether new fields of highly technical electronic equipment for industry.

In order to give industrialists the confidence and the means to build such an industry, it may be necessary to favour engineering—in all its branches, including aeronautical enginering— above other industries; to make capital more easily available to it; even to favour it in taxation. If it is sensible to discriminate in the national interest by allocating steel when steel is scarce, it is equally sensible to discriminate in other means of production.

Our Lead in Aircraft

Among capital goods industries that should be deliberately fostered, the aircraft industry is an outstanding example. If we throw away our lead in jet aircraft design, we shall deserve to suffer for it. The possibilities ahead are perhaps greater than even its most enthusiastic advocates can imagine. In 1911, for instance, what motoring enthusiast would have believed that in forty years the motor industry would be three times as important in world trade as railways and shipping combined?

Agricultural equipment is another example. In the last fifty years it has more than tripled its share of world trade. As the world shortage of food becomes more acute, and as prices rise, a steady demand for all the means of increasing food output—implements and machinery, prefabricated storage units, weed- and pest-killers, all shapes and sizes of tractors—is certain to grow.

These are the main lines along which we should seek to modernise our exports. A more exhaustive survey would find other industries which we must either allow to fade or must proceed to encourage. A rapidly developing chemical industry, for example, will be essential to our economic future. both as a vast research laboratory for finding ways to save and substitute raw materials, and as an export industry of the more complex chemical derivatives of coal and oil.

An Expanding Market

The same story could be told of other industries. Even with motor cars it seems more than likely that the safest and the most profitable future for exports lies less in the small and relatively easily assembled passenger car than in motor-car engines, and in heavy commercial lorries and trucks.

The guiding aim in all our export industries should be to concentrate on exploiting to the utmost the technical skill, mechanical inventiveness and experience which are our heritage as the oldest industrialised of the nations of the world. This does not mean concentrating on luxury markets which can have only a limited scope. It does—on the whole, though not exclusively—mean concentrating on serving the foreign producer rather than the foreign consumer. To earn our national living we shall have from now on to sell not consumer goods, but the means to make them. It may seem that this is a once-for-all business. But it is a reasonable hope that this is, over the whole world, an expanding rather than a shrinking market. And there are always replacements and new fittings to supply in the course of time.

Although this should be our aim, we are not making the best of our opportunities in this new field. The market for industrial equipment has more than doubled its share of world trade in the last fifty years, and shows every sign of growing still more. But between 1949 and 1951. America's exports of machinery and electrical apparatus (excluding vehicles) rose by 45 per cent, while Britain's rose by only 28 per cent. German engineering exports, meanwhile rose from £35 million in 1949 to £126 million in 1950, and to nearly £250 million in 1951.

At the moment, to be sure, the defence programme and a temporary shortage of steel are delaying some of our engineering exports. Again and again we hear of orders being lost, usually to the Germans, because of long delivery dates. Even so, much more could be done if the need to do it were recognised as a national priority.

Our eventual aim should be to build an engineering industry capable of meeting every demand as promptly and as cheaply as technical knowledge allows. We should aim, now, at having even a surplus capacity; at having too many trained men rather than too few. It would be less costly than propping up declining textile mills.

13 Exports and the balance of payments, 1960

(Political and Economic Planning, *Growth in the British economy: a study of economic problems and policies in contemporary Britain* (Allen & Unwin, 1960), pp. 18–19, 171–2)

The problem which has been in the forefront of discussion since the [1939–45] war has been the balance of payments. ... It was realised immediately after the war that a great expansion of exports would be needed if the country was to pay its way in the world, when it had lost overseas sources of income and had piled up debts overseas, and when it was facing worsened terms of trade. The targets fixed at that time were achieved and passed. To this extent the post-war export drive has been a great success. The increase which has already taken place in the volume of exports during the post-war period has been well in excess of the increase in the value of imports. For many years after the war it was easy to export because of the immense backlog of demand all over the world and the virtual absence from world markets of two main competitors, Western Germany and Japan. Since trading conditions became more normal in the 1950s, the development of British exports, in relation to those of other countries, especially Western Germany, has not been encouraging. ... Despite a quite considerable achievement, it has to be concluded that insofar as a balance of payments crisis can always be solved by increasing exports, exports have been inadequate.

*   *   *

In 1938 only 74% of Britain's imports of goods and services were financed by exports of goods and services. In 1957, 100% were so financed. ... Between 1938 and 1957 the value of United Kingdom exports of goods and services increased by over 500%, while the value of imports rose by some 350%. Yet apparently even this great improvement in the situation over the last twenty years has not been enough to put an end to worries about the balance of payments. Much has been done, but more is still needed.

23*

7

The workman

Nowhere are the fruits of economic progress to be seen more clearly than in the improved position of the workman. Hours of labour have been drastically shortened (*A1*, *C4*, *10*); wages are settled by negotiation between employers and unions either nationally or locally, where once they were much more a matter for individual bargaining (section B *passim*); industrial discipline is less severe and takes new forms (*C2*, *9*); and children no longer enter full-time employment at the early ages common in the nineteenth century. These gains have not come without effort, for employers until the First World War resisted the claims of unions to speak on behalf of labour. Nor does it follow that trade union strength can be automatically converted into the power to raise wages. Domestic service, the least organised of trades, shared in the general advance of wages in the later nineteenth century (*C8*), while the printers' unions, strong as they were, had to submit to mechanisation (*B11*). National unions have not been able to prevent the payment of relatively low wages in parts of the country where economic activity is at a low level; national rates tend to become minimum wages (*B15*). With the general acceptance by employers of the trade union as an unavoidable, indeed a stabilising, institution in the economy has gone an acknowledgment by trade union leaders of the need to promote industrial efficiency (*B14*). Not much has come of this change of heart, but it does represent a substantial theoretical shift in the attitude of trade unions compared to the sturdy refusal to look beyond narrow self-interest that characterised them until after the Second World War (*B7*). Trade unions, like businessmen, welcome state aid but frown on state intervention. Freed from restrictive legislation by acts of 1825, 1871, 1875 and 1906 trade unions enjoyed until 1966 the right to make what bargains they could with employers. The very recent attempts by government to renew legal restraints upon trade union activity perhaps represent a swing of the pendulum back towards a more coercive era. But the practical results of incomes policy since 1945, whether voluntary or otherwise, have so far been slight.

A. RECRUITMENT OF LABOUR

1 Indentured service, 1788-91

(Frances Collier, *The family economy of the working classes in the cotton industry, 1784-1833*, ed. Robert S. Fitton (Chetham Society, 3rd series, XII, 1965), pp. 55-6)

(i) *Child labour*, 1788

Be it remembered, It is this Day agreed by and between *Saml Greg* of *Manchester*, in the County of *Lancaster, Cotton Manufacturer* of the one Part, and *Thomas Smith, Hatters*, of *Heaton Norris in the County of Lancaster* of the other Part, as follows, That the said *Thos Smith Agreeath that Esther and Ann Smith* shall serve the said *Saml Greg* in his Cotton-Mills, in *Styall* as a just and honest Servant, *Thirteen* Hours in each of the six working Days, and to be at *theair* own Liberty at all other Times; the Commencement of the Hours to be fixed from Time to Time by the said *Saml Greg* for the Term of *Three* Years at the Wages of *one Penney per Week and Sufficient Meat Drink and Apparell Lodging washing and all other Things necessary and fit for a Servant.*

And that if the said *Esr and Ann Smith* shall absent *themselves* from the Service of the said *Saml Greg* in the said working Hours, during the said Term, that the said *Saml Greg* may not only abate the Wages proportionably, but also for the Damages sustained by such Absence. And that the said *Saml Greg* shall be at Liberty, during the Term, to discharge the Servant from his Service, for Misbehaviour, or want of Employ.

As Witness their Hands, this *Twenty Eight* Day of *Jany 1788*—

Witness   *By me Thomas Smith*
Mattw Fawkner

(ii) *Yearly hiring of adults*, 1791

Be it remembered, it is this Day agreed by and between *Samuel Greg*, of Styal, in the County of Chester, of the one Part, and *Wm Chadwick* of *Styall* of the other Part, as follows: That the said *Wm Chadwick* shall serve the said Samuel Greg in his Cotton-Mills, in Styal, in the County of Chester, as a just and honest Servant, *Twelve*

Hours in each of the six working Days, and to be at *his* own Liberty at all other Times; the Commencement of the Hours to be fixed from Time to Time by the said Samuel Greg, for the Term of *one* Year at the Wages of *foureteen Shilling per Week.*

And that if the said *Wm Chadwick* shall absent *him self* from the Service of the said Samuel Greg, in the said working Hours, during the said Term, without his Consent first obtained, that the said Samuel Greg may abate the Wages in a double Proportion for such Absence; and the said Samuel Greg shall be at Liberty, during the Term, to discharge the Servant from his Service, for Misbehaviour, or Want of Employ.

As witness their Hands, this *fiveth* Day of *Feby 1791.*

Witness   *William Chadwick*
Mattw Fawkner

2 Apprenticeship defended, 1817

(This petition from the watchmakers of London and Westminster is printed in the 'Report from the committee on the petitions of the watchmakers of Coventry', *B.P.P.*, 1817, VI, pp. 46-8)

1 That the obvious intention of our ancestors, in enacting the statute of the 5 Elizabeth, cap. 4, was to produce and maintain a competent number and perpetual succession of masters and journeymen, of practical experience, to promote, secure, and render permanent the prosperity of the national arts and manufactures, honestly wrought by their ability and talents, inculcated by a mechanical education, called a seven years' apprenticeship; whereby according to the memorable words of the statute itself 'it will come to pass, that the same law (being duly executed) should banish idleness, advance husbandry, and yield unto the hired person, both in time of scarcity and in time of plenty, a convenient proportion of wages'.

2 That it is by apprenticeships, that the practitioners in the arts and manufactures attain the high degree of perfection, whereby British productions have arrived at the great estimation in which they were heretofore held in foreign markets. ...

8 That the apprenticed artisans have, collectively and individually, an unquestionable right to expect the most extended protection from the Legislature, in the quiet and exclusive use and enjoyment of their several and respective arts and trades, which the law has already conferred upon them as a property, as much as it has secured the pro-

perty of the stockholder in the public funds; and it is clearly unjust to take the whole of the ancient established property and rights of any one class of the community, unless, at the same time, the rights and property of the whole commonwealth should be dissolved, and parcelled out anew for the public good. . . .

10 That in consequence of too minute a division of labour, injudiciously allowed in several manufactures, the workmen employed are not enabled to make throughout any one article however simple, or even to maintain themselves by their industry.

11 That the unlimited or promiscuous introduction of various descriptions of persons without apprenticeship into the manufactures occasions a surplus of manufacturing poor, and an unneccessary competition, ruinous to the commercial capital and industry of the nation; because the overflow of goods causes all the productions of the manufacturies to fall in price, and be sold to foreigners for less money than they cost in making; which deficiencies are necessarily made up by the ruin of the master manufacturers, bankruptcies, and dividends to creditors; and are the cause of increased parochial and other rates, thus necessarily created, for the support of the poor workmen, who are deprived of the fair price of their honest labour. . . .

17 That the system of apprenticeships, whether considered in a religious, political or moral point of view, is highly beneficial to the State, and from the neglect thereof is to be attributed the great defalcation of public morals, the numerous frauds committed in trade, the increased numbers of juvenile criminals, public trials and executions.

18 That the pretensions to the allowance of universal uncontrolled freedom of action to every individual founded upon the same delusive theoretical principles which fostered the French Revolution, are wholly inapplicable to the insular situation of this Kingdom, and if allowed to prevail, will hasten the destruction of the social system so happily arranged in the existing form and substance of the British constitution, established by law.

19 That the meeting highly approves the proceedings of the 62,875 masters and journeymen, who have already presented petitions, to the House of Commons, praying for leave to bring a Bill into Parliament to amend, extend and make more effectual the statute of apprenticeship, 5 Elizabeth, chap. 4. . . .

21 That the most effectual preventive against and check upon combinations of journeymen, as also of masters in any trade, is for the persons engaged in such trades to take apprentices as required by law.

3 Domestic service, 1837

(William Tayler, *Diary of William Tayler, footman, 1837*, ed. Dorothy Wise (St. Marylebone Society, 1962), pp. 33, 48)

It's surpriseing to see the number of servants that are walking about the streets out of place. I have taken an account of the number of servants that have advertised for places in one newspaper during the last week; the number is three hundred and eighty. This was the Times paper. There is considered to be as many advertisements in that as there is in all the rest of the papers put together, therefor I mite sopose there to be 380 advertisements in the rest. I am sertain half the servants in London do not advertise at all. Now, soposing seven hundred and sixty to of advertised and the same number not to of advertised, there must be at least one thousand, five hundred and twenty servants out of place at one time in London, and if I had reckoned servants of all work—that is, tradespeoples' servants—it would of amounted to many hundreds more. I am sertain I have under-rated this number. Servants are so plentifull that gentlefolk will only have those that are tall, upright, respectable-looking young people and must bare the very best character, and mechanics are so very numerous that most tradespeople sends their sons and daughters out to servise rather than put them to a trade. By that reason, London and every other tound is over run with servants.

*   *   *

16*th* This day has been spent about the same as most of my others. The first thing I do in the morning is to get up at half past six, goes to the water's side, stays until eight, comes home, haves my breakfast gets theirs ready at nine upstairs, then cleans the knives, fetches their breakfast down at ten, does a fiew other little jobs, and then goes out for a walk a little before eleven, and comes home a little before one. Gits their lunch ready and haves my own dinner by two, rests myself until three, then goes for a ride with the ladies until four, comes home, haves my tea, gets their dinner things ready at five, waits on them at dinner, brings it down and clears my part of it away by half past six, taks a walk or sits down and reads until eight, then takes up their tea, brings it down a little after eight, goes for another walk by the water's side for half an hour, then comes home and haves my supper. Goes to bed a little before eleven. In this way I goe on every day and so I mean to continue as long as I am here [Brighton] as I shall not get such good air when I get to London, therefore I get all

I can now. Very few servants go out so much as I do. Many have not an opertunity, some would rather stay at home and sleep, others would rather go to the publick house and get drunk, but I like pure air to either of this.

4 Sub-contract in the coal industry, 1842

(H. Scott, 'Colliers' wages in Shropshire, *c.* 1830–50', *Transactions of Shropshire Archaeological Society*, LIII, 1949–50, pp. 20–22)

This indenture, made the ———— day of ————, between the Butties of the one part, and the masters of the other part, whereas the said masters are possessed of a colliery in the parish of ————, and it has been agreed between them and the said butties that the latter parties shall work, get, and raise the unwrought mines of thick coal from the pair of pits at the said colliery, distinguished as ———— at the charter, and upon the terms and under and subject to the provisions, covenants, stipulations and agreements hereinafter expressed and contained. Now this indenture witnesseth that in consideration of the charter and payments and of the covenant of agreements hereinafter made payable and contained on the part of the parties hereto of the latter part, they, the said butties, for themselves jointly and severally, and for their respective executors and administrators, hereby covenant and agree to and with the said masters, their executors and administrators, in manner following; that is to say, that they the said butties, shall, and will immediately proceed at their own cost and charges in all things to work, get, and raise and undergo the unwrought mines of the thick coal from the aforesaid pits of the said parties hereto of the latter part, and shall and will deliver such coal at least 70 yards from the pits' mouth in the usual way and according to the best and most approved method. Nevertheless, under the inspection, control, directions and orders of the ground bailiff of the said masters who shall for the time being be employed at the said colliery, and also that they the said butties will, at their own expense, make and maintain all requisite gateroads, airways and headways, and do all necessary dead work, except dams and driving through faults, in and about the said pits, and will in all respects manage and carry on the same in a sure and regular course of mining, as practised in the neighbourhood of ————, with as much diligence, industry and despatch, as can be reasonably expected, and so as to obtain as great a weekly produce as the state of the mines will allow. And also shall and will employ a sufficient number of competent miners and workmen, and provide at their own expense all necessary tools and implements (except such things, tools and implements as are hereinafter mentioned), and covenanted to be provided by the parties hereto of the latter part; and will also provide slack for the engine without charging any charter for the same, and deliver such slack at the slack hole at the said colliery, and also will pay one half part of the expenses of the sick workmen and surgeon's charges, and also that they the said butties will not, during the subsistence of this agreement, engage with or work for any other coal master, or any other person whatsoever, but will, on the contrary, wholly and individually work for and attend to the interest of the said parties hereto of the latter part at the aforesaid pits, according to the best of their ability and judgment, and also that they the said butties will not do or commit any act, matter or thing whereby the said pits or the property in or about the same, may or can be in any manner injured or deteriorated in value. Also that they the said butties will pay unto the said masters, their executors, administrators or assigns, the sum of £—, being the balance paid by the parties hereto of the latter part to the late charter masters, on relinquishing the working of the said pits with interest after the rate of £5 per cent per annum thereon, by regular equal fortnightly payments of £—, at the least, and that the same shall be deducted and allowed from the charter prices at each reckoning, and that the said master, shall and may retain the same accordingly, and they shall have a security by way of lien on all the property of the said butties at the said pits, in addition to the other remedies hereinafter provided; and further, that they the said butties shall and will from time to time keep and maintain the pit workings, stock and property of equal and adequate value, with the amount of the valuation remaining due and unpaid, and in consideration of the covenants and agreements hereinbefore contained on the part of the said butties, they, the said masters, for themselves, their executors and administrators, hereby covenant and agree to and with the said butties, their executors and administrators, in manner following, that is to say: that they the said parties hereto of the latter part will pay and allow the said butties the following charter prices for each and every parcel of coals, lumps and slack, to be worked, raised, and delivered by them according to the true intent and meaning of these present; that is to say for every parcel of best coals, —; common coal, —; lumps and slack (except such slack as may be consumed in working the engine,—; fine slack,—. And for all cokes for the use of the furnace—per barrow, such cokes to be made three parts of coal and one part of round slack, each such parcel to be of the weight of—; such charter to include all the monies paid for wages; tools and implements to be provided by the said butties as aforesaid, and all dead work, save dams and driving through faults, for which the said butties are to be compensated under the clause of reference hereinafter contained; and such charter prices are to be subject to advance and reduction according to the fluctuations of workmen's wages, as hereinafter provided and also that they the said persons, parties hereto of the latter part, will make, find and provide all necessary dams, and all pit timber, rails, sleepers, pit ropes and trolley waggons, that may be necessary, and will also work the engine (subject to the supply of slack) and find water ease, and make reasonable compensation for driving through faults, and will pay one half part of the expenses of sick and surgeons' charges, and will pay the usual allowance to the widow of any workman who may be killed in the pit, and also pay and allow the said butties—per yard for gate-roading and—per yard for air heading. Provided always and it is, expressly declared and agreed by and between the parties hereto, that so long as any part of the said amount of valuation or sum of £—and interest shall be due and owing by the said butties, or if they shall not work the said pits in a regular and proper manner to the satisfaction of the said parties hereto of the second part, or if they the said butties shall suffer the pits and property to be lessened in value below the amount of the valuation from the time to time remaining unpaid, it shall be lawful for the said masters and the survivor or survivors of them, and the executors and administrators of such survivor, without giving any notice and without any further authority than such as is hereby given, to determine and put an end to this agreement if they think proper, and to seize and take possession of all the property in and about the said pits belonging to them the said parties hereto of the latter part and of all the horses, carriages, tools, implements and other effects of the said butties in and about the said colliery, and to sell and dispose of the same in such a manner as they think fit in order to satisfy the amount of valuation remaining unpaid, anything herein contained to the contrary notwithstanding. Provided also that (subject to the next preceding clause) if the said masters on their part or the said butties on their part shall be desirous of putting an end to this agreement, it shall be lawful for either of the said parties to do so on giving to the other parties one calendar month's notice in writing of their intention on that behalf. But it is hereby expressly understood and agreed that no such notice shall be given by the said butties unless for some reasonable cause to be ascertained and determined under the clause of reference hereinafter contained, and that after the expiration of such notice these presents and every covenant stipulation, and agreement herein contained shall cease and be void except as to any breaches by the said butties of the covenants herein contained on their part, and as to any claims by the said parties hereto of the latter part in respect of the said sum of £—and interest, or any part hereof. And the said butties shall, subject as aforesaid, be entitled to a valuation or remuneration for the gateroads, headways, undergone coals and deadwork. Provided always, and it is hereby also agreed and declared, that if there shall be any advance or reduction in the wages to be from time to time paid to the colliers and workmen employed in getting the said mines, then there shall be a proportionate advance or reduction of the charter prices to be fixed by the referees under the clause hereinafter contained. Provided also, and it is hereby lastly declared and agreed, that all matters to be determined by reference, according to the true intent and meaning of these presents shall be decided by two disinterested competent ground bailiffs, one to be chosen by the said masters, and the other by the said butties or by an umpire to be chosen by such ground bailiffs.

In witness whereof the said parties to these presents have hereunto set their hands and seals the day and year first written above.

5 Casual labour at the docks, 1891

(Royal commission on labour, *Minutes of Evidence*, Group B, Vol. I, *B.P.P.* 1892, XXXV, qq. 4587–93)

Earl of Derby

Is there a considerable variation in the number of men employed? —Yes, the joint committee[†] employed sometimes over 7,000 and sometimes less than 5,000. The last year we were employing from 8,000 to 9,000 men. I should be glad to give you some particulars about the trades in which the variations are the greatest. The goods in which we have the largest volume is wool, and wool is sold at five series of sales in the year, each of them lasting from three to five weeks. Between those sales about 300 to 400 labourers are sufficient for the ordinary work of the department, but during the sale this number

---

† The witness was W. E. Hubbard, chairman of the joint committee of management of the London and St Katherine and East India Docks.

will be increased to 1,000 or 1,100. I have made some particular inquiries as to where the surplus came to the wool department from, and we find that of the 700 or 800 that came there for these sales, which last nearly 20 weeks in the year, a very considerable number come from the other departments of the docks. We find that the number of men applying at the other departments is considerably diminished during the wool sales. We have identified some 200 out of the 700 employed at the wool sales which have just been closed as being labourers of the other departments of the London Docks. We have also made inquiries at warehouses and wharves in connexion with our docks, and we find there that all agree that men came from one to the other place, at which work is offering.

Then the tea trade has been mentioned as a trade which fluctuates largely?—It does not fluctuate now so much as it used to when the China tea was the principal kind coming into London. Then there were great fluctuations, because the season was concluded in about six weeks commencing from July. Now we are getting tea in a much larger degree from India and Ceylon, and it comes in much more regularly, arriving more or less all the year round. We have at our tea warehouses 540 men, of whom 230 are permanent, and the other 200 get about 10 months' employmet, and the last 120 about two-thirds—about eig ht months' employment in the year. Then a third point is regularity or irregularity of shipping. This is getting more regular, naturally, than it was in past times, chiefly owing to the substitution of steam for sail, and the steamers are getting more and more into large lines, which are advertised to arrive and depart at fixed dates, and that in the first instance provides more regular work, but at the same time it causes a pressure of work by a steamer being detained, and then making additional efforts to leave again at her fixed time.

Now, taking the maximum of men employed at something over 7,000, which is the figure you have given us, how many of these find full employment?—The number of our permanent labourers is 1,732, and they have full work.

All the year round?—All the year round. 1,015 are weekly labourers, which constitute our Class A., and they practically have also full employment; 1,750 we calculate get four-fifths of full employment and 876, Class C., get two-thirds; and the remainder perhaps only, one-half employment, that is to say, at the department where they are usually employed. They may get other work at the other departments, or may get other work from other employment when we are not employing them.

You spoke of Classes A., B., and C.; what is the distinction between them?—Class A. are men employed by the week, and have a wage of 24s. during the summer, and 21s. during the winter when the hours are shorter, guaranteed for them for each week that they are employed, and as a matter of fact they are usually employed. Class B. corresponds very much to what we used to call preference men in the old days. They get the preference of employment after the regular weekly servants are employed. Class C. follows after them.

Professor Marshall

As to the number for C. you gave 1,500 and afterwards 700. I have got down 1,500 in your first figure?—Then I was wrong, that 876 that I have just mentioned was the whole of Class C. You may take it that there are men from the bottom of Class C. who get less than two-thirds employment. I may say one cause of their getting less employment is that they do not apply with constant regularity. We found that allowing 5 per cent for sickness, which I think is a liberal supply, that 10 per cent of the men in Class B. do not apply every morning, and from 30 to 40 per cent of Class C. But I would also like to say that those classes are of quite recent formation, and it is somewhat difficult to get the men and some of the officers to take at once to a new system, but I hope, before long, to get the matter into better working order, and then doubtless the labour will be much more regular than it is now. It is our intention, and it is our wish, to make the work as regular as we can.

Earl of Derby

I understand that the joint committee desire to reduce the number of casual labourers and get as much permanent work as possible—It is so. We have issued from time to time instructions enjoining that matter, and we are endeavouring to carry it out. On the 30th December the joint committee passed this resolution, which only followed previous instructions given to the superintendents and heads of departments. The resolution is as follows: That it is very desirable that superintendents should use all endeavours to maintain, as far as possible, a regular staff of labourers for the work at their respective docks, and give them full employment in preference to the partial employment of a larger number, and that the employment of casual labour should be discouraged, except in cases of actual necessity.

6 Labour exchanges after twenty years, 1930

(William H. Beveridge, *Unemployment: a problem of industry, 1909 to 1930* (Longmans, 1930), pp. 322–3)

They have not abolished the hawking of labour. There has even in some occupations since the war been a development of Exchange service by trade unions for their members. Four out of every five engagements of insured workpeople are made otherwise than through the national system of Employment Exchanges.

This limitation of their achievement is not to be laid at the door of the Exchanges themselves. They had no compulsory powers and could rely only on gradual voluntary conversion of employers to their use; their promising early start in this direction was broken by the war, and since then, till within the last year or two, they have always had other work to put before placing. Nor have employers at any time proved easy to convert. The tendency to get labour in the old ways normally and try the Exchanges in emergencies only has been common. Those who have then found the Exchanges unable to meet the emergency have seldom thought of applying to their case the commonplace observation that a supplier of labour, as of any other article on whom one wants to rely in emergencies should also be given one's daily custom; only in this way can he keep a stock at call. The statistics given above of the growing proportion of Exchange work that involves the co-operation of two Exchanges in finding work for men or men for employers in districts other than their own, is in itself a sign of their being used by employers too much for emergencies and too little by way of daily routine. The achievement of the Exchanges, however, must not be underestimated. Their achievement in fact is normally in excess of the credit given to them and is more than the mere statistics of placing suggest. In so far as they are used disproportionately for emergencies rather than as daily sources of supply, their 5,000 daily placings include a relatively large proportion of difficult cases—of finding men for employers from a distance and of slowly draining the stagnant pools of mining labour; the placings in the insured occupations which they do not make include probably a disproportionately large proportion of cases where workmen laid off for a few days or weeks return automatically to their regular employer. If the Exchanges have not yet substituted an organised for an unorganised labour market, if their success has been least when it would have been most striking and is still most needed—in relation to casual labour—they have proved themselves in peace as in war an indispensable piece of national machinery.

B. COMBINATION AMONG MASTERS AND MEN

1 A dispute in the Northumberland and Durham coal industry, 1765

(*Newcastle Chronicle*, 21 September 1765)

Whereas several scandalous and false reports have been and still continue to be spread abroad in the country concerning the Pitmen in the Counties of Durham and Northumberland absenting from their respective employments before the expiration of their Bonds; This is therefore to inform the Public that most of the Pitmen in the aforesaid Counties of Durham and Northumberland were bound the latter end of August, and the remainder of them were bound the beginning of September, 1764, and they served till the 24th or 25th of August, 1765, which they expect is the due time of their servitude; but the honourable Gentlemen in the Coal Trade will not let them be free till the 11th of November, 1765, which, instead of 11 months and 15 days, the respective time of their Bonds, is upwards of 14 months. So they leave the most censorious to judge whether they be right or wrong. For they are of opinion that they are free from any Bond wherein they were bound.—And an advertisement appearing in the newspapers last week commanding all persons not to employ any Pitmen whatever for the support of themselves and families, it is confidently believed that they who were the authors of the said advertisement are designed to reduce the industrious poor of the aforesaid counties to the greatest misery: as all the necessaries of Life are at such exorbitant prices, that it is impossible for them to support their families without using some other lawful means, which they will and are determined to do, as the said advertisement has caused the people whom they were employed under to discharge them from their service:—Likewise the said honourable Gentlemen have agreed and signed an Article, not to employ any Pitman that has served in any other colliery the year before; which will reduce them to still greater hardships, as they will be obliged to serve in the same colliery for life; which they conjecture will take away the ancient character of this Kingdom as being a free nation.—So the Pitman are not designed to work for or serve any of the said Gentlemen, in any of their collieries, till they be fully satisfied that the said Article is dissolved, and new Bonds and Agreements made and entered into for the year ensuing.

2 Combination Act, 1800

(Statute 39 and 40 Geo. III, c. 106)

An Act to repeal an Act, passed in the last session of Parliament, intituled, An Act to prevent unlawful combinations of workmen; and to substitute other provisions in lieu thereof.

(All contracts heretofore entered into for obtaining an advance of wages, altering the usual time of working, decreasing the quantity of work, etc. (except contracts between masters and men) shall be void.)

II And be it further enacted, that no journeyman, workman, or other person shall at any time after the passing of this act make or enter into, or be concerned in the making of or entering into any such contract, covenant, or agreement, in writing or not in writing, as is herein-before declared to be an illegal covenant, contract, or agreement; and every journeyman and workman or other person who, after the passing of this act, shall be guilty of any of the said offences, being thereof lawfully convicted, within three calendar months next after the offence shall have been committed, shall by order of such justices, be committed to and confined in the common gaol, within his or their jurisdiction, for any time not exceeding three calendar months, or at the discretion of such justices shall be committed to some house of correction within the same jurisdiction, there to remain and to be kept to hard labour for any time not exceeding two calendar months.

III And be it further enacted, that every journeyman or workman, or other person, who shall at any time after the passing of this act enter into any combination to obtain an advance of wages, or to lessen or alter the hours or duration of the time of working, or to decrease the quantity of work, or for any other purpose contrary to this act, or who shall, by giving money, or by persuasion, solicitation, or intimidation, or any other means, wilfully and maliciously endeavour to prevent any unhired or unemployed journeyman or workman, or other person, in any manufacture, trade, or business, or any other person wanting employment in such manufacture, trade, or business, from hiring himself to any manufacturer or tradesman, or person conducting any manufacture, trade, or business, or who shall, for the purpose of obtaining an advance of wages, or for any other purpose contrary to the provisions of this act, wilfully and maliciously decoy, persuade, solicit, intimidate, influence, or prevail, or attempt or endeavour to prevail, on any journeyman or workman or other person hired or employed, or to be hired or employed in any such manufacture, trade, or business, to quit or leave his work, service, or employment, or who shall wilfully and maliciously hinder or prevent any manufacturer or tradesman, or other person, from employing in his or her manufacture, trade, or business, such journeymen, workmen, and other persons as he or she shall think proper, or who, being hired or employed, shall, without any just or reasonable cause, refuse to work with any other journeyman or workman employed or hired to work therein, and who shall be lawfully convicted of any of the said offences, shall, by order of such justices, be committed to and be confined in the common gaol, within his or their jurisdiction, for any time not exceeding three calendar months; or otherwise be committed to some house of correction within the same jurisdiction, there to remain and to be kept to hard labour for any time not exceeding two calendar months.

IV And for the more effectual suppression of all combinations amongst journeymen, workmen, and other persons employed in any manufacture, trade or business, be it further enacted, that all and every persons and person whomsoever, (whether employed in any such manufacture, trade, or business, or not), who shall attend any meeting had or held for the purpose of making or entering into any contract, covenant, or agreement, by this act declared to be illegal, or of entering into, supporting, maintaining, continuing, or carrying on any combination for any purpose by this act declared to be illegal, or who shall summons, give notice to, call upon, persuade, entice, solicit, or by intimidation, or any other means, endeavour to induce any journeyman, workman, or other person employed in any manufacture, trade, or business, to attend any such meeting, or who shall collect, demand, ask, or receive any sum of money from any such journeyman, workman, or other person, for any of the purposes aforesaid, or who shall persuade, entice, solicit, or by intimidation, or any other means, endeavour to induce any such journeyman, workman, or other person to enter into or be concerned in any such combination, or who shall pay any sum of money, or make or enter into any subscription or contribution, for or towards the support or encouragement of any such illegal meeting or combination, and who shall be lawfully convicted of any of the said offences, within three calendar months next after the offence shall have been committed, shall, by order of such justices, be committed to and confined in the common gaol within his or their jurisdiction, for any time not exceeding three calendar months, or otherwise be committed to some house of correction within the same jurisdiction, there to remain and be kept to hard labour for any time not exceeding two calendar months. . . .

VI And be it further enacted, that all sums of money which at any time heretofore have been paid or given as a subscription or contribution for or towards any of the purposes prohibited by this act, and shall, for the space of three calendar months next after the passing of this act, remain undivided in the hands of any treasurer, collector, receiver, trustee, agent, or other person, or placed out at interest, and all sums of money which shall at any time after the passing of this act, be paid or given as a subscription or contribution for or towards any of the purposes prohibited by this act, shall be forfeited, one moiety thereof to his Majesty, and the other moiety to such person as will sue for the same in any of his Majesty's courts of record at Westminster; and any treasurer, collector, receiver, trustee, agent, or other person in whose hands or in whose name any such sum of money shall be or shall be placed out or unto whom the same shall have been paid or given, shall and may be sued for the same as forfeited as aforesaid.

[All contracts between masters or other persons for reducing the wages of workmen or for altering the hours of work or for increasing the quantity of work, are to be void. Masters convicted of such agreements, shall be fined 20*l*: half to go to the Crown, half to the informer and the poor of the parish.]

XVIII And whereas it will be a great convenience and advantage to masters and workmen engaged in manufactures, that a cheap and summary mode be established for settling all disputes that may arise between them respecting wages and work; be it further enacted by the authority aforesaid, that, from and after the first day of August in the year of our Lord one thousand eight hundred, in all cases that shall or may arise within that part of Great Britain called England, where the masters and workmen cannot agree respecting the price or prices to be paid for work actually done in any manufacture, or any injury or damage done or alleged to have been done by the workmen to the work, or respecting any delay or supposed delay on the part of the workmen in finishing the work, or the not finishing such work in a good and workmanlike manner, or according to any contract; and in all cases of dispute or difference, touching any contract or agreement for work or wages between masters and workmen in any trade or manufacture, which cannot be otherwise mutually adjusted and settled by and between them, it shall and may be, and it is hereby declared to be lawful for such masters and workmen between whom such dispute or difference shall arise as aforesaid, or either of them, to demand and have an arbitration or reference of such matter or mat-

24*

ters in dispute; and each of them is hereby authorized and empowered forthwith to nominate and appoint an arbitrator for and on his respective part and behalf, to arbitrate and determine such matter or matters in dispute as aforesaid by writing, subscribed by him in the presence of and attested by one witness, in the form expressed in the second schedule to this Act; and to deliver the same personally to the other party, or to leave the same for him at his usual place of abode, and to require the other party to name an arbitrator in like manner within two days after such reference to arbitrators shall have been so demanded; and such arbitrators so appointed as aforesaid, after they shall have accepted and taken upon them the business of the said arbitration, are hereby authorised and required to summon before them, and examine upon oath the parties and their witnesses, (which oath the said arbitrators are hereby authorised and required to administer according to the form set forth in the second schedule to this act), and forthwith to proceed to hear and determine the complaints of the parties, and the matter or matters in dispute between them; and the award to be made by such arbitrators within the time being after limited, shall in all cases be final and conclusive between the parties; but in case such arbitrators so appointed shall not agree to decide such matter or matters in dispute, so to be referred to them as aforesaid, and shall not make and sign their award within the space of three days after the signing of the submission to their award by both parties, that then it shall be lawful for the parties or either of them to require such arbitrators forthwith and without delay to go before and attend upon one of his Majesty's justices of the peace acting in and for the county, riding, city, liberty, division, or place where such dispute shall happen and be referred, and state to such justice the points in difference between them the said arbitrators, which points in difference the said justice shall and is hereby authorised and required to hear and determine and for that purpose to examine the parties and their witnesses upon oath, if he shall think fit.

3 A combination of Bolton machine-makers, 1831

(G. W. Daniels, 'A turn-out of Bolton machine-makers in 1831', *Economic History*, 4 (1929), 591)

Gentlemen,
    We, your workmen wish to inform you that we Request a Reduction of our time to work for the future as follows viz. from Six O'Clock in the Morning to Six O'Clock at Night allowing half an hour for

Breakfast and one hour for Dinner and half an hour for Tea and to leave work on Saturdays at four O'Clock Respecting lads that the[y] be Reduced in proportion to the Men allowing one Lad to four men and the lads to serve five years before Twenty one years of age and that by Legal Indenture

May 21st 1831

We your workmen wish you to un[der]stand that We are giving you a fortnight notice from the Date hereof—for the above mentioned times,

We Remain Yours Respectfully

Your Servants

[Here follow the names of 134 workmen.]

4 The Tolpuddle martyrs, 1834

(*The Times*, 20 March 1834)

Spring Assizes, Western Circuit, Dorchester. Monday, March 17. Crown Court (before Baron Williams). Administering unlawful oaths.

James Lovelace, George Lovelace, Thomas Stanfield, John Stanfield, James Hammet, and James Brine were indicted for administering ... a certain unlawful oath and engagement, purporting to bind the person taking the same not to inform or give evidence against any associate, and not to reveal or discover any such unlawful combination[1]...

John Lock. I live at Half Puddle. I went to Toll Puddle a fortnight before Christmas. I know the prisoner James Brine. I saw him that evening at John Woolley's. He called me out and I went with him. He took me to Thomas Stanfield's, and asked me if I would go in with him. I refused and went away. I saw him in about a fortnight afterwards in a barn. He asked me if I would go to Toll Puddle with him. I agreed to do so. James Hammet was then with him. Edward Legg, Richard Peary, Henry Courtney, and Elias Riggs were with us. They joined us as we were going along. One of them asked if there would not be something to pay, and one said there would be 1s to pay on entering, and 1d a week after. We all went into Thomas Stanfield's house into a room upstairs. John Stanfield came to the door of the room. I saw James Lovelace and George Lovelace go along the passage. One of the men asked if we were ready. We said, yes. One of

---

[1] The indictment was framed on 37 Geo. III, 123, against seditious and illegal confederacies.

them said, 'Then bind your eyes', and we took out handkerchiefs and bound over our eyes. They then led us into another room on the same floor. Someone then read a paper, but I don't know what the meaning of it was. After that we were asked to kneel down, which we did. Then there was some more reading; I don't know what it was about. It seemed to be out of some part of the Bible. Then we got up and took off the bandages from our eyes. I had then seen James Love-lace and John Stanfield in the room. Some one read again, but I don't know what it was, and then we were told to kiss the book, when our eyes were unblinded, and I saw the book, which looked like a little Bible. I then saw all the prisoners there. James Lovelace had on a white dress, it was not a smock-frock. They told us the rules, that we should have to pay 1s then, and a 1d a week afterwards, to support the men when they were standing out from their work. They said we were as brothers; that when we were to stop for wages we should not tell our masters ourselves, but that the masters would have a note or a letter sent to them.

∗   ∗   ∗

Mrs Francis Wetham. I am the wife of a painter in the town. In October, last year, James Lovelace and another person came to our shop; he said he wanted something painted from a design he had brought; he had two papers with him, on one was a representation of a skull, and on the other a skeleton arm extended with a scythe; he said it was to be painted on canvas, a complete skeleton on a dark ground, six feet high; over the head, 'Remember thine end'. I asked him what it was for, whether a flag or a sign; he told me it was a secret for a society, and he would tell me no more; if I wanted further information I was to send to him, 'J. Lovelace, Toll Puddle'.

∗   ∗   ∗

The following letter was then put in and read:

Bere Heath, Feb. 1, 1834.

Brother,

We met this evening for the purpose of forming our committee. There was 16 present, of whom 10 was chosen—namely, a president, vice-president, secretary, treasurer, warden, conductor, three outside guardians and one inside guardian. All seemed united in heart, and expressed his approval of the meeting. Father and Hallett wished very much to join us, but wish it not to be known. I advised them to come Tuesday evening at 6 o'clock, and I would send for you to come at that time, if possible, and enter them, that they may be gone before the company come. I received a note this morning which gave me great encouragement, and I am led to acknowledge the force of union.

(Signed by the secretary.)

The following rules were then put in and read:

General Rules

1 That this Society be called the Friendly Society of Agricultural Labourers.

*   *   *

20 That if any master attempts to reduce the wages of his workmen, if they are members of this order, they shall instantly communicate the same to the corresponding secretary, in order that they may receive the support of the grand lodge; and in the meantime they shall use their utmost endeavours to finish the work they may have in hand, if any, and shall assist each other, so that they may all leave the place together, and with as much promptitude as possible.

21 That if any member of this society ... solely on account of his taking an active part in the affairs of this order ... shall be discharged from his employment... then the whole body of men at that place shall instantly leave that place, and no member of this society shall be allowed to take work at such place until such member be reinstated in his situation.

[22 If a member divulge any secret of the society, members throughout the country shall refuse to work with him.]

23 That the object of this society can never be promoted by any act or acts of violence, but, on the contrary, all such proceedings must tend to injure the cause and destroy the society itself. This order therefore will not countenance any violation of the laws.[1]

---

[1] The prisoners were found guilty. On 19 March they were sentenced to seven years' transportation. On 16 April, Lord Howick, in answer to a question in Parliament, said that he believed that their ship had already sailed. The remainder of their sentence was remitted in 1836.

5 Combination of engineering employers, 1852

(Thomas Hughes, *Account of the lock-out of engineers* ... *in 1851–2* (Cambridge and London: National Association for the Promotion of Social Science, 1860), pp. 44–7)

Employers of Operative Engineers and the Amalgamated Society (Private and Confidential)
For Members only

Central Association of Employers of Operative Engineers, &c. Offices, 30, Bucklersbury

London, 24th January, 1852

Dear Sir,—I beg leave to subjoin an extract from the minutes of the Conference held here this day, on the subject of the constitution and rules of this Association.

It is felt by the Conference that the time has now arrived when the members of the Association should finally determine upon their future course of action; and at the same time that its measures should be perfectly effectual for securing the defensive objects for which they are associated.

After two days' lengthened discussion, and the advantage of much information to which they have access, the Conference have come to the unanimous resolution, to recommend the plan of operations developed in the subjoined Minute of their proceedings, for the adoption of their Members.

I am, therefore, directed to intimate to you that you are respectfully solicited, *without delay*, to enclose in the envelope herewith sent, your approval of or dissent from the recommendations of the Conference; that a meeting of this Committee will be held on Thursday next, to receive and consider the collective returns from the members; and that at that meeting the whole mode, terms, and time of re-opening the various establishments will be finally determined; or suggestions for these objects prepared for the consideration of an aggregate meeting of the whole members of the Association.

(By order of the Conference)
I remain, dear Sir
Yours faithfully,

Sidney Smith, *Secretary*

At a Meeting of the Executive Committee of the Central Association of Employers of Operative Engineers, to confer with a Deputation from the Lancashire Committee, held on the 24th of January, 1852.

The Conference proceeded to take into consideration the steps to be recommended for adoption by the members of the Association, in consequence of the perseverance of the Amalgamated Society and other Trades' Unions in adhering to rules and practices equally inimical to the free action and just rights of the artisan and labouring classes, and to the exercise, by employers, of the fair control which every master is entitled to maintain over his own establishment.

The Conference further in consideration that the demands of the members and executive council of the Amalgamated Society and of other Trades' Unions, had not been withdrawn within the time limited by the Committee; and that the various unjustifiable practices of these combinations had not been disavowed by any body of the artisans lately in the employment of the members of the Association; determined that it was indispensably necessary—for the protection of the operative and labouring classes from the annoyances, intimidation, personal violence, and exclusion from employment, to which they had been habitually subjected by the influence, organization, and ascendancy in the various shops of members of these Unions, and for the relief of employers from the interference and dictation thereby inflicted on them —to advise the members of the Association universally to adopt the following measures of self-defence and self-vindication.

First. That no member of this Association shall engage, admit into (or, after he shall have become cognizant of the same), continue in his service or employment, in any capacity whatever, any member of any Trades' Union or Trades' Society, which takes cognizance of, professes to control, or practises interference with, the regulations of any establishment, the hours or terms of labour, the contracts or agreements of employers or employed, or the qualification or terms of service.

Second. That no deputations of workmen, of Trades' Unions, committees, or other bodies, with reference to any objects referred to in Article 1st, be received by any member of this Association on any account whatever; but that any person forming part of, instigating, or causing such deputation, shall be dismissed forthwith; it being still perfectly open to any workman, individually, to apply on such subject to his employer; who is recommended to be at all times open and accessible to any personal representation of his individual operatives.

Third. That employers be especially solicited, *as much as possible*, to avoid the delegation of the engagement or contract of their workmen to others, and to take a more *personal* superintendence of control, or engagements with their hands—and in the most especial manner, that they impress upon every person engaged by them their anxiety that, in case of any molestation, annoyance, or obstruction in pursuing their avocations, or procuring employment, they should at once apply and complain to the principals of the establishment, who should sift such complaint to the bottom, and to dismiss all persons who had been proved to have offered or abetted such molestation or obstruction.

Fourth. That no member of this Association shall engage or continue in his employment any person whatsoever, until he has read, in presence of one witness at least, to such person the rules, if any, of his establishment, and also the following.

DECLARATION, *by the undersigned, on engaging in the employment of* [here insert name, address, and trade of employer] *I*, A. B. [here insert, Christian and Surname of person declaring] *do hereby honestly, and in its simplest sense and plainest meaning, declare, that I am neither now, nor will, while in your employment become a member or contributor, or otherwise belong to or support any Trades' Union or Society, which, directly or indirectly, by its Rules, or in its meetings or transactions of its business, or by means of its officers or funds, takes cognizance of, professes to control or interferes with the arrangements or regulations of this or any other manufacturing or trading establishment, the hours or terms of labour, the contracts or agreements of employers or employed, or the qualifications or period of service. I do also further declare, that I have no purpose or intention to call in question the right of any man to follow any calling in which he may desire to engage, or to make what arrangements, and engage what workmen he pleases, upon whatever terms they choose mutually to agree.*

(*Signed*)

Dated the   day of   185

Signed

Witness

Fifth. That no member of this Association shall engage any workman who has been previously in employment elsewhere, without ascertaining from what establishment he was discharged, and whether the cause of his leaving had any reference to an infringement of the objects of the foregoing declaration.

Sixth. That no member of this Association shall, on any pretext whatever, permit or submit to dictation, interference, or direct or indirect tampering with the management of his establishment, or the engagement or conditions of the service of his workmen; but that whenever any attempts are made to abrogate or compromise the free operation of the foregoing provisions, such member shall at once apply, if he requires it, for the advice, award, and assistance of the Executive Committee, who shall be bound to afford him every assistance and support called for by the circumstances of the particular case.

Seventh. That, in the event of a strike or turn-out occurring in the establishment of any member of this Association, for reasons or from causes which shall, in the opinion of the Executive Committee, entitle the employer so assailed to its countenance and support, it is hereby and shall continue to be distinctly understood, that all the members of the Association shall sustain, according to their power and ability, such member in upholding the objects of the Association; it being expressly understood and declared, that no acts shall warrant the interference of this Committee, except such as it is the declared object of the foregoing provisions to prevent.

Eighth. That, in order as far as possible lies in the power of this Association, to obviate any inconvenience which may arise to meritorious workmen, for being deprived of any advantages they may fancy they derive from the legitimate objects from which existing Trades' Unions or Societies have been diverted, this Association gives full power and authority to the Executive Committee, to submit for its sanction a plan for the establishment of a new, sound, and legitimate Benefit Society.

6 Miners on strike, 1858

(Report by the Chief Constable of Staffordshire, in Staffordshire Record Office. Mr E. R. Lloyd, inspector of schools, kindly supplied a transcript of this document.)

Notes of a Meeting of Colliers, held at Horsley Heath, Staffordshire, August 30, 1858:
Present, in the open-air 800 Colliers; with Col. Hogg, Chief Constable of Staffordshire, with a strong body of police assisted by Capt. Seagrave Chief Constable of Wolverhampton, and a detachment of men under him.—Time, from 10 a.m. till 1.20.

Joseph Linney was called to the chair. He gave out a Hymn. When it had been sung he said that their employers compelled them to strike. Before the strike the work they had would not support their wives and their children, and now their masters wanted to reduce one shilling. He had looked at it it in all its various shapes and forms, and a more barbarous action never could be acted, than was now attempted by the masters. From Xmas last to the time that the strike took place, the average wages of the coal miners was not more than 15/- a week. (Voices: It was not so much as that.) Out of the 15/- they had to pay 1s for drink and 6d for 'sick' money before they left the 'field'; and when they got home there was the rent to pay, and also the taxes. After that how much had they got to support their families? In general not more than 11s a week; and yet their employers were so 'hard hearted' as to treat them in the way they had. He did not wish to injure their employers, but it was the duty of their masters to give them a fair day's wages for a fair day's work; and he thought they would all agree with him when he said with the auctioneer—'We ask no more and we will take no less'. (*Ld. app.*) The speaker said he was a teetotaller and a Sunday School teacher. Because he would not drink he had been turned out of a pit. ... When the policemen saw the dangers that the miners worked in they said—'Why we would not work in such places for a pound a day'. (*Ap.*) And the colliers might get a pound a day if they liked—for the coal was the spring of all commerce and industry, and a pound of it was worth more than a pound of gold. They had to sell their labour, and it was their duty to sell it at the highest price. If all the colliers in the Kingdom were to lay down their tools and demand a high price for their labour they could get it. It was because of the extravagance of their masters in a time of prosperity that they now that trade was bad wanted to reduce their wages. They wanted even now to spend on themselves what they got out of the wages of their workmen. But should they reduce their wages? (*Cries of No*!!) He was glad to hear them say so. They had gone on comfortably till Oldbury Wake; and if the meeting intended to go on till West Bromwich Wake, which was held at the end of November, they would then take a fresh start and remain on strike till Christmas. (*Laughter and applause*) In this great struggle colliers had been working with their brains and had turned poets. They had been asking other districts to join them. He held a letter in his hand from one of these districts. It was from Brierley Hill and Kingswinford. That letter asked the miners of that (the Tipton, Oldbury and West Bromwich) district to come amongst them and have a good meeting, and then the writer said—'We will lay down our tools to a man and will work no more till you have got your wages and we have got ours'. (*Ld. applause.*) [The letter was dated from Brierley Hill, incorrectly spelled, and signed 'A Miner'.] The agitation was spreading far and wide. He told them (the meeting) that they would do it, and they were doing it. (*Applause.*) In a fortnight they should get their money. . . .

Job Radford, a brickmaker, of Oldbury, then spoke at some length encouraging the miners to form a trades association, similar to that of the brickmakers of Manchester—in consequence of which association the brickmaker now got twice the wages that he used to get. . . .

Job Radford again spoke inculcating temperance; after which it was determined that a meeting should be held at Brierley Hill on that day week at 11 o'clock; and on that evening (Monday) at Tipton; after which the proceedings terminated.

7 The engineers, 1851-67

(Royal commission on trade unions, *Minutes of evidence B.P.P.* 1867, XXXII, qq. 572–82, 597–617, 665–71, 827–36, 841–61, 868–84, 895–912, 920–27).

Mr William Allan examined.

(*Chairman,*) You are secretary of the Amalgamated Society of Engineers, are you not?—Yes.

When did that society begin?—It was established on the lst of January 1851. But perhaps it might be as well for me to explain that it was established out of a number of societies that previously existed.

Out of several societies which you call branch societies?—No, independent societies.

Independent societies joined this in 1851?—Yes.

How many members does it comprise at present?—33,600.

And how many branches?—308 branches.

Is the number of members increasing?—It increases at the rate of between 2,000 and 3,000 per year.

Have you got the increase for 1865 and 1866; were 3,000 members added in 1865 and 3,300 in 1866?—Yes.

Are those individual members, or were they unions that came in? —They were new members.

Have you established any new branches this year?—Yes; a few this year, and a number last year.

And have you branches in England, Wales, Scotland, Ireland, the colonies, and the United States, and one in France?—Yes. For the information of the Commissioners I should say that in England and Wales there are 238 branches, numbering 27,856 members; in Scotland there are 33 branches, and 3,218 members; in Ireland 11 branches, with 1,371 members. In the British colonies there are 14, that includes Australia, Canada, Malta, New Zealand, and Queensland, and those 14 branches contain 626 members. The United States have 11 branches, with 498 members. In France we have only one branch, which is in Croix, in the north of France, and numbers 30 members. Altogether there are 308 branches, with 33,599 members.

How are your funds obtained?—By a subscription of 1s per week from each member.

Will you state about what your fund is at present?—In round numbers at the present time it is 140,000*l*.

Is that in the funds?—Not in the funds,—in different banks. That is to say, the accumulated fund is about 140,000*l*.

And what is about your present annual income. 86,885*l*, is it not? —Yes; that would be for 1865. Our 1866 report is not out yet.

(*Earl of Lichfield*) You say that the subscription is 1s a week, and the number of members 33,000; Where does the rest of the money come from?—From admissions. There is a certain amount that every candidate has to pay, not less than 15s, as an entrance fee, and, according to age, it may amount to 3*l* 10s. You will see, therefore, that we derive a large amount of revenue in the way of admissions into the society, the entrance fees being, as I have said, according to age.

(*Chairman*) What was your expenditure in 1865?—The expenditure in that year was 49,172*l*.

What were the heads under which that 49,000*l* was distributed?—To members out of employment there was 14,076*l*; to sick members, 13,785*l* 14s 9d; and to superannuated members (that is to say, members who were too old to gain the ordinary rate of wages at the trade—we allow such from 7s to 9s per week each) 5,184*l* 17s 4d. Then, in funerals on members' deaths or their wives' deaths we paid 4,887*l*.

Is any allowance to widows included in that last amount?—At the death of a member we pay the widow 12*l*, and that is included in the amount. All expenses, so far as the funeral is concerned, are included. Then we paid 1,800*l* in cases of accident, that is, where members got disabled from following their employment in consequence of some accident occurring to them, for instance, under the loss of an arm or anything of that kind.

(*Sir D. Gooch*) Ceasing to be able to follow the trade?—Yes; we had 18 claimants and they received 100*l* each.

(*Lord Elcho*) Is that all the privilege they can derive?—No, they have afterwards the privilege of paying 6*d* a week and deriving sick benefit, and also the funeral money at death, and, generally speaking, five out of every six, I should say, retain their position in the society by the payment of 6*d* per week. Then we have a benevolent fund, through which we have paid away 820*l*; that is a fund that is in addition to the ordinary subscription of 1*s* per week; it is to relieve cases of extraordinary distress that may exist in consequence of men being long out of employment or in sickness.

(*Mr Hughes*) That does not come out of the 1*s* per week?—No, that does not come out of the 1*s* per week.

(*Lord Elcho*) Is that voluntary?—No, it is compulsory; we find that voluntary subscriptions do not come in so well as compulsory ones.

(*Mr Hughes*) Is that a levy?—It is a levy, but is agreed to by a majority of votes; that is to say, it is put to the members whether they will agree to it or not.

(*Lord Elcho*) Is it put to them annually?—No; only when the fund gets so low that we require it to be increased.

(*Mr Booth*) Is that levied on all branches?—On every member throughout the society, no matter where they are.

Does one levy a year suffice?—Yes; but during the cotton famine we found that two levies were necessary in order to meet the distress in Lancashire and other districts. Then to the branch officers during 1865 we paid 4,337*l*; that is to say, to the staff throughout the whole of the 300 and odd branches.

(*Chairman*) That expense for the staff extends to America, therefore? —To all the branches; it is a complete return from the whole of the branches. Then there are several small items, but which are scarcely worth troubling the Commissioners with. I will, however, put the report in so that the Commissioners will be able to refer to it at any time. While we are on the subject of the benefits, I may just state that since the commencement of the society for 15 years the average rate of our expenditure to men out of employment has amounted to 18,656*l*. The sick benefit has averaged 7,675*l*. The superannuation, that is to say, the payments to the old members, has averaged 1,795*l*. The accident benefit amounted to 826*l* upon an average, and the funerals to 2,306*l*. The benevolent grants, that is the fund I have referred to, have averaged for 12 years (it has only been 12 years in existence), 533*l* per year, and assistance to other trades that were engaged in disputes with their employers, or locked out, amounted to 724*l* per year, taking the average of 13 years.

With regard to the 18,000*l*, was that paid merely for the time that the men were out of employ, or towards the expenses in sending them to places where they could get employment, and other expenses of that sort?—The 18,656*l* was paid to members out of employment, and for their railway fares from one town to another in going to different situations.

(*Mr Harrison*) Out of that 18,000*l* what proportion has been paid to men out of employment owing to a strike, lock-out, or some labour dispute; I mean where work was to be had, but where it was a dispute about the terms?—We have only had one dispute which you may call important in our trade since the commencement of the society, and that was in 1852; in the first six months of that year we expended 40,000*l* on a lock-out, but it was not our fault that we were out of employment; it was the fault of the employers who locked us out.

Irrespective of that affair what is the proportion?—We have not kept a separate account of the amount spent under this head, but leaving that 40,000*l* out I should say it does not exceed 10 per cent, as far as any strikes with our employers are concerned.

(*Mr Hughes*) That is to say 10 per cent, on 18,000*l*?—Yes.

(*Earl of Lichfield*) Have the operations of your society had the effect of equalizing wages in different districts where you have branches?—They have had the effect of equalizing wages to a very considerable extent, but have not made a uniform rate of wages.

Let us take for instance, any two large towns in Lancashire; has the society had the effect there of equalizing wages between the two places?—I would ask what am I to understand by equalizing; does that mean a uniform rate?

I do not refer to a uniform rate in every place; but to an equalizing of the rate of wages between a place like Manchester and a place like Bolton?—The wages in Manchester are a considerable deal higher than the wages in Bolton.

But has the action of the society tended to equalize the rate of wages in those different places?—Yes, it has decidedly tended to bring them nearer to a certain rate.

Do they vary very much in different towns of the same district, say Lancashire, for instance?—Yes.

To what extent do they vary?—1*s*, 2*s*, and 3*s* per week.

(*Chairman*) Do you mean more than the expense of living would account for?— I believe that the cost of living is pretty near the same, go where you will, if you live upon the same diet. I know that if I go anywhere out of London I have to pay as much as people in London.

The only advantage in reality possessed by Manchester and those places is, that coals and house rent are much cheaper; but if they buy the same quality of goods as we do in London, I do not see there would be much difference so far as the actual living is concerned. . .

Do you find that the possession of very large funds, and the fact that they belong to a very powerful organization, such as your society is, tends generally to make the members of your society disposed to enter into such a dispute, or the contrary. I am not asking now with regard to the council but the members?—I should say that the members generally are decidedly opposed to strikes, and that the fact of our having a large accumulated fund tends to encourage that feeling amongst them. They wish to conserve what they have got, as I have heard it put here, the man who has not got a shilling in his pocket has not much to be afraid of, but with a large fund such as we possess, we are led to be exceedingly careful not to expend it wastefully, and we believe that all strikes are a complete waste of money, not only in relation to the workmen but also to the employers.

Have you found by experience that your society has done anything to promote the same feeling or the same practice in other trade societies?—Many of the societies (the Amalgamated Carpenters and others I could mention) have taken in fact our constitution and our mode of management as their guide.

Has your society in recent times ever interfered in trade questions with a view of bringing about a settlement?—Yes.

Has it ever interfered to put a stop to or dissuade, or discountenance, a threatened strike or an actual strike?—Decidedly; we have recommended that no strike should take place, at least in 20 cases in as many months.

In your own trade or in other trades do you mean?—In our own trade. The question in dispute has generally arisen from the fact of the high rate of provisions and other necessaries, and our members have consequently wished to have an advance of wages.

In your trade, within the last 10 or 12 years, have wages fluctuated much, or have they been more or less stationary?—They have fluctuated little or nothing till within the last 12 months, when there has been an upward tendency. It must be borne in mind that I now speak of London and Manchester. Away in the north of England and in Scotland the wages have improved vastly within these last few years— year after year almost; but that has arisen from the fact of the large amount of shipbuilding going on at Newcastle and different parts round there as well as on the Clyde.

How often within the last 10 years has your society supported or maintained a strike directly for the purpose of raising wages?—I think there have been only some three or four; two at Blackburn and one at Preston, where the strike has been directly for wages, and there has been one at Keighley, but certainly not more than six of that description.

And for a reduction of hours, how many strikes have there been?— There has not been any strike of any importance for that object, with the exception of one at Glasgow, and that did not originate with our society.

What have been the grounds or causes of such disputes as you have had in the trade within the last 10 years?—They have principally arisen from piecework and the large number of boys employed.

They have been, then, I understand, for the purpose of enforcing or maintaining trade rules?—Yes, regulating the trade.

Have they been for the most part to introduce new rules or to maintain those already existing?—To maintain existing rules and customs.

Speaking generally, what number of trade disputes have your society been engaged in over the last 15 years (taking you back to the lock-out) the sole purpose of which was, not to maintain something which was pre-existing, but to introduce a change?—I could not speak positively as to the exact number, but they have been very few indeed. We have had disputes with reference to doing away with piece-work, but very few disputes indeed (perhaps two or three only) of that description to which you refer.

(*Mr Roebuck*) Will you explain what the dispute was in reference to boys being employed?—That they were introduced in a larger proportion than what the trade recognizes.

What do you mean when you speak of what the trade recognizes in the introduction of boys?—It depends on the class of work. For instance, in an engine factory very few boys are employed in comparison with the number of men. In machine shops and tool-making establishments there are a large number employed, and in some instances they have introduced boys to what we think an alarming **extent,** and the result is that the men have objected to it.

Could the boys do the work of the men?—They might be capable of doing a portion of it.

And did the masters think they could?—I have no doubt they did, or else they would not have employed them.

Why should you prevent a master from employing boys who can

do the work?—We have a perfect right to say to him 'If you employ a certain number of boys beyond what we conceive to be a proper number, we will not work for you.'

A proper number means the number that you like?—What the men think right.

(*Mr Harrison*) With regard to the wages and trade customs which the men have recognized and from time to time attempted to maintain, are they wages or trade customs which have been introduced as novelties by the society, or is the society engaged in maintaining that which is recognized amongst workmen generally, whether members of the society or not?—We have little dependence to place on non-society men; we generally regulate our own affairs, and if the non-society men fall in with our views well and good. Occasionally, when these disputes do occur there is a meeting held of society and non-society men at which the question is discussed and they come to an understanding one way or the other, and, generally speaking, the non-society men, if there is anything in the nature of a strike, leave their employment as well as the society men.

(*Mr Roebuck*) I think you said that a great number of the strikes arose from the resistance of a lowering of wages?—Yes.

When provisions rise I suppose you sometimes ask for a rise of wages?—It has only been, as I have said, within these last 12 months, that any material change has taken place, except in the north of England and in Scotland.

But when the masters endeavoured to lower the wages did they ever state that provisions had fallen?—No.

They never gave that as a reason for wishing to give a less rate of wages?—As a rule they assign no reason at all beyond the simple fact that they want to reduce the wages, inasmuch as the orders are not coming in so rapidly as they should; that trade is bad in fact.

Then the reason they give is slackness of trade?—Yes.

But you have never heard that the masters have said that as the wages were raised when provisions rose, they thought they ought to be reduced when provisions fell?—No; I have heard that employers say "When trade was good we have given you an advance of wages, and when trade is bad you ought to submit to a reduction," but not assigning it to the fact of the price of provisions in any shape or form.

But that position of the masters has been resisted?—It has been resisted and is being resisted now at Blackburn. They propose there to reduce the wages 2*s* per week, and the men are all on strike in consequence of refusing.

At Blackburn has there ever been a rise of wages in consequence of a briskness of trade?—Yes.

And the men, being pleased with that, have not yielded to the design of the masters to lower the wage when the trade is slack?—Precisely so.

So that you have one rule for yourselves and another for your master?—We keep what we can get, as a general rule.

*　　*　　*

What are the principal complaints that the executive council have had to deal with as coming from members of the society,—have there been complaints of conduct on the part of employers, or foremen, or overseers?—Complaints I think generally arise when changes take place in different factories, when a new superintendent or manager is introduced; on the principle that new brooms sweep clean they begin to alter the regulations and to discharge men and a great deal of ill-feeling is created in consequence of that. For instance, the gentleman who succeeded Sir Daniel Gooch as superintendent of the Great Western Railway has endeavoured to reduce the rate of wages, and we have endeavoured to prevent his doing it. His line of argument, I suppose, is that there is not a dividend paid to the shareholders. Our mode of arguing the question is that it is not a matter of any importance to us whether there is a dividend or not, that if you enter into a bad contract it is your business, not ours.

Have you ever found that those who have taken a prominent part in trade disputes have met with any difficulty in obtaining work?—In some instances I have known cases of that kind, where they could not obtain employment, and have been obliged to go into some other line of business.

And have you found in practice that in any works the society men are refused employment?—Yes, at some places they object to employ them if they know that they are society men.

(*Mr Roebuck*) Do you know that at this very time the great firm of John Brown & Co., at Sheffield, have issued an advertisement that no union man will be employed?—I am not aware that a document of that description has been issued, but I am aware that we pay no attention to it.

Do you know that that factory is full of men now?—I have no doubt it is; am I to understand that that question is meant to convey that Mr. Brown's establishment is at work without any society men.

Yes?—Then I say I very much doubt it.

(*Mr Hughes*) Do you know of any lists having been published by masters, or by men, with a view of marking men?—Yes.

Have you ever seen such lists?—I have one or two with me.

Can you produce them?—Yes; I now hold in my hand a list which, in 1855, was published by the firm of Sharp, Stewart & Co., of Manchester, at the Atlas Works; we had a dispute there in consequence of the number of boys employed. Our members, and in fact the non-society men as well, left the employment. Directly after they left the employment, Messrs. Sharp, Stewart & Co. published this list, and sent it to the employers, announcing to them that these men had struck work. Now it is quite clear that the object was to prevent those men from getting employment; then here is an amended list (*producing it*) containing a few more names; they were not exactly satisfied with the first list, but they got hold of some other names, and they published a second edition of the list, as you might call it. Then our members who were employed in that shop at the time thought that what was good for the goose would not be at all bad for the gander, and they published a list of those that went in.

In your experience have those sorts of lists been common?—They are not common, but it is unusual to publish them; for instance, in the great dispute of 1852 we did not publish any lists, and we only published this list from the fact that the employers had shown us the example.

(*Mr Mathews*) It was a sort or retaliation?—Yes.

You were speaking about the admission of boys into shops, and I think you stated that the union generally set their face against the admission of boys beyond a certain number?—Yes.

In point of fact if the masters encouraged the admission of boys beyond the number dictated by the men, the result would be that the men would withdraw from that particular shop?—It might end in that.

You set your face against competition in labour by the free admission of boys?—We endeavour to prevent an overplus of labour in our market by the admission of boys.

So that while keen competition is brought to bear on the operation of masters in every department of trade, the workmen contrive by this means to shut out the competition of labour?—No, I think it generally regulates itself.

But would not the effect of the unions be to interfere with that self-regulation?—Decidedly so; we do interfere as regards the society men.

(*Mr Roebuck*) And if any evil followed from that interference, the evil would fall upon the poor?—I do not know about whom it would fall on, that is such a broad question.

(*Sir Daniel Gooch*) How do you deal with machine men, and what proportion do they form of your society?—Generally speaking at the railway factories, and shops and engine factories in the country, the machine men are what we call labourers, and we do not admit them at all, because in the first place they are not in receipt of what we call the ordinary rate of wages, and you will see by the heading of that proposition form which I handed round that our members must be in receipt of the ordinary rate of wages.

When you have a strike and advise men to go out, you force the machine men and others of that class to go out as well?—Yes, as a rule.

The class of mechanics form a minority, do they not, in machine factories, and so on?—The machine men form a small portion of the whole.

In a shop like Sharp's where there is so much of the work done by machine there are more machine men, are there not?—It depends upon the class of machines. In large engine shops, such as Maudsley's and Penn's, there are not so many machines. In the locomotive shops there will be a larger number of men employed at machines than in marine shops, as we call them.

And your action does throw these men out of work?—Yes.

And brings hardship, therefore, upon men unconnected with the dispute?—Yes; if the men in one portion of the works come out on strike, as a rule the works are closed, because the employer cannot go on unless he has his complement. I may mention that in the great lock-out of 1852, for instance, the employers never took into account any of those differences, but they locked them all out.

Do the society ever do anything for the benefit of those people whom they turn out?—We endeavour to get subscriptions for them, but we do not pay them anything out of our funds.

(*Mr Roebuck*) That is to say that other people may be charitable to them?—Our own members.

You are not charitable out of your own funds I understand you to say?—We do not seek charity from others. For instance, in 1852, when those labourers were thrown out of employment in consequence of the masters locking them out, through no fault of theirs, we went to our members and asked them to subscribe to those out of employment one day's wages per week, so that the charity was altogether on our side.

(*Mr Mathews*) But I understand that in the case specified all your men were not out of employment?—There were only about 3,000 out of employment of our men, and 9,000 in employment, and it was from them that we asked the one day's wages a week.

(*Sir D. Gooch*) Your rules do not seek to regulate any scale as to the number of apprentices that may be employed?—No.

How is that regulated?—It is more a trade custom.

What do you mean by that?—I should say that here in London, for instance, they will not average one apprentice to three journeymen.

That is low, is it not?—It is low. Then in some machine shops they will have as many as two apprentices, perhaps, to one journeyman. It all depends upon the class of work.

(*Mr Mathews*) Would your union consent in any case to apprentices doing the work that men do?—Yes, they do it now in particular districts.

In the higher skilled work?—No, the employers do not find it to their advantage.

But supposing they did?—We would endeavour to prevent it.

(*Sir D. Gooch*) Supposing that English employers were competing with foreign locomotive builders, what chance have English houses to compete with foreign houses if the manufacturer has no power in deciding the mode of carrying on his business. If he can by the employment of boys get his work done cheaper why should he not, and what interest is it to your society of engineers to shut out boys from the works, in order that foreign competition may come in by reason of a regulation of that kind?—I should say that it is not to their interest to let any of the orders go away if they can possibly avoid it. But it must be first shown that the orders have gone away in consequence of their not being able to execute the works here as cheap. Take Belgium, for instance, in the iron trade, there they were reducing the wages 10 per cent in order to compete with the English, while at the same time the English were reducing theirs 10 per cent in order to compete with the Belgians. . . .

(*Sir D. Gooch*) It might be, might it not, that those are cases of competition with French houses, and that they are making those engines now without a profit, all that they get by those engines are going to the workmen, and the workmen not assisting the masters in getting a fair profit. If that is so, that will not last long?—I can scarcely understand an employer manufacturing 20 engines for less than a remunerative price. But the fact is, that at that very price the wages are higher now than ever they were before at Beyers and Peacock's, who are making some of those engines.

It may be an advantage, I suppose, and is an advantage in point of fact for one of these large establishments to be kept going, even without a shilling profit; the manufacturer would rather keep his works going than let them stand?—That is not a rule, I think, with Manchester men or Scotchmen.

Do you not think that work has been done at Manchester now in the engineering branch without any profit?—I can scarcely conceive of such a contract as that taken being done without profit.

(*Mr Mathews*) Would it not be better for large manufacturers such as these you allude to, to take work occasionally without a profit, rather than to let their works stand?—We have no objection to the employers taking a contract without profit; but what we say is, if you do, you ought not to make us pay for it.

(*Sir D. Gooch*) But the interest of the employer and the employed is to work together, is it not?—There I differ. Every day of the week I hear that the interests are identical. I scarcely see how that can be, while we are in a state of society which recognizes the principle of buying in the cheapest and selling in the dearest market. It is their interest to get the labour done at as low a rate as they possibly can, and it is ours to get as high a rate of wages as possible, and you never can reconcile those two things.

You prevent the master by your rules from employing that kind of labour which he finds to be the most profitable to him; for instance, that of boys. If he found that he could dispense with nine-tenths of the mechanics and employ boys at a lower rate, why should he not be as free to do that on his own premises and works, and using his own capital, as the men are to do the work on the terms that they like? —Then we say, You are perfectly at liberty to employ as many boys as you like, but we object to the system and will not work for you.

(*Mr Mathews*) That is another way of coercing the masters, is it not?—We leave them to themselves. There is another reason why we object to the system of apprentices. Generally speaking, workmen like to get their own sons into the trade, but in many large establishments (take Swindon, Crewe, and Wolverton) apprentices are brought in who are the sons of farmers say, and supersede our own sons, and we are obliged to move out of the district to get our sons in elsewhere.

(*Mr Booth*) What is your object in limiting the number of apprentices?—To keep the wages up; no question about it.

8 The summons to the first Trades Union Congress, 1868

(Sidney and Beatrice Webb, *The history of trade unionism* (2nd ed. Longmans, 1896), pp. 487–8)

Manchester, April 16th, 1868

Sir:

You are requested to lay the following before your Society. The vital *interests* involved, it is conceived, will justify the officials in convening a special meeting for the consideration thereof.

The Manchester and Salford Trades' Council having recently taken into their serious consideration the present aspect of Trades Unions, and the profound ignorance which prevails in the public mind with reference to their operations and principles, together with the probability of an attempt being made by the Legislature, during the present Session of Parliament, to introduce a measure which might prove detrimental to the interests of such Societies *unless some prompt and decisive action be taken by the working classes' themselves*, beg most respectfully to intimate that it has been decided to hold in Manchester, as the main centre of industry in the provinces, a Congress of the representatives of Trades' Councils, Federations of Trades, and Trade Societies in general.

The Congress will assume the character of the Annual Meetings of the Social Science Association in the transactions of which Society the artisan class is almost excluded; and papers previously carefully prepared by such Societies as elect to do so, will be laid before the Congress on the various subjects which at the present time affect the Trade Societies, each paper to be followed by discussion on the points advanced, with a view of the merits and demerits of each question being thoroughly ventilated through the medium of the public press. It is further decided that the subjects treated upon shall include the following:

1 Trade Unions an absolute necessity.
2 Trade Unions and Political Economy.
3 The effect of Trade Unions on foreign competition.
4 Regulation of the hours of labour.
5 Limitation of apprentices.
6 Technical Education.
7 Courts of Arbitration and Conciliation.
8 Co-operation.
9 The present inequality of the law in regard to conspiracy, intimidation, picketing, coercion, etc.

10 Factory Acts Extension Bill, 1867: the necessity of compulsory inspection and its application to all places where women and children are employed.

11 The present Royal Commission on Trades' Unions—how far worthy of the confidence of the Trade Union interests.

12 Legalization of Trade Societies.

13 The necessity of an Annual Congress of Trade Representatives from the various centres of industry.

All Trades' Councils, Federations of Trades, and Trade Societies generally, are respectfully solicited to intimate their adhesion to this project on or before the 12th of May next, together with a notification of the subject of the paper that each body will undertake to prepare, and the number of delegates by whom they will be respectively represented; after which date all information as to the place of meeting, etc., will be supplied.

It is not imperative that all Societies should prepare papers, it being anticipated that the subjects will be taken up by those most capable of expounding the principles sought to be maintained. Several have already adhered to the project, and have signified their intention of taking up the subjects, Nos. 1, 4, 6, and 7.

The Congress will be held on Whit-Tuesday, the 2nd of June next, its duration not to exceed five days; and all expenses in connection therewith, which will be very small, and as economical as possible, will be equalized amongst those Societies sending delegates, and will not extend beyond their sittings.

Communications to be addressed to Mr W. H. Wood, Typographical Institute, 29 Water Street, Manchester.

By order of the Manchester & Salford Trades' Council.

S. C. Nicholson, *President*

W. H. Wood, *Secretary*

9 The trade union as friendly society, 1879

(A. F. J. Brown, ed. *English history from Essex sources, 1750–1900* (Chelmsford: Essex County Council, 1952), p. 209. The document is quoted from the *Colchester Mercury*, 29 March 1879)

Each member subscribed 1*s* per week and for that he received 10*s* per week when out of employment and 12*s* per week when sick; and during the time such relief was paid, the weekly subscription was not called for, a feature, he believed, novel in Benefit Societies. Any member meeting with an accident, which permanently incapacitated him from following his trade as a carpenter and joiner, received £100; and for partial disablement, a sum of £50 was paid. The subscription also covered insurance of tools; and on the occasion of the fire at Mr Dobson's premises at Colchester, some men who were members had their tools replaced at a cost of £50. Any person after being a member 19 years was entitled to a superannuation allowance of 7s per week, and after 25 years membership to 8s per week for life ... On the death of a member his widow or relatives received £12, and should a member lose his wife he received £5 ....

With regard to the Colchester Branch of this Society, at the end of 1877 they had a balance of £233 15s 1½d. For out of work benefit they had paid during the year £3 10s. That was a light amount, but fortunately the depression which had raged throughout the country had not extended to Colchester, and the enterprise of Colchester business men had kept the building trade active. Their total expenditure for 1878 was £28 17s 8½d, including £14 2s paid for sick benefit. Their income for the year was £110 11s 11d, so they had a balance left of £82. Although there were something like 200 carpenters and joiners in the town, this Branch only numbered from 45 to 50 members.

10 New unionism, 1890

(Philip S. Bagwell, *The railwaymen: the history of the National Union of Railwaymen* (Allen & Unwin, 1963), p. 149, quoting the secretary's report to the annual general meeting of the Amalgamated Society of Railway Servants, 1890)

We have now, while still adhering to our old principles, adopted methods which are associated with robust and even aggressive trade unionism. ... We are a trade union with benefit funds, not a friendly society with a few mutual protection benefits, and this cannot be made too clear to members of the railway service.

11 Mechanisation and a craft union, 1893–6

(Ellic Howe, *The London compositor: documents relating to wages, working conditions and customs of the London printing trade, 1785–1900* (Bibliographical Society, 1947), pp. 490–1, 494–5. These documents are extracts from reports or circulars issued by the Typographical Association, the union for provincial compositors (*i*), and the London Society of Compositors (*ii*).)

(i) Piecework and machinery, 1893

In London they had not had much experience of piecework. In fact they had practically none at present; but they had not been able to resist the demand of the employers, where these machines had been introduced, the demand to formulate a piece scale. The first office with which they had negotiations was the *Sportsman*, where the Thorne machines were tried. They were started on stab[†]—a very low stab, he was sorry to say—but all were entirely in the dark as to what the machines were capable of doing, and met them rather too fairly. After they had been running twelve months, the proprietors of the *Sportsman* made a claim for the piece system. Their association resisted it, but the proprietors put up their backs, and insinuated that the reason they declined to frame a piece scale was that they were antagonistic to the machines, and would not allow the men to show what they were capable of doing. A piece scale was framed which did not meet with approval; a split took place, and the house was closed to the Society. The next case was the *Daily News*. There they had the most liberal 'stab terms for fifteen or sixteen months. At the end of that time they were met with the same demand for a piece scale. They framed a scale as high as possible, but not so high that the employers could very well reject it. But after the scale had been agreed to between the Executive and the proprietors, the men stepped in and declined to accept it, alleging that the rate was too low. The Executive were, perhaps, not very sorry that they did decline. There was a certain amount of friction caused by this, but at length it was tried. This scale had only been in operation for a very short time, so they were not able to give any very definite opinion on the subject.

(ii) Inevitability of technical progress, 1895–6

No subject of greater importance could well engage the attention of a trade organization than that which has necessitated the convening of the present meeting—namely the attempt to substitute machine for hand labour. During the past fifty years rumour has from time to time been busily engaged in magnifying the capabilities of the various inventions which have been designed ostensibly for the purpose of supplanting the hand compositor; and looking at the number and variety of machines which have during that period been introduced, and the signal failure which has attended the various and costly experi-

---

† Established or standard rate of wages, as opposed to piecework.

ments, it is, perhaps, excusable that members should have allowed themselves to be lulled into a false sense of security against the prospective competition of machinery, and to feel that, come what may, they have nothing to fear from its attacks. Now, however, and for the first time in the history of the Society, the general Trade is confronted with the same problem that other industries have had to solve, a demand having been made for the fixing of a Scale of Prices and Conditions of Working to govern machine work, which shall be applicable to the Trade as a whole. Happily, as far as can be judged at present, machinery can only be partially used, and for certain classes of work, although looking at the marvellous developments of recent years, it is impossible to say what the future may bring forth.

For some time, however, an unmistakable desire has been evinced by the employers to try machinery, although it is held in many quarters that, when the cost of the machine or the royalty (if on hire), with the various incidental expenses are taken into account, the saving in comparison with hand composition is *nil*. Whether this be so or not, the fact remains that machines are being introduced into different offices, and, once introduced, have a disagreeable habit of remaining. Therefore, by common consent, it will be agreed that the time has arrived when a workable Scale should be fixed, enabling the operator to earn a good wage, while at the same time protecting as far as possible the interests of the case hand.

12 Restrictive practices in wartime, 1917

(Sidney Webb, *The restoration of trade union conditions* (Nisbet & Co. 1917), pp. 33–9)

We shall realize better what is the character of the 'Trade Union conditions' that have been suspended if we state the changes in the organization and management of the factory that their abrogation has permitted. During these fateful two years the employers in practically all industries have, to a greater or less degree—

(*i*) Changed the processes of manufacture, notably so as to enable work formerly done by skilled craftsmen to be done by women or labourers;

(*ii*) Introduced new and additional machinery with the same object;

(*iii*) Engaged in work or on processes formerly done by skilled craftsmen, boys, women, and unapprenticed men;

(*iv*) Increased the proportion of boys to men;

(*v*) Substituted piecework and bonus systems for time wages; and that without any printed and collectively-agreed-to piecework list of prices, or other protection against a future cutting of rates;

(*vi*) Increased the hours of labour, sometimes refusing also any satisfactory addition for overtime, night duty, and Sunday work;

(*vii*) Speeded up production, getting rid of all customary understandings among the workers of what constituted a fair day's work, or what times should be taken for particular jobs;

(*viii*) Suppressed demarcation disputes and ignored all claims, whether to kinds of work or particular jobs, of particular unions, particular grades, particular sets of craftsmen, or a particular sex.

*   *   *

A large section of British industry has at last learned by experience as it had long admitted in theory, the lesson of the economic advantage of a large output, of production for a continuous demand, of standardization and long runs, of the use of automatic machinery for the separate production of each component part, of team-work and specialization among the operatives, of universalizing piece-work speed and of not grudging to the workers the larger earnings brought by piece-work effort. We do not think it is any exaggeration to say that the 15,000 or 20,000 establishments, large or small, in every conceivable industry, with which the Ministry of Munitions, the Board of Trade, the War Trade Department, and the Admiralty have been in touch, are now turning out, on an average, more than twice the product per operative employed that they did before the war; whilst, assuming the same standard rates of wages, grade by grade, the labour-cost works out considerably lower than under the old system.

13 The general strike, 1926

('Sir John Simon on the General Strike: a speech delivered in the House of Commons on Thursday May 6th, 1926' in *Liberal leaders on the General Strike: speeches by the Rt Hon the Earl of Oxford & Asquith and Rt Hon Sir John Simon M.P.* (Macmillan, 1926), Appendix I, pp. 57–9. The directive here quoted was issued on 1 May 1926. The general strike began on 3 May and ended nine days later.)

The Trades Union Congress General Council and the Miners' Federation of Great Britain having been unable to obtain a satisfactory settlement of the matters in dispute in the coal-mining industry, and the Government and the mine-owners having forced a lock-out, the General Council, in view of the need for co-ordinated action on the part of affiliated unions in defence of the policy laid down by the General Council of the Trades Union Congress, directs as follows:

TRADES AND UNDERTAKINGS TO CEASE WORK

Except as hereinafter provided, the following trades and undertakings shall cease work as and when required by the General Council:

Transport, including all affiliated unions connected with transport, i.e. railways, sea transport, docks, wharves, harbours, canals, road transport, railway repair shops and contractors for railways, and all unions connected with the maintenance of, or equipment, manufacturing, repairs, and groundsmen employed in connection with air transport.

Printing Trades, including the Press.

Productive Industries

(a) *Iron and Steel.*

(b) *Metal and Heavy Chemicals Group.*—Including all metal workers, and other workers who are engaged, or may be engaged, in installing alternative plant to take the place of coal.

Building Trade.—All workers engaged on building, except such as are employed definitely on housing and hospital work, together with all workers engaged in the supply of equipment to the building industry, shall cease work.

Electricity and Gas.—The General Council recommend that the Trade Unions connected with the supply of electricity and gas shall co-operate with the object of ceasing to supply power. The Council request that the executives of the Trade Unions concerned shall meet at once with a view to formulating common policy.

Sanitary Services.—The General Council direct that sanitary services be continued.

Health and Food Services.—The General Council recommend that there should be no interference in regard to these, and that the Trade Unions concerned should do everything in their power to organise the distribution of milk and food to the whole of the population.

With regard to hospitals, clinics, convalescent homes, sanatoria, infant welfare centres, maternity homes, nursing homes, schools, the General Council direct that affiliated Unions take every opportunity to ensure that food, milk, medical and surgical supplies shall be efficiently provided.

14 Trade unions and industrial efficiency, 1951

(Trades Union Congress, *Trade unions and productivity: report of a team of British trade union officials who investigated the role of unions in increasing productivity in the U.S.A.* (T.U.C., 1951), pp. 5, 58–9)

British trade unionism, to repeat what almost amounts nowadays to a platitude, is standing on the threshold of a new social, economic and industrial order—a situation which has been created in part by the trade unions themselves. The way has been long and arduous, but 'mass' and 'hard core' unemployment and social insecurity, characteristics of social injustice, have, we hope, disappeared for good.

British Productivity

But what lies beyond this threshold of Labour Movement achievement? Trade unionists are looking for the answer with the same sense of responsibility as that exercised in their contribution to the reconstruction and stabilising of the national economy. By and large the answer has been found. It is to seek a rising standard of life for all, achieved through increasing industrial productivity or output per man hour. This then is the real problem confronting trade unions: to find ways and means of increasing productivity—a problem concerned mainly with industrial policy and actions as distinct from the political pressure to achieve full employment and economic stability.

Trade Union Role

Productivity has shown a long-term tendency to rise but trade unions have not in the past claimed any direct responsibility. During and since the war, however, trade unions and the Trades Union Congress have sought, through joint consultative and advisory machinery from national to shop floor levels, to increase productivity.

*   *   *

British trade unions can learn from American trade union experience. Equally, we are convinced that American unions could learn something by studying our methods and attitudes, particularly in the field of politics and working class education.

One of the difficulties of utilising American experiences, however, is the considerable difference in the industrial and economic environments in which the two trade union movements operate. Many differences will have become apparent in the reading of this report.

Different Circumstances

The United States of America has not experienced a post-war economic situation calling for a policy of wage restraint on the part of trade unions. Nor is there full employment, as we understand the phrase, in the United States. Moreover, owing to the size of the home market and the comparatively small percentage of American industrial capacity employed on export production, there is not the same urgency as in Britain to keep prices down—provided wages are not left behind. Britain has to keep prices down, not only to maintain a high standard of living but to compete effectively with other countries in order to secure imports of foodstuffs and raw materials on which full employment and the standard of living depend.

British trade unions, in voluntarily adopting a policy of wage restraint, accepted obligations and responsibilities far beyond anything contemplated in America. By securing wage increases, American unions compel management to improve industrial efficiency. British unions, as things are at present, expect increased wages, or lower consumer prices, to follow increased productivity. For too long, too great a section of British management has not been willing to face up to the need to become thoroughly efficient and progressive.

America's Advantages

We do not ignore management's problems. The scattered and small-scale nature of Britain's foreign markets often prevents integration with home production and does not permit the long production runs enjoyed by American industry which encourages greater specialisation and the breaking down of jobs into simple unskilled functions. Americans also enjoy, almost without limit, varied and high quality raw materials obtained without the difficulties created by foreign politics and trade agreements, quotas, tariffs and exchange rates.

An added advantage to American industry is, that their operatives are provided with mechanical power amounting to two or three times that available to their British colleagues. This factor alone does much to explain the differences between American and British rates of productivity and is largely a reflection on past British management. It is a problem which can be overcome but the speed with which it can be tackled is limited by present shortages of power, capital equipment and manpower. Under conditions of full employment, increased capital goods production can be achieved only at the expense of the production of consumers' goods or by increasing output per man hour. There is, therefore, the utmost urgency to make the most effective use of existing manpower.

Wages and Prices

Within conditions of full employment it is very doubtful if wage increases secured by British unions (presumably on an industry-wide basis) would necessarily lead to an increased rate of mechanisation and hence productive efficiency. A higher output per man hour might be achieved in some factories through redeployment. But if industry-wide wage increases were secured irrespective of whether individual managements carried out redeployment it would be difficult to avoid a rise in prices unless unions were prepared to force inefficient firms into bankruptcy.

The adoption of the wage-restraint policy indicated clearly enough that most British unions were well aware of the nature of the problems confronting industry.

Social Obligations

The increasing willingness to accept incentive schemes can be taken as a sign of recognition of the fact that wage restraint did not mean a restriction of earnings or puchasing power. A wage policy in which earnings are related to output and factory efficiency is therefore the most obvious one for unions to pursue. This does not, of course, take into account the anomalies and inequalities of the existing national wage structure. The extent, however, to which any wages policy should be pursued by individual unions without regard to the benefits which can accrue to the whole community from reduced prices will be a test of the social consciences of unions. Some workers have fewer opportunities than others to increase productivity but that is no reason why they should not share the benefits of increased productivity through lower consumer prices.

New Industrial Role

The need in industry for decisive trade union action in which unions must accept their responsibilities as well as claim their rights is perfectly clear. Where managements are progressive and seeking to use 'scientific management' techniques in a reasonable manner to step up production, unions should be prepared to co-operate. If managements try to be aggressive the need for effective trade union action is accentuated—not to the point of resisting new development but to see that abuses are eliminated and that the inaccuracies of 'scientific management' are not exploited at the expense of workpeople. Where managements are not sufficiently enterprising and progressive, are unwilling to step up efficiency or extend markets through lower prices, then unions must press them to do so.

15 The breakdown of national agreements, 1938–67

(Royal Commission on trade unions and employers' associations, *Report* (Cmnd. 3623, 1968), pp. 13–15)

56 Substantive industry-wide agreements lay down the length of the normal working week, regulations for overtime, weekend and shift working, and for statutory and annual holidays. These provisions are usually observed fairly closely in the industries to which they apply. They also regulate pay. In this respect their provisions show a great deal of variety. Some fix only two time rates, one for skilled workers and another for unskilled, leaving individual firms to deal with intermediate and other grades. Others prescribe a list of different rates for a catalogue of different grades, with in addition a series of special additional payments for special duties or conditions of work. In industries in which a substantial number of women are employed, women are usually treated as forming a separate grade with rates of pay lower than those for skilled men. Some agreements make no provision for payment by results: others do so, but in different degrees of detail. Some describe their rates as minimum rates, others as standard rates.

57 The figures in Tables A [and B] show that over the last thirty years there has been a decline in the extent to which industry-wide agreements determine actual pay. In 1938 there was only a modest 'gap' between the rates which they laid down for a normal working week and the average earnings which men actually received. By 1967 the two sets of figures had moved far apart. . . . Together the tables record a remarkable transfer of authority in collective bargaining in this group of industries. Before the war it was generally assumed that industry-wide agreements could provide almost all the joint regulation that was needed, leaving only minor issues to be settled by individual managers. Today the consequences of bargaining within the factory can be more momentous than those of industry-wide agreements.

26*

Table A Earnings and time rates of men in October 1938

Industry	Time rates for a normal week			Average earnings of adult male manual workers in last pay week October 1938
Engineering	Fitters:	£3	7 2	£3 13 8
	Labourers:	£2	10 4	(General engin-eering, and iron and steel found· ing)
Shipbuilding and repairing	Platers:	£3	8 0	£3 10 1
	Labourers:	£2	9 0	
Building	Bricklayers:	£3	13 2	£3 6 6
	Labourers:	£2	15 0	(Building and decorating)
Civil Engineering	Labourers:	£2	4 0 to	£3 2 10
		£3	2 0	(Public Works Contracting)

58 The three major elements in the 'gap' between agreed rates and average earnings are piecework or incentive earnings, company or factory additions to basic rates, and overtime earnings. Most industry-wide agreements give no more guidance on piecework or incentive earnings than to say that, for the 'average' worker, there should be a given minimum level. The actual prices or incentive 'values' are usually fixed within the factory and earnings now generally exceed the minimum by a generous margin. Additions to basic rates include 'lieu' payments to those on timework to compensate them for not having the opportunity to raise their earnings through piecework, and job-rates which may be settled on some system of job evaluation. Both are usually fixed within the factory or the company. The length of the normal working week and the rate of overtime pay, are both generally settled by industry-wide agreements, but the decision to work overtime is a matter for the factory, and this governs the volume of overtime earnings.

Table B Earnings and time rates of men in certain industries in October 1967

Industry	Times rates for normal week of 50 hours (national or provincial rate)	Average weekly earnings of adult male manual workers for industry (Second pay-week in Oct. 67)
Engineering Fitter:	£11 1 8	£21 7 9
Labourer:	£ 9 7 4	to £24 8 5
Building Craftsmen:	£14 13 4	£21 13 8
Labourer:	£12 11 8	
Shipbuilding and repairing		
Fully skilled:	£11 1 4	£21 17 8
Labourer:	£ 9 6 0	
Dock labourers	£15 0 0	£22 16 6
Cocoa, chocolate and sugar confectionery	£10 15 6	£21 7 5
Electrical cable making	£11 8 4 to £13 5 6	£23 9 4
Furniture manufacturing	£13 0 0 to £14 3 4	£22 5 4
Motor vehicle retail and repairing trade	£11 0 0 to £13 10 0	£18 10 4
Soap, candle and edible fat manufacturing	£10 6 6 to £11 7 0	£23 10 5
Footwear manufacturing	£11 12 6	£19 14 4

*   *   *

In summary it may be said that, on the basis of the evidence available, about half the workers covered by industry-wide agreements or statutory wage regulations are employed in industries where the rates specified are generally exceeded and most of the rest work in jobs not covered by any form of wage fixing at industry level. Even where rates fixed at industry level are fairly closely followed, more often than not they are supplemented by high levels of overtime earnings. On any reasonable estimate the effective regulation of pay levels by industry-wide agreement is now very much the exception rather than the rule in Britain and is largely confined to the public sector.

C. WAGES AND WORKING CONDITIONS

1 Borrowing from the employer, 1789–96

(T. S. Ashton, *An eighteenth-century industrialist: Peter Stubs of Warrington, 1756–1806* (Manchester University Press, 1939), pp. 32–3)

		£	s	d
	Jacob Cooper Dr.			
1789 April 18	To Cash at sundry times left in arear when settled for small beer	1	4	3
	Benjamin Jolley Dr.			
1788 Sept. 13th	To Cash for Bedstocks	2	5	0
1790 Octr. 30th	To Cash lent to pay for your Cloaths	1	1	0
	Thomas Gad Dr.			
1790 June 26	To paid Mr. Thomas Skitt for Rent due to Mrs Hind & an Ale score	3	8	4
Nov. 8	To Cash to your wife to pay your rent with	1	10	0

*   *   *

February the 6 1796.

Sir I am-tould that you ar verrey Busey: If you ar I should be verrey glad to work for you. If you will be so good as to lend me four or five pound I will come on Saturday and be hired to you and live in Warrington and you may stop so much a week from me. Please to send it letter against wednesday to the Black horse and rainbow Liverpool to Jonathan Cranage file Cutter, Aughton. That money would pay my debts and oblige your humble servant

Jonathan Cranage

2 Industrial discipline, 1797 and 1817

(i) 1797

(George Unwin and others, *Samuel Oldknow and the Arkwrights: the industrial revolution at Stockport and Marple* (Manchester University Press, 1924), p. 198)

WHEREAS

The horrid and impious Vice of profane CURSING and SWEARING —and the Habits of Losing Time—and DRUNKENNESS—are become so frequent and notorious; that unless speedily checked, they may justly provoke the Divine Vengeance to increase the Calamities these Nations now labour under.

NOTICE is hereby given,

That all the Hands in the Service of

SAMUEL OLDKNOW

working in his Mill, or elsewhere, must be subject to the following

RULE:

That when any person, either Man, Woman or Child, is heard to CURSE or SWEAR, the same shall forfeit One Shilling—And when any Hand is absent from Work, (unless unavoidably detained by Sickness, or Leave being first obtained), the same shall forfeit as many Hours of Work as have been lost; and if by the Job or Piece, after the Rate of 2*s* 6*d* per Day,—Such Forfeitures to be put into a Box, and distributed to the Sick and Necessitous, at the discretion of their Employer.

MELLOR, 1*st December*, 1797

(*ii*) 1817

(*The Observer*, 21 September 1817)

To Master Tradesmen, Manufacturers, etc.:

Mr Editor—By inserting this in your paper, you will very much oblige your humble servant—An Observer. Gentlemen, I am much surprised that a custom now almost universal, and which is attended with great inconvenience and hardship to a very industrious and respectable class of people, should be continued. I mean the weekly payment of your men, which is now done on a Saturday night, and generally so late that it is impossible for them to procure necessaries for themselves and their families for Sunday, the only day on which they can have any rest or enjoyment.

Some people think that were you to pay them on a Friday, you would lose the benefit of the men's work on the Saturday, but this, I apprehend, would only happen to those men who invariably keep St Monday as a holiday, and whether it was St Monday or St Saturday, it would make no difference in the end.

3 Employment of children in factories

(*i*) *Early anxiety in Manchester*, 1796

('Report of the committee on the state of children in manufactories', *B.P.P.*, 1816, III, pp. 139–40)

Resolutions for the consideration of the Manchester Board of Health, by Dr Perceval, January 25, 1796

It has already been stated that the objects of the present institution are to prevent the generation of diseases; to obviate the spreading of them by contagion; and to shorten the duration of those which exist, by affording the necessary aids and comforts to the sick. In the prosecution of this interesting undertaking, the Board have had their attention particularly directed to the large cotton factories established in the town and neighbourhood of Manchester; and they feel it a duty incumbent on them to lay before the public the result of their inquiries:

1. It appears that the children and others who work in the large factories, are peculiarly disposed to be affected by the contagion of fever, and that when such infection is received, it is rapidly propagated,

not only amongst those who are crowded together in the same apart-
ments, but in the families and neighbourhoods to which they belong.

2. The large factories are generally injurious to the constitution of
those employed in them, even where no particular diseases prevail,
from the close confinement which is enjoined, from the debilitating
effects of hot or impure air, and from the want of the active exercises
which nature points out as essential in childhood and youth, to invi-
gorate the system, and to fit our species for the employments and for
the duties of manhood.

3. The untimely labour of the night, and the protracted labour of
the day, with respect to children, not only tends to diminish future
expectations as to the general sum of life and industry, by impairing
the strength and destroying the vital stamina of the rising generation,
but it too often gives encouragement to idleness, extravagance and
profligacy in the parents, who, contrary to the order of nature, subsist
by the oppression of their offspring.

4. It appears that the children employed in factories are generally
debarred from all opportunities of education, and from moral or
religious instruction.

5. From the excellent regulations which subsist in several cotton
factories, it appears that many of these evils may, in a considerable
degree, be obviated; we are therefore warranted by experience, and are
assured we shall have the support of the liberal proprietors of these
factories, in proposing an application for Parliamentary aid (if other
methods appear not likely to effect the purpose), to establish a general
system of laws for the wise, humane, and equal government of all
such works.

(*ii*) *Parliamentary opinions*, 1816

(*Hansard's Parliamentary Debates*, 1st series, 33 (3 April 1816),
cols. 884–6)

State of Children in Manufactories

Sir Robert Peel rose, in pursuance of a previous notice, to submit a
motion to the House respecting the state of children employed in the
cotton manufactories. The object of his motion was altogether
national, as it affected the health and morals of the rising generation,
and went to determine, whether the introduction of machinery into
our manufactories was really a benefit. The principal business in our
cotton manufactures was now performed by machinery, and, of course,

interrupted the division of work suitable to the respective ages, which formerly was practised in private houses. The consequence was, that little children of very tender age were employed with grown persons at the machinery, and those poor little creatures, torn from their beds, were compelled to work, even at the age of six years, from early morn till late at night—a space of perhaps 15 or 16 hours! He allowed that many masters had humanely turned their attention to the regulation of this practice; but too frequently the love of gain predominated, inducing them to employ all their hands to the greatest possible advantage. Some time ago he had introduced a bill into the House, for regulating the work of apprentices, which was attended with the happiest results, and their time was limited; but children were still subjected to all the hardships to which carelessness or avidity might expose them. The House were well aware of the many evils that resulted from the want of education in the lowest classes. One object of his present bill was to enable manufacturing children to devote some of their time to the acquirement of a little useful simple knowledge, such as plain reading and writing. He hoped those poor children would experience the protection of the House, for if it were not extended to them, all our excellent machinery would be productive of injury. It might, perhaps, be said, that free labour should not be subjected to any control; but surely it could not be inconsistent with our constitution, to protect the interests of those helpless children. The hon. baronet concluded with moving, 'That a committee be appointed to take into consideration the state of the children employed in the different manufactories of the united kingdom; and to report the same, together with their observations thereupon to the House'.

Mr Finlay said, he had no intention to oppose the motion of the hon. baronet, but he could not concur in the calumny which he had uttered against all the manufactories of the united kingdom (cries of No, no! from all parts of the House); by stating, that children of a tender age were torn from their beds at an early hour, and compelled to work during a space of 15 or 16 hours. To such a statement he must give a flat denial. No doubt there might be some manufactories where those practices were permitted, but he could state that in the manufactories of Scotland, and especially that part of Scotland with which he was connected they did not exist. In fact, it was a general rule in the cotton manufactories of that country, not to employ any children under ten years of age; and as to the general healthiness and convenience of those establishments, he would venture to affirm that the cotton mills of Glasgow were not only situated most advantageously for health, but were conducted upon the most liberal plan. With respect to the object of the hon. baronet's motion, he believed it to be a very laudable one, and likely to produce much good; but he could not help feeling some degree of surprise that the hon. baronet should have deferred such a measure till now, when he was about to quit the concern in which he had been so many years engaged. Those abuses, of which he now complained, must have existed for many years, and equally in his own establishments, as in others, yet he had never before thought it necessary to propose any measures of relief.

Sir Robert Peel hoped the House would permit him to say a few words in reply to what had fallen from the hon. member who spoke last. He had accused him of calumniating all the manufactories of the United Kingdom, but he trusted no one would suspect him of calumniating a body for which he felt so much respect. With regard to another imputation cast upon him by the hon. member, that he had not brought the present measure forward till he was about to quit the concern in which he was engaged, he would remind the hon. member that twelve years ago, when he was deeply interested in it, he brought in a bill for regulating the apprentices in the cotton trade.

Mr Lyttleton said, he was persuaded that it would be found impossible, when the committee went into the business, to apply such a bill as was brought forward last session, to all the manufactories in the kingdom.

Mr Curwen took that opportunity of protesting against the principle of legislating for the regulation of the authority of parents over their children, who must be best aware of the quantity of work those children were able to bear, and who must undoubtedly feel most for their distresses. Such a proceeding was a libel on the humanity of parents. He stated that in all the manufactories of which he knew any thing, regulations had been established, which considerably ameliorated the condition of the persons employed, and the air of all of them had been of late so much improved, that labour was not half so distressing as it had been previously. He however had no objection to a committee.

Mr Courtenay was well convinced that the hon. baronet could be influenced by no other motive than that of humanity, but he thought that there appeared no reason for imputing a system of cruelty to the great body of the manufacturers of the kingdom. Regulations of the best description had been generally established by them, which would be sufficiently manifested to the committee, to which he had no objection.

Mr R. Gordon expressed bis gratitude to the hon. baronet, for having so zealously and uniformly exerted himself in support of the interests of humanity. Much good had been already accomplished, and much might be anticipated from the appointment of a committee. The House might remember a circumstance, which by a similar inquiry, had transpired the last session. It had appeared, that overseers of parishes in London, were in the habit of contracting with the manufacturers in the north for the disposal of children; and those manufacturers agreed to take one idiot for every nineteen sane children. In this manner, waggon loads of those little creatures were sent down, to be at the perfect disposal of their new masters. The fact alone was sufficient to justify an inquiry.

Mr Wrottesley also spoke in favour of a committee, conceiving that the establishment of good regulations in individual manufactories, was not sufficient cause to prevent parliament from compelling their more general adoption. If there were 99 manufactories well regulated, while the 100th was in a different state, that alone would authorize the exertion of parliament.

4 The beginning of effective factory legislation, 1833

(i) The Factory Act, 1833

(3 and 4 Will. IV, c. 103)

An Act to regulate the Labour of Children and young Persons in the Mills and Factories of the United Kingdom.

... no person under eighteen years of age shall be allowed to work in the night, that is to say between the hours of half-past eight o'clock in the evening and half-past five o'clock in the morning, except as hereinafter provided, in or about any cotton, woollen, worsted, hemp, flax, tow, linen, or silk mill or factory. ...

II And be it further enacted, that no person under the age of eighteen years shall be employed in any such mill or factory in such description of work as aforesaid more than twelve hours in any one day, nor more than sixty-nine hours in any one week, except as hereinafter provided.

...

VI And be it further enacted, that there shall be allowed in the course of every day not less than one and a half hours for meals to every such person restricted as hereinbefore provided to the performance of twelve hours work daily.

VII And be it enacted, that from and after the first day of January one thousand eight hundred and thirty-four it shall not be lawful for any person whatsoever to employ in any factory or mill as aforesaid, except in mills for the manufacture of silk, any child who shall not have completed his or her ninth year of age.

VIII And be it further enacted, that from and after the expiration of six months after the passing of this act, it shall not be lawful for any person whatsoever to employ, keep, or allow to remain in any factory or mill as aforesaid for a longer time than forty-eight hours in any one week, nor for a longer time than nine hours in any one day, except as herein provided, any child who shall not have completed his or her eleventh year of age, or after the expiration of eighteen months from the passing of this act any child who shall not have completed his or her twelfth year of age, or after the expiration of thirty months from the passing of this act any child who shall not have completed his or her thirteenth year of age: Provided nevertheless, that in mills for the manufacture of silk, children under the age of thirteen years shall be allowed to work ten hours in any one day. ...

[XI No child under thirteen to be employed without a certificate that the child is of normal strength and appearance.] ...

XVII ... it shall be lawful for His Majesty by Warrant under his Sign Manual to appoint during His Majesty's pleasure four persons to be Inspectors of factories and places where the labour of children and young persons under eighteen years of age is employed, ... and such Inspectors or any of them are hereby empowered to enter any factory or mill, and any school attached or belonging thereto, at all times and seasons by day or by night, when such mills or factories are at work. ...

XVIII And be it further enacted, that the said Inspectors or any of them shall have power and are hereby required to make all such rules, regulations, and orders as may be necessary for the due execution of this act, which rules, regulations, and orders shall be binding on all persons subject to the provisions of this act; and such inspectors are also hereby authorised and required to enforce the attendance at school of this act. ...

XX And be it further enacted, that from and after the expiration of six months from the passing of this act, every child hereinbefore restricted to the performance of forty-eight hours of labour in any one week shall, so long as such child shall be within the said restricted age attend some school. ...

(*ii*) *A factory inspector's half-yearly report*, 1836

(Reports of the factory inspectors, *B.P.P.*, 1836, XLV, pp. 10–11. This report was made by the inspector whose district included the towns of Manchester, Leeds and Huddersfield—and 2300 mills.)

When in my former Report of the 25th August 1835, I adverted to a very general concurrence which had been expressed to me by mill-owners and operatives in restraining the moving power, I had also an eye to the limited number of superintendents with which I was furnished for so extensive a division, and the possibility that on principles of economy more would not be allowed. As far as regards mills in which steam-power is alone employed, I believe still that concurrence would be generally given to this restraint; but there are a numerous class of mills, particularly in remoter parts of the division, where water-power is alone used, and where steam could not be applied for want of the necessary supply of coals. Here some consideration is in justice due to the situation of these mill-owners; and if from this cause, and a regard to other principles, the question of restraint shall be deemed finally objectionable, a remedy for the evil of over-working may be found in the appointment of a sufficient number of superintendents to keep a vigilant watch over the proceedings of the refractory, without which over-working by the latter will *inevitably* be practised, to the misery of children so employed, and the great annoyance of the community at large. The activity which has hitherto been displayed by the superintendents, in the execution of this part of their duty, is the best warrant of their future success; but it must be kept in mind, that this success has mainly depended on the superintendents being allowed to enter the interior of mills, where alone it is possible to detect offences of this description. The well-disposed and more respectable part of the community, who conscientiously act up to the spirit and letter of the law, and who have therefore nothing to conceal, can have no objection to these occasional visitations by well behaved and perfectly conciliatory officers; but many of the refractory, who are bent on the infraction or evasion of the law, and who have already suffered for their offences, are averse to these visitations, and have given notice to the superintendents, that they shall not be allowed to enter their mills in future: but these are the very persons requiring to be controlled, and who will most assuredly continue to transgress, if some effective check upon their actions be not imposed.

Notwithstanding the numerous convictions and heavy penalties already inflicted for offences of this description, I continue to receive reports from the very quarter where these convictions have lately occurred, that several of the most determined transgressors, after concurring with their neighbours in a resolution to work only 12 hours per diem, or 69 hours per week, have again reverted to 13 or 14 hours per day; and one gentleman, from that quarter, who has always been perfectly willing to adopt the 12 hours limit, writes that be is again surrounded by persons who violate the law, and that if these proceedings cannot be restrained, he himself will be compelled to become an over-worker in his own defence.

These are some of the difficulties which, as I have before frequently remarked, the inspector and his superintendents will most assuredly have to encounter in their endeavours to extend that protection to children which it is the great object of the law to grant, and which, without the aid of an efficient law, they as assuredly, in such cases as are here adverted to, will not receive.

Amidst all this it is highly gratifying to be enabled to report, that the manufacturing districts, in all their branches of industry, are now in a state of very general and great prosperity. This prosperity, and the uninterrupted demand for goods, may be, and probably is, one cause of the over-working so much complained of; but the demand for our goods is not likely, in my opinion, to cease or even to be diminished. New mills are now erecting in various parts of the country, and many old ones being at the same time enlarged or improved, more and more hands will consequently be wanted, the demand for children will proportionally increase, and if sufficient numbers of legal age cannot be procured, every artifice and evasion will be attempted to smuggle into the mills infants, altogether unfitted for the assigned work. The business of the inspector and his superintendents may thus too be so magnified in its difficulties as to render it impracticable of effectual execution with the present limited establishment.

5 Town and country wages, 1836

(L. Marion Springall, *Labouring life in Norfolk villages, 1834–1914* (Allen & Unwin, 1936), p. 141. The writer of this letter had migrated to Leeds from Norfolk under a poor-law scheme for which Mr Baker was the Yorkshire agent.)

Leeds

Mr Baker, Sir,   *24 July, 1836*

I hope you will excuse my writing to you, as I do it to let you know the situation our family is in, and to settle the reports of some idle men who came down and could get plenty of work and good wages, but

were too idle to work or even to seek for it; they went back and said
they had put us in the workhouse and told many other stories, which
were all lies, to screen themselves. I have to return you many thanks
for your kindness to myself and family for getting us employment. I
will let you know what we earned in Norfolk, and what we now earn
in Leeds, then you will see whether we have bettered ourselves or not.
My family and myself never earned more than 10s a week, except for
about five weeks in harvest, then we might earn about £1 a week,
and that had to keep me and my wife and seven children; and for three
months my eldest son when he was out of work. The wages we receive
in Leeds are, my wages are 18s and sometimes a guinea, a week. My
eldest daughter is in a situation, at £6 10s a year, and the rest of the
family bring me in 15s a week. We have received better treatment since
we came to Leeds than we did in our own part of the country, where
people are so poor they are fit to eat one another up alive.

J. Clarke

6 Fluctuations in wages, 1847–69

(Thomas Brassey, *On work and wages* (3rd ed. 1872), pp. 37–42)

The fall in wages, which follows a commercial panic, when production is diminished and employment is scarce, proves how closely the
rate of wages depends upon alterations in the relation between supply
and demand. When the panic took place in the railway world in 1847–8,
even the common labourers, employed on the Eastern Union Railway,
accepted lower wages.

In 1849, on the Royston and Hitchin Railway, labour was cheaper
than it has ever been since. The reduction was a direct consequence
of the depression, caused by the collapse of railway enterprise in
1847–8. Men who, on the North Staffordshire line, shortly before the
panic, had been paid 3s 6d a day only earned half-a-crown on the Royston and Hitchin line.

A member of my father's staff [Mr Mackay] informs me that at
the same period and from the same cause the wages of the navvies,
which in the inflation of the railway mania in 1846 had been advanced
in some cases to 6s a day, in the collapse following on the panic were
so much reduced that on the Cheshire Junction line, the cost of the
work was in consequence diminished by not less than fifteen per cent.

The following statement gives the weekly wages earned by men
employed on railway works from 1843 to 1869. The notes furnish a
comparative statement of the cost of work represented by the different

rates of wages, and contain a short explanation of the extent of the demand for labour at the different periods included in the Return:

Periods

	1843	1846	1849	1851	1855	1857	1860	1863	1866	1869
Masons	21/	33/	24/	21/	25/6	24/	22/6	24/	27/	27/
Bricklayers	21/	30/	24/	21/	25/6	22/6	22/6	24/	27/	25/6
Carpenters and Blacksmiths	21/	30/	22/6	21/	24/	22/6	22/6	24/	25/6	24/
Navvies, Getters (Pickmen)	16/6	24/	18/	15/	19/	18/	17/	19/	20/	18/
Fitters (Shovellers)	15/	22/6	16/6	14/	17/	17/	16/	17/	18/	17/
Cost of labour only, per cube yard:										
Of Brickwork	2/3	3/9	2/9	2/3	2/6	2/6	2/4	2/6	2/9	2/6
Of Earthwork	/4$\frac{1}{2}$	/7$\frac{1}{2}$	/5	/4	/5$\frac{1}{2}$	/5$\frac{1}{4}$	/5	/5$\frac{1}{2}$	/5$\frac{3}{4}$	/5$\frac{1}{2}$

Gloucester and Bristol Railway, period of general depression, provisions for men and horses very cheap. Men plentiful, excellent workmen. Clay cuttings, on the Gloucester to Stonehouse line, taken out at 6d a yard, inclusive of horse labour. [1843]

Lancaster and Carlisle, Caledonian, Trent Valley, North Staffordshire, Eastern Union Railways in construction. Height of the railway mania. Demand for labour excessive, very much in excess of supply. Beer given to men as well as wages. Look-outs placed on the roads to intercept men tramping, and take them to the nearest beershop to be treated and induced to start work. Very much less work done in the same time by the same power. Work going on night and day, even the same men working continuously for several days and nights. Instances recorded of men being paid for forty-seven days in one lunar month. Provisions dear. Excessively high wages, excessive work, excessive drinking, indifferent lodgings caused great demoralisation, and gave the death-blow to the good old navvy already on the decline. He died out a few years after this period.[1] [1846]

Great Northern, Oxford, Worcester and Wolverhampton, Oxford and Birmingham, Chester and Holyhead Railways, in construction. Great reduction in wages caused by the financial embarrassments in October 1847, and political turmoils and revolutions in 1848 on the

---

[1] Other experienced contractors do not admit that good navvies can no longer be obtained.

Continent and at home. General distrust, aggravated by the unsettled state of affairs abroad. Works stopped in 1847, partially resumed in 1848. The 1846 contract not yet completed. In 1849 work comparatively plentiful. Provisions moderate in price. [1849]

Shrewsbury and Hereford, North Devon, in construction. Contracts taken in 1846 now all completed. Great depression in the labour market.. But little work going on. Political affairs on the Continent unsettled. Provisions very cheap. [1851]

Leicester and Hitchin, Leominster and Kingston Railways in construction. Work still very slack during this period. Best men gone to France, Spain, Belgium, Switzerland, and Italy to Mr Brassey's works. Crimean War, militia all called under arms. These circumstance tended to raise wages. Provisions dear, horse provender excessively high, costing 5s a day each horse. [1855]

Shrewsbury and Crewe, Leominster and Kingston Railways in construction. Work still very slack; the effects of the Crimean War had not wholly passed away [1857].

Knighton and Craven arms, Woofferton and Tenbury, widening of Shrewsbury and Hereford, Severn Valley Railway Works in construction. Men plentiful, provisions cheap [1860].

Tenbury and Bewdley, South Staffordshire, Ludlow and Clee Hill, Wenlock, Nantwich and Drayton, widening of Shrewsbury and Hereford, Worm Valley drainage, Letton Valley drainage in construction. Men plentiful, provisions rather dear [1863].

Wellington and Drayton, widening of Nantwich and Drayton, Hereford Loop, Hooton and Parkgate, Wenlock and Craven arms, Ebbw Vale in construction [1866].

Silverdale and Drayton, Sirhowy, widening of Abergavenny, and Merthyr Railways, and London drainage works in construction. Provisions rather dear. [1869]

The explanatory memorandum does not exhaust the list of Mr Brassey's contracts in progress at the several dates mentioned. Those contracts only are included which happened to be in the recollection of the writer, whose immediate field of observation was necessarily limited to a few contracts in the Midland Counties.

7 Children's employment, 1862

(Children's employment commission, *First Report*, with appendix,
B.P.P. 1863, pp. xciii–xciv, 13, 199, 301, 322)

(*i*) *General recommendation of the commissioners*

639. The branches of manufacture to which the recommendations
of this Report apply, employ the following number of children and
young persons:

The Potteries	11,000
The Lucifer Match Manufacture	1,613
The Percussion Cap Manufacture	150
The Paper-staining Manufacture	1,150
Finishing and "Hooking"	2,300
Fustian cutting	1,563
Total children and young persons employed ..	17,776

640. If Parliament should think fit to adopt our recommendations
with regard to the above-named manufactures, the considerable
number of upwards of seventeen thousand more children and young
persons will be placed under the protection, and be benefited by the
privileges of the Factory Act.

(*ii*) *The potteries*

*Mr Geo. L. Ashworth and Brothers' Earthenware Manufactory,
Hanley*

John Lawton, manager of Messrs. Ashworth's manufactory—I have
been 20 years manager and foreman here. I was a hollow-ware presser
27 years before that. I was a mould runner for about one year. I was
not 9 when I began to work. I learnt reading, writing, and arithmetic
at the Sabbath school and at night school. I went to Sabbath school
when I was 5 years of age, and I began to attend night school when
I was about 14 or 15.

Boys are principally employed in running moulds and turning the
jigger. The average age for boys beginning to turn jigger and run
moulds would be 10 or 11. Some would begin a year or so earlier.

I think a child of 9 or 10 is old enough for this work; nor have I
ever found that children have been injured by it.

The children all work for the men, and are paid by the men.

The wages of a child of 9 or 10 would be 2*s* or 2*s* 6*d*.

Jigger turning is turning the wheel which turns the whirler on which the potter makes plates, saucers, and cups.

Mould running is carrying the plates, saucers, and cups, and sometimes dishes, on the moulds, into the drying house or stove, and there placing it on the shelves to dry. They also have to enter the stoves to turn the ware.

I never heard of children or their parents complaining either of their being worked too hard or paid too little. Perhaps, but very rarely, I have heard of children who happen to have a bad master being badly treated; but in such a case of course we should interfere.

There is a great improvement in this respect to what it was some years ago. The lads are now themselves so independent, that some of the men dare not ill-use them. The lads would go off altogether.

Lads do not work now as they did. The men generally give over at an earlier hour, that is, at 6 o'clock; beyond this hour men seldom work now.

The lads always have their half hours for breakfast and their hour for dinner, which is spent in play. I do not know a single case where a man has prevented his children having their hours for their food and play. If they did work a little beyond the regular time for meals, their lads would still have the same time.

The mould runner also assists in wedging the clay for the man. It is throwing or beating the clay to drive out the air. It requires strength in proportion to the quantity of clay the boy lifts. A boy may please himself as to the quantity of clay he lifts at a time. I consider it a healthful exercise. The children come in the morning to light the fires. They seldom come before 6. If children were overworked at mould running, it would weaken them, but we seldom find boys are injured by the work, although the stoves are hot.

These lads generally go to the trade at which they have been assisting. When lads get about 12 or 14, they begin to apply for places as apprentices. We don't take any over 14, because they would not be bound beyond 21. We generally take them between 12 and 14, according to their size and aptness. We very often take them from the mould runners employed at our own works. All our apprentices are bound by a stamped agreement.

The stamp is only 2s 6d now. The master pays for the stamp. As a rule, we always keep our apprentices to the expiration of their time. This is the general rule when apprentices are bound.

Boys are also employed by the dippers. Their age would generally be from 12 to 14. Their work is carrying baskets full of ware, handing the ware to the dipper, and then carrying it away. Little boys are not so much employed in this work, because care is required. If the ware was broken the master would suffer. In the case of mould running the man only would be the loser; and, besides, the ware in the dipping house has been already fired, and, therefore, of more value than the clay on the mould which the mould runner carries.

✳   ✳   ✳

Memorial of Employers in the Potteries

To the Right Honourable Sir George Grey, Bart., Her Majesty's Principal Secretary of State for the Home Department.

We, whose signatures are subjoined being manufacturers in the district of Staffordshire, commonly known as the Staffordshire Potteries, having under our consideration the following facts with regard to the employment of children in this district:

I. That children are employed in the potteries at a very early age, and in a way to interfere injuriously with their education.

II. That the majority of the children appear to be taken from school in this district before they are 10 years old. As we are led to believe from the fact that the average age of the first class in 15 out of the 21 national schools of this district is only 10·2 years, and that four-fifths of the children leave before they reach the first class.

III. That in 22 out of about 116 manufactories in the district where inquiry was made, it was found that there were 177 children under 10 years of age, and 576 others under 13 years of age; which leads us to believe that it is the employment in the potteries which causes the early removal of the children from school before noticed.

IV. That this state of things is the cause of various moral and physical evils to the youthful population of this district;

(a) A vast amount of ignorance, as is evidenced by the fact that out of 670 working children questioned on the subject, 185 (or 27·6 per cent.) professed themselves unable to read;

(b) That the employment of children at so tender an age is injurious to their health, stunts their growth, and causes in many cases a tendency to consumption, and distortion of the spine, etc., as we have the evidence of competent medical men to testify.

V. That much as we deplore the evils before mentioned it would not be possible to prevent them by any scheme of agreement between the manufacturers as to the employment of children; as a portion only of the employers could be brought to consent to such agreement.

Taking all these points into consideration, we have come to the conviction that some legislative enactment is wanted to prevent children from being employed at so early an age, and to secure to them at any rate a minimum of education; and we would respectfully urge upon the Legislature, the advisableness of appointing a commissioner to inquire into the matter, and consult as to the best means of remedying the evils complained of.

(Signed)

Minton & Co.	Stoke-upon-Trent.	Samuel Elkin	Longton
John Dimmock & Co.	Hanley.	Sampson, Bridgwood, & Son	Ditto.
Edwd. Jno. Ridgway	Ditto.	Charles Allerton & Sons	Ditto.
J. W. Parkhurst	Ditto.	Pinder, Bourne & Co.	Burslem.
Wilkinson & Richard	Ditto.	Hope & Carter	Ditto.
Joseph Clementson	Ditto.	Cook, Edye, & H. Calken	Ditto.
Josiah Wedgwood & Sons	Etruria.	John S. Hill	Ditto.
Geo. W. Ashworth & Bros.	Hanley.	Wm. Adams	Turnstall.
William Webberley	Longton.	Beech & Hancock	Ditto.
Frederick Chetham	Ditto.	Edwd. Challoner	Ditto.
John Lockett	Ditto.	Wm. Adams & Sons	Stoke.
Thomas Betbury	Ditto.	Elijah Hughes	Cobridge.
James Broadhurst	Ditto.	Richard Edwards	Dutchall.

(*iii*) *Lace-finishing*, *Nottingham*

Ellen Cresswell, age 13.—Frame clips. Has done so at two mistresses and been at a warehouse also. At the mistresses' the hours were called from 8 o'clock to 8, but at one she has gone at 7 a.m. and stayed till 11 and 11½ p.m., and one night till 12. There were about six girls at each place, none much younger than herself, i.e. from 9 upwards, and all stayed when she did. At the other mistress's she never stayed later than 9 or 9½. Her hours here are 8 till 7, with an hour for dinner and half an hour for tea (1 and 5). Gets 5s. Has been to school on Sunday for some years, and to a week day school for half a year before she went out to work. Can read, but not very well. (Spells slowly.) Cannot write or sum. (Cannot read the number '28' in large print.) [The overlooker said that this girl was slow. A quick worker at the same work could get 9s or 10s.]

Elizabeth Leyland, age 15.—Began lace work with her mother at 9 years old, and went to a mistress at 11. The hours there were 8 to 8, and if busy till 9; not always, but 'off and on nights.' Was at a small warehouse in Pilchergate with 20 girls, where the hours were 8 to 8; but she generally took work home with her for mother, and helped her. Can read a verse without stopping (reads badly). Goes to a night school now to learn to write.

Elizabeth Bradley, age 13.—Went to a lace mistress at 7 years old. The regular hours were 8 to 8 or 9.

Emma Collier, age 12.—Went out to a lace mistress when 'going' 6. When busy, which was often, and in two or three different seasons of the year, went at 6 a.m. and worked till 9 at night, but not later. Has done this for two or three weeks together. Was about 7 when she began to work these hours. There were eight girls, some about her own age, and all used to stay. At first had 1*s* a week, after about a year 2*s*.

Harriet Craig, age 18.—Went to a lace mistress at 7 years old, and to another at 10. At the first place there were four girls, some younger than herself. At the second there were 10 girls, eight of them about 8 or 9 years old. At this place the hours were reckoned from 8 a.m. to 9 p.m., but they worked till 10 quite as often as till 9, and on Saturdays generally till 11, but not always. Does not remember staying later than 11. All the girls stayed. Has been to a Sunday school, but to no other. Cannot read, except just a few quite short words.

(iv) Chimney sweeps

Examined by Mr F. D. Longe, December 10, 1862.

No. 8.

Mr Thomas Howgate, Nelson Street, Leeds.—I have been a chimney sweeper in Leeds for 24 years. I began as a climbing-boy. I worked as a climbing-boy until I was 17. I worked for my father. I learnt to climb by going up chimneys with other boys behind me. I was 8 years old when I began. I was not hurt. I have no marks on my knees now. Lads were ill-used by their masters. I never had a climbing boy myself. I would not be bothered with them. I can do best without them. Sometimes I am asked whether I will bring a boy, and I say, No. I have been refused because I had no boy. The chimneys where I have been refused can be swept with the machine just as well as by the boy. Some of the sweeps here have climbing boys. A man at Woodhouse has 3 boys; they are all his own children. Another man, whom we call ——, has a climbing boy; he is his own child too. I do not know of any other sweeps who have climbing boys. My son is about 10. He goes to school at Mr Wood's. I pay 9*d* a week, sometimes 1*s* for him. I like to have him at this school, because it is near, so we can get him when we want him. He is getting on very well. It would be cheaper to send him to the National School I know. The climbing boys were stopped here in 1843. There is no reason for having the boys except to save the sweep the trouble of sweeping himself. It is no hardship on sweeps enforcing the Act. We had an association here, but it has been broken up. Boys are always used here, I believe, at —— House. If they could stop boys being used in this house they might be stopped everywhere else about here.

8 Domestic service and its rewards, 1902

(Dorothy C. [Mrs C. S.] Peel, *How to Keep House* (Constable, 1902), pp. 133–6)

The young working-girl of to-day prefers to become a Board School mistress, a post-office clerk, a typewriter, a shop girl, or a worker in a factory—anything rather than enter domestic service; not because the work is lighter or the pay better, but because in these professions she has the full use of her hours of liberty, and, more important reason than all, she enjoys a higher social position: she is in point of fact a 'young lady'.

∗   ∗   ∗

Wages of Domestic Servants—In the last ten years wages of domestic servants have risen. They vary according to locality. In Scotland and Ireland they are lower than in England, and in small provincial towns they are lower than in London. In London, where there are so many flats, there is a demand for superior general servants, who ask and obtain from £18 to £24 a year. First-rate parlourmaids, in every way able to perform the duties of a good manservant, can obtain large wages. Good cooks also can demand almost any yearly sum, while, owing to the difficulty of finding footmen, laundry, kitchen, and scullery maids, their wages have risen considerably.

It is interesting to compare the amounts given in the original edition of Mrs Beeton with those which are offered in the *Morning Post* and

other papers and demanded by London registry office keepers. Mrs
Beeton's list is as follows:

Housekeeper	£20 to £45		Upper-housemaid	£12 to £20
Maid	12 to 25		Maid-of-all-work	9 to 14
Head-nurse	15 to 30		Kitchenmaid	9 to 14
Cook	14 to 30		Scullerymaid	5 to 9

And so on.
The following list is drawn up after inquiry at registry offices and
perusal of advertisements, and from personal experience:

Womenservants

Cook-housekeeper	£30 to £80
Professed cook	30 to 60
Good cook	25 to 35
Plain cook	18 to 25
Cook-general	16 to 25
General	10 to 25
Parlourmaid, head	30 to 35
Thorough parlourmaid	24 to 30
Parlourmaid	18 to 26
House-parlourmaid	16 to 26
Housemaid, head	24 to 30
Housemaid, thorough	22 to 26
Housemaid	18 to 22
Housemaid, under	16 to 20
Between maid	10 to 18
Schoolroom maid	14 to 18
Kitchenmaid	16 to 24
Scullerymaid	12 to 18
Young girl to help (or 3/- a week)	6 to 12
Nurse, superior upper	30 to 50
Good nurse	24 to 30
Nurse	20 to 25
Under-nurse	16 to 20
Nurserymaid	10 to 18
Maid, superior	30 to 60
Good maid	25 to 35
Useful maid	22 to 30
Young ladies' maid	16 to 26

Menservants

Butler	£50 to £100
Single-handed manservant	40 to 60

Footman, upper	30 to	50
Footman, under	20 to	35
Page	8 to	16
Hall boy	8 to	16

Job Servants

Cooks, 10/6 to £1 10/-a week, with food, washing, and beer

Parlourmaids, 10/6 to £1 a week, with food, washing, and beer

Housemaids, 10/6 to 15/6 a week, with food, washing, and beer

Workwomen, 2/6 to 3/6 a day, food and beer, in many cases railway fare.

Charwoman, 2/6 a day and food, generally beer

Boy for two or three hours a morning, 2/- to 5/- a week and breakfast

Waitress for dinners and afternoons, 5/6; for late parties, 7/6

Waiters, 7/6 to 10/6

Caretakers, 10/6 to 15/6 a week, with coals and light

Board Wages

10/6 to 15/6 per week per head, exclusive of firing and light.

9 Beginnings of 'scientific' management, 1914

(J. A. Hobson, *Work and wealth: a human valuation* (Macmillan, 1914), pp. 204–5)

Until lately the detailed organisation of labour and its utilisation for particular technical processes had received little attention in the great routine industries. Even such technical instruction as has been given to beginners in such trades as building, engineering, weaving, shoe-making, etc., has usually taken for granted the existing tools, the accepted methods of using them and the material to which they are applied. To make each sort of job the subject-matter of a close analysis and of elaborate experiment, so as to ascertain how it could be done most quickly and accurately and with the least expenditure of needless energy, comes as a novel contribution of business enterprise. To get the right man to use the right tools in the right way is a fair account of the object of Scientific Management. At present a man enters a particular trade partly by uninstructed choice, partly by chance, seldom because he is known by himself and his employer to have a natural or acquired aptitude for it. He handles the tools that are tradi-tional and are in general use, copying the ways in which others use them, receiving chance tips or suggestions from a comrade or a foreman, and learning from personal experience how to do the particular work in a way which appears to be least troublesome, dangerous, or exhausting. Both mode of work and pace are those of prevailing usage, more or less affected by machinery or other technical conditions.

The scientific manager discovers enormous wastes in this way of working. Part of the waste he finds due to improper tools and improper modes of working, arising from mere ignorance; part he attributes to systematic or habitual slacking, more or less conscious and intentional on the part of the workers. The natural disposition of the worker to 'take it easy' is supplemented by a belief that by working too hard he deprives some other worker of a job. Scientific Management, therefore, sets itself to work out by experiment the exact tool or machine appropriate to each action, the most economical and effective way by which a worker can work the tool or machine, and the best method of selecting workers for each job and of stimulating them to perform each action with the greatest accuracy and celerity. By means of strictly quantitative tests it works out standard tools, standard methods of work and standard tests for the selection, organisation, stimulation, and supervision of the workman.

10 Wages and hours, 1914-25

([Balfour] Committee on industry and trade, *Survey of Industrial relations* (H.M.S.O., 1926), pp. 10-11, 130-1)

If we attempted to measure roughly the change of 'real' wages for each of the above-mentioned classes of workmen by the simple expedient of dividing the estimated percentage increase in weekly rates of money wages by the index of cost of living, we should find that, taking the 1914 level as 100 in each case, the average 'real' rate of time wages for skilled workers at the end of 1924 would be round about 94 and that for unskilled workers about 106. Taking separately the exposed and 'sheltered' industries as described above, we should find that in the former group the average 'real' time rates of wages would be roughly represented by 91 and in the latter group by 114. Too much importance must not be attached to these precise figures; in view of the uncertainty of some of the data, they must be regarded rather as rough approximations than as ascertained percentages. But whatever qualifications may apply to the actual figures, it is a legitimate inference from the available data that, in industries in which time rate of wages prevail, skilled workers employed in industries directly exposed to foreign competition were in 1924 on the average less well off than before the war, while, on the other hand, unskilled workers generally, and workers both skilled and unskilled in the so-called 'sheltered' industries, have generally speaking, if with some exceptions, improved their average position as regards purchasing power.

Needless to say, the above conclusions, being based on weekly rates of wages, apply to workpeople in full employment, and take no account of the allowances to be made at the two dates 1914 and 1924 for unemployment, short time or overtime. The qualifications to be made on this account are discussed below.

We have now to consider how far the above inferences may be modified by such considerations as the prevalence and recent spread of piece-work or other methods of industrial remuneration dependent on output.

In 1906, the year to which the last comprehensive inquiry into earnings referred, the proportion of workers in industry on time rates of wages was found to be no less than 75 per cent of the whole number covered by the returns. There are no official statistics as to subsequent changes in the proportions, but... on the whole, recent tendencies have been towards an extension of piece work, including under this term all systems of payment by results. It is evident that the change of weekly piece-work earnings cannot be accurately represented by a mere comparison of piece-work prices, since it is directly affected by a number of other factors which there are no official data for determining. It is, however, a matter of common knowledge that, in trades in which piece and time rates are worked side by side by similar workers in similar classes of work for the same hours, piece-work earnings are usually higher than time rate earnings. Hence, in any trades in which piece-work has become materially more prevalent since 1914, the true average rise in weekly earnings will be somewhat greater than the recorded rise in time rates. In the absence of any official data which enable precise allowance to be made for this factor, we may be permitted, for the purposes of illustration, to refer to statistics collected from its members by the Engineering and Allied Employers' National Federation which show that in April, 1924, the general average rise of money earnings since July, 1914, for engineering workers, including both skilled and unskilled, was about 73 per cent when working on piece, and 65 per cent when working on time. The general average rise of earnings (piece and time), taking into account not only the increases of time and piece earnings but also the shift from time to piece, is also stated to have been about 73 per cent. While, therefore, as stated above, skilled workers at time rates in the more exposed trades in 1924 were on the average less well off than before the war, the same does not necessarily apply to skilled workers on piece-work, whose total number has tended to increase as compared with time workers.

✳   ✳   ✳

Hours of Labour, 1925

In a few cases the reduced hours agreed upon in 1919 and 1920 have since been slightly increased, but the general effect of the changes has been to establish a normal working week (apart from overtime) of 48 hours or less in nearly all the organised industries; though it is probably the case that in some of the less well organised trades a longer working week is often worked. The principal exceptions to the general rule in the organised trades are certain industries in which, owing to the nature of the process, work is carried on continuously by day and night, and throughout the week-end. For example, in pig iron manufacture, heavy chemical manufacture, lead and spelter manufacture, and in the gas and electricity supply industries in some districts, although 8-hour shifts are in operation, the shift-workers employed on or in connection with the actual processes of smelting or manufacture generally work an average of seven shifts per week, though in some such cases the average weekly hours actually worked are less than 56 owing to the employment of spare men who take turns of duty in the absence of the regular shift workers. It should also be observed that the 8-hour turns in these industries are inclusive of time taken for meals in the intervals which occur in waiting on the process of manufacture.

In Table 10 in Chapter VI, particulars are given showing, for each of the principal industries in respect of which information is available, the hours recognised by employers and workpeople as constituting a full ordinary week's work (exclusive of overtime) (a) at the end of 1918 (prior to the reductions which were generally made in 1919 and 1920), and (b) at June, 1925. In considering the information given in the table, it should be borne in mind that no general statistics as to hours of labour in all industries have been completed since 1906, and that the information given relates only to the hours specified in agreements signed by organisations of employers and workpeople, or in the absence of formal agreements, the hours known to be generally recognised by such organisations. As the hours agreed upon are not necessarily binding on non-associated employers employing nonunion workmen, cases will be found (especially in industries and districts in which the employers' and workpeople's organisations are not strongly represented) in which hours in excess of those agreed upon are actually worked. It should also be noted that, as already mentioned, in many of the less well organised industries, in which the hours of labour are not governed by collective agreements between employers and workpeople, the working week is probably longer, on the whole, than in the organised industries dealt with in the table.

8

The Poor Law and public health

By the middle of the eighteenth century the English Poor Law already had a considerable history behind it. Since the reign of Elizabeth I it had, in theory, provided a national minimum for the destitute so that none need die of starvation. In practice the Poor Law was a branch of local government administered according to no uniform pattern. It often gave help to the poor, not only to the destitute, and by the 1830s it was in the eyes of ratepayers a costly, and in the eyes of economists an inefficient, machine for the relief of destitution. The reforms of 1834 were based upon the findings of a royal commission that produced a celebrated but to modern investigators, a careless and inadequate analysis of the problems (3). Drastic administrative changes followed, but it may be doubted whether the lot of the pauper altered very much. There was nothing new about the stigma of pauperism in 1834; the new workhouses did not incarcerate more than a minority of paupers; the aged, the widows, the unemployed largely went on getting doles from the relieving officer. Rigorous insistence on the workhouse test was applied in relatively few unions and mostly not after 1834, but after 1870. It was the hope of tough-minded social workers that fear of the workhouse would call forth either a greater effort from the potential pauper, or charitable help from his friends, relatives, or patrons. Meanwhile, the study of social phenomena was adding to men's understanding of the causes of poverty. The new Poor Law administrators themselves helped to demonstrate the connection between sickness and pauperism and to initiate the public health movement in the 1840s (4). Private investigators later in the century and in the reign of Edward VII showed that poverty reflected more discredit on the organisation of the economic system than on the individual pauper. Low wages, unemployment, irregular employment as well as more personal causes such as large families and drink were held to explain poverty and these findings resulted in a shift in public policy (7–12). The Poor Law gave way to insurance against sickness and unemployment, to old age pensions, to family allowances, to municipal housing and ultimately to the creation of new towns. It is by now generally realised that this bundle of measures scarcely amounts to a 'welfare state' but it does reflect the better understanding of the causes of poverty that characterises the twentieth century and the greater willingness to make public provision for the weaker members of the society.

1 The working of the settlement laws, 1774–1824

(*i*) 1774

(S. A. Cutlack, ed. 'The Gnosall records, 1679 to 1837: poor law administration' in Staffordshire Record Society, *Collections for a history of Staffordshire* (Stafford, 1936), p. 73)

County of Stafford

The Examination of Francis Evans of the parish of Gnosall in the said County Labourer upon Oath before us John Williamson & John Turton Esq^rs two of his Majesty's Justices of the Peace in and for the said County, the 11th Day of June in the Year of our Lord, 1774 touching the Place of his last legal Settlement.

This Examinant saith, That he was born in the parish of Gnosall in the said County which was the place of his fathers settlement. That Twenty years since or thereabouts he was hired to Mr Butler of the parish of Claybrook in the county of Leicester for Eleven months which he served. That his master then hired him again for eight pounds a year and they agreed that this examinant should be absent from the service a week in some part of the year—to prevent his gaining a settlement—when his master could best spare him—That he continued with Mr Butler upon his farm in the parish of Claybrook aforesaid under the last mentioned hiring for two years except a week in each year which time he came to see his friends in Staffordshire and his master hired a Labourer in his room to do his work, and stopped such Labourers pay out of this examinants wages in each of the two last mentioned years. That since the said service in the parish of Claybrook he hath not to his knowledge & belief done any act whereby he could gain a settlement And that he hath a wife named Mary and three children namely Francis aged about seven years, Elizabeth aged about four years and Joseph about one year & three months old now residing in the parish of Gnosall aforesaid ————————

Sworn the day &   The mark of
year first before written
before us ————   X
    John Williamson   Francis Evans
    John Turton

(*ii*) 1824

(J. D. Chambers, *Nottinghamshire in the eighteenth century: a study of life and labour under the squirearchy* (P. S. King, 1932), p. 265)

Articles of agreement made and entered into this 22nd day of October 1824 between the occupiers of lands or tenements in Parish of Bleasby and Church Wardens and Overseers of Bleasby.

Whereas the number of persons applying for relief to Parish of Bleasby has greatly increased. ...

It is mutually agreed between the parties that if any of them hire any person to serve him for any period exceeding Fifty one weeks, or shall make any agreement by which any servant shall obtain a settlement in the Parish, then the person so hiring or making such agreement shall pay into the Churchwardens and Overseers the Sum of Ten Pounds to be used for the relief of the Poor.

2 Speenhamland, 1795

(*Reading Mercury*, 11 May 1795)

Berkshire, to wit

At a General Meeting of the Justices of this County, together with several discreet persons assembled by public advertisement,[1] on Wednesday the 6th day of May, 1795, at the Pelican Inn in Speenhamland (in pursuance of an order of the last Court of General Quarter Sessions) for the purpose of rating Husbandry Wages, by the day or week, if then approved of, [names of those present].

Resolved unanimously,

That the present state of the Poor does require further assistance than has been generally given them.

Resolved,

That it is not expedient for the Magistrates to grant that assistance by regulating the Wages of Day Labourers, according to the direc-

---

[1] *Reading Mercury*, May 4, contained an advertisement of a general meeting of justices 'to limit, direct, and appoint the wages of day labourers'.

tions of the Statutes of the 5th Elizabeth and 1st James: But the Magistrates very earnestly recommend to the Farmers and others throughout the county, to increase the pay of their Labourers in proportion to the present price of provisions; and agreeable thereto, the Magistrates now present, have unanimously resolved that they will, in their several divisions, make the following calculations and allowances for relief of all poor and industrious men and their families, who to the satisfaction of the Justices of their Parish, shall endeavour (as far as they can) for their own support and maintenance.

That is to say,

When the Gallon Loaf of Second Flour, weighing 8lb 11 ozs shall cost 1s.

Then every poor and industrious man shall have for his own support 3s weekly, either produced by his own or his family's labour, or an allowance from the poor rates, and for the support of his wife and every other of his family, 1s 6d.

When the Gallon Loaf shall cost 1s 4d.

Then every poor and industrious man shall have 4s weekly for his own, and 1s and 10d for the support of every other of his family.

And so in proportion, as the price of bread rise or falls (that is to say) 3d to the man, and 1d to every other of the family, on every 1d which the loaf rise above 1s.

By order of the Meeting,
W. Budd, Deputy Clerk of the Peace.[1]

3 The principles of 1834

(Report from his Majesty's commissioners for inquiring into the administration and practical operation of the poor laws, *B.P.P.*, 1834, XXVII, pp. 24–5, 127–8)

The Out-door Relief to the Impotent (using that word as comprehending all, except the able-bodied and their families) is subject to less abuse. The great source of Poor-Law mal-administration is, the desire of many of those who regulate the distribution of the parochial

---

[1] Simultaneously the Magistrates published a recommendation to overseers to grow potatoes, setting poor people to work and offering them one-third or one-fourth of the crop, and to sell at 1s a bushel; also to get in a stock of peat, faggots, furze etc, in the summer and to sell at a loss in the winter.

fund to extract from it a profit to themselves. The out-door relief
to the able-bodied, and all relief that is administered in the workhouse, afford ample opportunities for effecting this purpose: but no
use can be made of the labour of the aged and sick, and there is little
room for jobbing if their pensions are paid in money. Accordingly we
find, that even in places distinguished in general by the most wanton
parochial profusion, the allowances to the aged and infirm are moderate.

The out-door relief of the sick is usually effected by a contract with a
surgeon, which, however in general includes only those who are
parishioners. When non-parishioners become chargeable from illness,
an order for their removal is obtained, which is suspended until they
can perform the journey; in the mean time they are attended by the
local surgeon, but at the expense of the parish to which they belong.
This has been complained of as a source of great peculation; the surgeon
charging a far larger sum than he would have received for attending
an independent labourer or a pauper, in the place of his settlement.
On the whole, however, medical attendance seems in general to be
adequately supplied, and economically, if we consider only the price
and the amount of attendance.

The country is much indebted to Mr Smith, of Southam, for his exertions, to promote the establishment of dispensaries, for the purpose
of enabling the labouring classes to defray, from their own resources,
the expense of medical treatment. . . . It appears to us, that great good
has already been effected by these dispensaries, and that much more
may be effected by them; but we are not prepared to suggest any
legislative measures for their encouragement. . . .

We have dwelt at some length on out-door relief, because it appears
to be the relief which is now most extensively given, and because it
appears to contain in itself the elements of an almost indefinite extension; of an extension, in short, which may ultimately absorb the whole
fund out of which it arises. Among the elements of extension are the
constantly diminishing reluctance to claim an apparent benefit, the
receipt of which imposes no sacrifice, except a sensation of shame
quickly obliterated by habit, even if not prevented by example; the
difficulty often amounting to impossibility on the part of those who
administer and award relief, of ascertaining whether any and what
necessity for it exists; and the existence in many cases of positive
motives on their parts to grant it when unnecessary or themselves to
create the necessity. The first and third of these sources of mal-

28*

administration are common to the towns and to the country, the second, the difficulty of ascertaining the wants of the applicant, operates most strongly in the large towns.

✳   ✳   ✳

Remedial Measure

The most pressing of the evils which we have described are those connected with the relief of the Able-bodied. They are the evils, therefore, for which we shall first propose remedies.

If we believed the evils stated in the previous part of the Report, or evils resembling or even approaching them, to be necessarily incidental to the compulsory relief of the able-bodied, we should not hesitate in recommending its entire abolition. But we do not believe these evils to be its necessary consequences. We believe that, under strict regulations, adequately enforced, such relief may be afforded safely and even beneficially.

In all extensive communities, circumstances will occur in which an individual, by the failure of his means of subsistence, will be exposed to the danger of perishing. To refuse relief, and at the same time to punish mendicity when it cannot be proved that the offender could have obtained subsistence by labour, is repugnant to the common sentiments of mankind; it is repugnant to them to punish even depredation, apparently committed as the only resource against want.

In all extensive civilized communities, therefore, the occurrence of extreme necessity is prevented by alms giving, by public institutions supported by endowments or voluntary contributions, or by a provision partly voluntary and partly compulsory, or by a provision entirely compulsory, which may exclude the pretext of mendicancy.

But in no part of Europe except England has it been thought fit that the provision, whether compulsory or voluntary, should be applied to more than the relief of *indigence*, the state of a person unable to to labour, or unable to obtain, in return for his labour, the means of subsistence. It has never been deemed expedient that the provision should extend to the relief of *poverty*; that is, the state of one, who in order to obtain a mere subsistence, is forced to have recourse to labour.

From the evidence collected under this Commission, we are induced to believe that a compulsory provision for the relief of the indigent can be generally administered on a sound and well defined principle; and that under the operation of this principle, the assurance that no one need perish from want may be rendered more complete than at present, and the mendicant and vagrant repressed by disarming them of their weapon, the plea of impending starvation.

It may be assumed, that in the administration of relief, the public is warranted in imposing such conditions on the individual relieved, as are conducive to the benefit either of the individual himself, or of the country at large, at whose expense he is to be relieved.

The first and most essential of all conditions, a principle which we find universally admitted, even by those whose practice is at variance with it, is, that his situation on the whole shall not be made really or apparently so eligible as the situation of the independent labourer of the lowest class. Throughout the evidence it is shown, that in proportion as the condition of any pauper class is elevated above the condition of independent labourers, the condition of the independent class is depressed; their industry is impaired, their employment becomes unsteady, and its remuneration in wages is diminished. Such persons, therefore, are under the strongest inducements to quit the less eligible class of labourers and enter the more eligible class of paupers. The converse is the effect when the pauper class is placed in its proper position, below the condition of the independent labourer. Every penny bestowed, that tends to render the condition of the pauper more eligible than that of the independent labourer, is a bounty on indolence and vice. We have found, that as the poor's rates are at present administered, they operate as bounties of this description, to the amount of several millions annually.

The standard, therefore, to which reference must be made in fixing the condition of those who are to be maintained by the public, is the condition of those who are maintained by their own exertions. But the evidence shows how loosely and imperfectly the situation of the independent labourer has been inquired into, and how little is really known of it by those who award or distribute relief. It shows also that so little has their situation been made a standard for the supply of commodities, that the diet of the workhouse almost always exceeds that of the cottage, and the diet of the gaol is generally more profuse than even that of the workhouse. It shows also, that this standard has been so little referred to in the exaction of labour, that commonly the work required from the pauper is inferior to that performed by the labourers and servants of those who have prescribed it: So much and so generally inferior as to create a prevalent notion among the agricultural paupers that they have a right to be exempted from the amount of work which is performed and indeed sought for by the independent labourer.

We can state, as the result of the extensive inquiries made under this Commission into the circumstances of the labouring classes, that the agricultural labourers when in employment, in common with the other classes of labourers throughout the country, have greatly advanced in condition; that their wages will now produce to them more of the necessaries and comforts of life than at any former period. These results appear to be confirmed by the evidence collected by the Committees of the House of Commons appointed to inquire into the condition of the agricultural and manufacturing classes, and also by that collected by the Factory Commissioners. No body of men save money whilst they are in want of what they deem absolute necessaries. No common man will put by a shilling whilst he is in need of a loaf, or will save whilst he has a pressing want unsatisfied. The circumstance of there being nearly fourteen millions in the savings banks, and the fact that, according to the last returns, upwards of 29,000 of the depositors were agricultural labourers, who, there is reason to believe, are usually the heads of families, and also the fact of the reduction of the general average of mortality, justify the conclusion that a condition worse than that of the independent agricultural labourer, may nevertheless be a condition above that in which the great body of English labourers have lived in times that have always been considered prosperous. Even if the condition of the independent labourer were to remain as it now is, and the pauper were to be reduced avowedly below that condition, he might still be adequately supplied with the necessaries of life.

But it will be seen that the process of dispauperizing the able-bodied is in its ultimate effects a process which elevates the condition of the great mass of society.

In all the instances which we have met with, where parishes have been dispauperized, the effect appears to have been produced by the practical application of the principle which we have set forth as the main principle of a good Poor Law administration, namely, the restoration of the pauper to a position below that of the independent labourer.

The principle adopted in the parish of Cookham, Berks, is thus stated:

'As regards the able-bodied labourers who apply for relief, giving them hard work at low wages by the piece, and exacting more work at a lower price than is paid for any other labour in the parish. In short, to adopt the maxim of Mr Whately, to let the labourer find that the parish is the hardest task master and the worst paymaster he can find, and thus induce him to make his application to the parish his last and not his first resource'.

In Swallowfield, Berks, labour was given 'a little below the farmers' prices'.

The principle adopted by the Marquis of Salisbury, in Hatfield, Herts, is set forth in the following rules:

'All persons, except women, employed by the parish, under the age of fifty, shall be employed in task work. The value of the work done by them shall be calculated at five-sixths of the common rate of wages for such work. Persons above the age of fifty may be employed in such work as is not capable of being measured, but the wages of their labour shall be *one-sixth* below the common rate of wages'.

4 Pauperism and public health, 1842

(Poor Law Commissioners, Report on the sanitary conditions of the labouring population of Great Britain, *B.P.P.*, House of Lords, 1842, XXVI, pp. 369–70)

First, as to the extent and operation of the evils which are the subject of the inquiry:

That the various forms of epidemic, and other disease caused, or aggravated, or propagated chiefly amongst the labouring classes by atmospheric impurities produced by decomposing animal and vegetable substances, by damp and filth, and close and overcrowded dwellings prevail amongst the population in every part of the kingdom, whether dwelling in separate houses, in rural villages, in small towns, in the larger towns—as they have been found to prevail in the lowest districts of the metropolis.

That such disease, wherever its attacks are frequent, is always found in connexion with the physical circumstances above specified, and that where those circumstances are removed by drainage, proper cleansing, better ventilation, and other means of diminishing atmospheric impurity, the frequency and intensity of such disease is abated; and where the removal of the noxious agencies appears to be complete, such disease almost entirely disappears.

That high prosperity in respect to employment and wages, and various and abundant food, have afforded to the labouring classes no exemptions from attacks of epidemic disease, which have been as frequent and as fatal in periods of commercial and manufacturing prosperity as in any others.

That the formation of all habits of cleanliness is obstructed by defective supplies of water.

That the annual loss of life from filth and bad ventilation are greater than the loss from death or wounds in any wars in which the country has been engaged in modern times.

That of the 43,000 cases of widowhood, and 112,000 cases of destitute orphanage relieved from the poor's rates in England and Wales alone, it appears that the greatest proportion of deaths of the heads of families occurred from the above specified and other removable causes; that their ages were under 45 years; that is to say, 13 years below the natural probabilities of life as shown by the experience of the whole population of Sweden.

That the public loss from the premature deaths of the heads of families is greater than can be represented by any enumeration of the pecuniary burdens consequent upon their sickness and death.

That, measuring the loss of working ability amongst large classes by the instances of gain, even from incomplete arrangements for the removal of noxious influences from places of work or from abodes, that this loss cannot be less than eight or ten years.

That the ravages of epidemics and other diseases do not diminish but tend to increase the pressure of population.

That in the districts where the mortality is the greatest the births are not only sufficient to replace the numbers removed by death, but to add to the population.

That the younger population, bred up under noxious physical agencies, is inferior in physical organization and general health to a population preserved from the presence of such agencies.

That the population so exposed is less susceptible of moral influences, and the effects of education are more transient than with a healthy population.

That these adverse circumstances tend to produce an adult population short-lived, improvident, reckless, and intemperate, and with habitual avidity for sensual gratifications.

That these habits lead to the abandonment of all the conveniences and decencies of life, and especially lead to the overcrowding of their homes, which is destructive to the morality as well as the health of large classes of both sexes.

That defective town cleansing fosters habits of the most abject degradation and tends to the demoralization of large numbers of human beings, who subsist by means of what they find amidst the noxious filth accumulated in neglected streets and bye-places.

That the expenses of local public works are in general unequally and unfairly assessed, oppressively and uneconomically collected, by separate collections, wastefully expended in separate and inefficient operations by unskilled and practically irresponsible officers.

That the existing law for the protection of the public health and the constitutional machinery for reclaiming its execution, such as the Courts Leet, have fallen into desuetude, and are in the state indicated by the prevalence of the evils they were intended to prevent.

5 A programme of sanitary reform, 1845

(Second report of the commissioners on the state of large towns and populous districts, *B.P.P.*, 1845, XVIII, pp. 13–68)

That in all cases the local administrative body appointed for the purpose have the special charge and direction of all the works required for sanitary purposes, but that the Crown possess a general power of supervision.

That before the adoption of any general measure for drainage a plan and survey upon a proper scale, including all necessary details, be obtained, and submitted for approval to a competent authority.

That the Crown be empowered to define and to enlarge from time to time the area for drainage included within the jurisdiction of the local administrative body.

That, upon representation being made by the municipal or other authority, or by a certain number of the inhabitants of any town or district, or part thereof, setting forth defects in the condition of such place, as to drainage, sewerage, paving, cleansing, or other sanitary matters, the Crown appoint a competent person to inspect and report upon the state of the defects, and, if satisfied of the necessity, have power to enforce upon the local administrative body the due execution of the law.

That the management of the drainage of the entire area, as defined for each district, be placed under one jurisdiction.

That the construction of sewers, branch sewers, and house drains, be entrusted to the local administrative body.

That the duty of providing the funds necessary to be imposed upon the local administrative body, and that the cost of making the main and branch sewers be equitably distributed among the owners of the properties benefited; and that the expense of making the house-drains be charged upon the owners of the house, to which the drains are attached, etc.

That some restriction be placed on the proportionate rates in the pound to be levied in one year, but if the local administrative body finds that there is need for larger funds, for the immediate execution of works for sanitary measures, than can be provided by such rates, it be empowered to raise, by loan on security of the rates, subject to the approval of the Crown, such sums as may be requisite for effecting the objects in view.

That provision always be made for the gradual liquidation of such debts, within a given number of years.

That the whole of the paving, and the construction of the surface of all streets, courts and alleys be placed under the management of the same authority as the drainage.

That the provisions in local Acts, vesting the right to all the dust, ashes, and street refuse in the local administrative body, be made general; and that the cleansing of all privies and cess-pools at proper times, and on due notice, be exclusively entrusted to it.

That it be rendered imperative on the local administrative body, charged with the management of the sewerage and drainage, to procure a supply of water in sufficient quantities not only for the domestic needs of the inhabitants, but also for cleansing the streets, scouring the sewers and drains, and the extinction of fire. . . .

That measures be adopted for promoting a proper system of ventilation in all edifices for public assemblage and resort, especially those for the education of youth.

That, on complaint of the parish medical or other authorised officer, that any house or premises are in such a filthy and unwholesome state as to endanger the health of the public, and an infectious disorder exists therein, the local administrative body have power to require the landlord to cleanse it properly, without delay; and in case of his neglect or inability, to do so by its own officers, and recover the expense from the landlord.

That the local administrative body have power to appoint, subject to the approval of the Crown, a medical officer properly qualified to inspect and report periodically upon the sanitary condition of the town or district, to ascertain the true causes of disease and death, more especially of epidemics increasing the rates of mortality, and the circumstances which originate and maintain such disease, and injuriously affect the public health of such town or populous district.

[Recommendations for abating factory exhalations and nuisances; for regulating the width of new courts, the accommodation of cellar-dwellings and the sanitation of new houses; for power to buy out new water companies at the end of a term of years; for controlling odging-houses; for providing public spaces and walks.]†

6 Avoidable mortality, 1861-71

([W. Farr,] Supplement to the Registrar General's thirty-fifth report, *B.P.P.*, 1875, XVIII, Part II, pp. viii-ix, xiv-xv, xxviii-xxxiii, xli-xlii)

In the last twenty years the towns of England have increased from five hundred and eighty to nine hundred and thirty-eight; their population from nine to fourteen millions; and the health of the whole population of the country has remained stationary.

Breeders reject weakly animals from their stock, and thus achieve success. By the care now taken of the humblest member of the human race the weakly, it is said, survive; they marry and propagate, and thus, as some contend, the proportion of inferior organizations is raised. The imbecile, the drunkard, the lunatic, the criminal, the idle, and all tainted natures were once allowed to perish in fields, asylums, or gaols, if they were not directly put to death, but these classes and their offspring now figure in large numbers in the population.

2. Probable Decrease of Mortality

Such are samples of the many obstacles to the sanitary progress of a nation, and it is evident that at present they can only be overcome in part; but there is no ground for despair. There has been progress. The mean lifetime of sovereigns and peers is prolonged; it was in past ages much shorter than the lifetime of the unhealthy labourers in the cities of to-day. The mortality of the city of London was at the rate of 80 per 1000 in the latter half of the seventeenth century, 50 in the eighteenth, against 24 in the present day. The mortality in the liberties of the city of London within and without the walls was in the four plague years 1593, 1625, 1636, 1665, at the rate of 24, 31, 13, and 43 per cent. In the city alone 90,472 persons died of plague in the four epidemics, and 53,604 of other diseases. The enumerated population of the city was 130,178 in 1631. In the cholera epidemic year of 1849 the mortality from all causes in the metropolis was only 3 per cent. And in the last two epidemics there was a further decline. Thus it is as certain that the high mortality can be reduced by hygienic appliances down to a certain limit as it is that human life can be sacrificed.

---

† The first general Public Health Act (1848) was based on this report and that of the Select Committee on the Health of Towns, *B.P.P.*, 1840, XI.

The analysis of the causes of the mortality renders it still further certain that the actual mortality of the country can be reduced. Many of the destroyers are visible, and can be controlled by individuals, by companies, and by corporate bodies, such as explosions in coal mines, drowning in crazy ships, railway collisions, poisonings, impurities of water, pernicious dirts, floating dusts, zymotic contagions, crowdings in lodgings, mismanagements of children, neglects of the sick, and abandonments of the helpless or of the aged poor.

Furthermore, including the London district of Hampstead, there are fifty-four large tracts of England and Wales which actually experience a mortality at the rate of only seventeen per 1000—less by *five* than the average mortality per 1000 of the whole country, less by ten than in nine districts, and less by *twenty-two* than the mortality reigning for ten years in Liverpool. Now the healthy districts have a salubrious soil, and supply the inhabitants with waters generally free from organic impurities. The people are by no means wealthy; the great mass of them are labourers and workpeople on low wages, whose families get few luxuries, and very rarely taste animal food. Their cottages are clean, but are sometimes crowded and impurities abound; the sanitary short-comings are palpable.

It will not, therefore, be pitching the standard of health too high to assert that any excess of mortality in English districts over *17 annual deaths* to every 1000 living is an excess not due to the mortality incident to human nature, but to foreign causes to be repelled, and by hygienic expedients conquered.

It is right to state that the real is greater than the apparent mortality of these districts; they are increasing, and contain an undue proportion of population at the younger healthiest ages, so that a correction for this makes the mortality *20* instead of *17*. That is the rate of their stationary mortality if the population were stationary, if births equalled deaths, and there were no migration.

The mean annual deaths at the rate of 22.4 in the ten years 1861–70 were 479,450 in England; and had the rate of mortality been 17 the annual deaths would not have exceeded 363,617; so the overplus due to the operation of causes existing, but less destructive in the healthier districts was 115,833. The hope of saving any number of these 115,833 lives annually by hygienic measures, is enough to fire the ambition of every good man who believes in human progress.

✳   ✳   ✳

... It will be interesting now to give a few illustrative cases of the changes in the marriage, birth, and death rates (1) where a new industrial enterprise has been suddenly developed, and (2) where a branch of industry is declining. As instances of the former, take the districts of Ulverston, Guisbrough, and Stockton; of the latter St. Austell and Redruth in Cornwall, where the works and the population decline. Ulverston contains Barrow-on-Furness. It owes the increase to a cause thus referred to in the Quarterly Return of the Registrar-General:

'The mortality often augments with the increased prosperity of a district; and this is curiously illustrated by Ulverston, a romantic district extending from Morecambe Bay to Lake Windermere. Ulverston, in the years 1841–50, was one of the healthiest districts of England; the mortality did not exceed 18 in 1000. A change took place, and in the ten years 1851–60 the mortality rose to 20 in 1000. The deaths in the June quarter of 1864 were considerably above the average of previous years, caused, says the registrar of Dalton, 'in part by the increase of the population, and in part by the prevalence of scarlatina and measles.' He adds: 'but there is no distress; work is plentiful, wages good, and provisions cheap. Labourers are earning 3s 6d a day; artisans 4s 3d and upwards.'

The population of many of the townships and parishes of the Ulverston district, at the feet of its fells, and round the shores of its meres, is stationary, and in some instances has declined; it is an old iron district, which had seen its works decay when coal came into use for smelting, but of late a pure haematite ore has been discovered in the carboniferous limestone of Dalton-in-Furness, for which there is a great demand. The population of the parish rose from 4683 to 9152 in the interval of the two last censuses, and, with the parishes in its vicinity, gave the increase which raised the population of Ulverston district from 30,556 in 1851 to 35,738 in 1861.

The mortality of the district of Ulverston, exclusive of Dalton, in the two last quarters, was at the rate of 26 and 23 in 1000; while that of Dalton was at the rate of 42 and 31; and it is in this sub-district that the spectacle is presented of "work plentiful, wages good, provisions cheap," and "the prevalence of destructive epidemics." This coincidence is reproduced over and over again. And it must not be supposed on that account that work, good wages, and cheap provisions are in themselves bad things; for they are as salutary as they are attractive to the masses of mankind. But our industrial armies are cut down by the camp diseases which are generated by the inadequate house accommodation, and by the want of sanitary arrangements, which are never carried out in the neighbourhood of new works. Impure water, impure air, their own exhalations, kill men, women, and children on the spot, and breed the leaven which devastates the towns and valleys in the vicinity. For the sins of a parish are often visited on its neighbours in thousands round. Thus

South Wales has been rendered prosperous by the mines, and unhealthy by the negligence of the people. The mining population appears to be even less careful of life than the manufacturing population.'

The increase of population went on; for the excess of births exceeded the excess of deaths.

This timely warning was not lost on the energetic authorities of Barrow-on-Furness; and though the mortality in the three decenniads that ended in 1870 increased and was 18, 20, 21, it has gone no further; sanitary measures have been undertaken and are still proceeding. The marriage-rate and the birth-rate, evoked to a higher pitch by the prosperity of the place, rose more rapidly than the death-rate. In cases of this kind of rapid concentration of population the high birth-rate is not the cause of the high death-rate; the first is caused by the prosperity, the second by the defective sanitary provisions. All the rates had been low in the first healthiest decenniad, quite in conformity with the general law.

Guisbrough and Stockton, including Middlesborough, exhibit a similar series of phenomena; the population became thicker and the mortality increased; the marriages and births also increased.

With a declining copper and tin industry in St. Austell and in Redruth, Cornwall, the mortality slightly declined; but to nothing like the same extent as the marriage and the birth-rates.

*   *   *

March of an English Generation through Life

It is possible, by means of Tables so constructed, to follow any large number of people through the whole of their ages, and to point out the casualties under which they will probably succumb.

Age 0–5. The first thing to observe is, that the fatality children encounter is primarily due to the changes in themselves. Thus 1,000,000 children just born are alive, but some of them have been born prematurely; they are feeble; they are unfinished; the molecules and fibres of brain, muscle, bone are loosely strung together; the heart and the blood, on which life depends, have undergone a complete revolution; the lungs are only just called into play. The baby is helpless; for his food and all his wants he depends on others. It is not surprising then that a certain number of infants should die; but in England the actual deaths in the first year of age are 149,493, including premature births, deaths by debility and atrophy; diseases of the nervous system 30,637, and of the respiratory organs 21,995. To convulsions, diarrhoea, pneumonia, bronchitis, chiefly their deaths are ascribed; little is positively known; and this implies little more than that the brain and spinal marrow, nerves, muscles, lungs, and bowels fail to execute their functions with the exact rhythm of life. The first two are said by pathologists to be often rather symptoms of diseases unknown than diseases in themselves. The total dying by miasmatic diseases is 31,266; but it is quite possible that several of the children dying of convulsions die in the early stages of some unrevealed zymotic disease, whose symptoms have not had time for development. Convulsion is a frequent precursor in children of measles, whooping cough, scarlet fever, fever; indeed Dr. C. B. Radcliffe well remarks 'in the fevers of infancy and early childhood, especially in the exanthematous forms of these disorders, convulsion not unfrequently takes the place occupied by rigor in the fevers of youth and riper years.' Many of the cases of pneumonia may also in like manner be whooping-coughs and other latent zymotic diseases. In the second year of life pneumonia, bronchitis, and convulsions are still the prevalent, and most fatal diseases; many also die then of measles, whooping-cough, scarlatina, and diarrhoea. Scarlet fever asserts its supremacy in the second, third, fourth, and fifth years of age. Whooping-cough is at its maximum in the first year, measles in the second, scarlatina in the third and fourth years. Thus these diseases take up their attacks on life in succession and follow it onwards.

The deaths from all causes under the age of five years are 263,182. The number ascribed to infanticide is very few; but the deaths by suffocation (overlaying, etc.) are more numerous; and so are the deaths directly referred to the 'want of breast-milk.' The total deaths by burns, injuries, drowning, and all other kinds of violence are 5175.

By a physiological law 511,745 boys are born in England to 488,255 girls; and by another law 141,387 boys and 121,795 girls die in the first five years of life; so that at the end of five years the original disparity in the numbers of the two sexes is so much reduced that at the age of five years the boys only slightly exceed the girls in number. The greater mortality of boys is due to difference of organisation, for the external conditions are substantially the same in which boys and girls are placed.

Great as is the influence of organisation itself, the difference of external circumstances and sanitary condition exercise a very real influence on life, disease, and death in childhood.

Thus, even in the Healthy districts of the country, out of 1,000,000 born, 175,410 children die in the first five years of life; but in Liverpool

District, which serves to represent the most unfavourable sanitary conditions, out of the same number born, 460,370, nearly half the number born, die in the five years following their birth. This is 284,960 in excess of the deaths in the Healthy Districts.

Out of 1,000,000 Children born alive (1) in the Healthy Districts, (2) in all England, and (3) in the District of Liverpool, the Numbers dying under Five Years of Age by Nineteen Groups of Causes

	Healthy Districts	England	Liverpool District
Deaths from all Causes	175,410	263,182	460,370
Total Zymotic Diseases	49,761	87,099	171,009
of which			
Small-pox	602	3,331	5,175
Measles	5,257	11,507	25,514
Scarlatina	11,373	17,959	26,818
Diphtheria	4,194	2,425	3,395
Whooping-cough	9,650	14,424	32,551
Typhus (with Enteric and Common			
Fever)	2,807	5,401	9,297
Diarrhoea, Dysentery	9,354	20,344	51,911
Cholera	399	1,129	4,255
Other Zymotic Diseases	6,135	10,579	12,093
Cancer	110	71	62
Scrofula and Tabes	5,335	8,115	11,694
Phthisis	2,650	4,469	5,116
Hydrocephalus	6,604	9,296	14,972
Diseases of the Brain	22,692	40,065	49,840
Diseases of the Heart, and Dropsy	1,304	1,507	2,038
Diseases of the Lungs	27,884	41,476	79,893
Diseases of the Stomach and			
Liver	4,431	4,778	4,874
Violent Deaths	4,232	5,175	17,107
Other causes	50,401	61,131	103,765

The above Table shows how many children die from the several groups of causes (1) in the healthy districts, (2) in all England, and (3) in the Liverpool District. There is a greater increase in Liverpool from smallpox and measles than from scarlet fever; and diphtheria was more fatal in the healthy districts than in all England. Diarrhoea and cholera were greatly aggravated in the other districts of England; so were whooping-cough, and typhus, under which were registered typhus, typhoid, infantile remittent, and relapsing fever. The diseases of the lungs were more fatal to children in Liverpool than diseases of the brain.

The children of Norway fare better than the children of sunny Italy; to which it may well be still an *officina gentium*. Out of 100 children born alive the deaths in the first five years of life are in Norway 17, Denmark 20, Sweden 20, England 26, Belgium 27, France 29, Prussia 32, Holland 33, Austria 36, Spain 36, Russia 38, Italy 39. Russia is almost as fatal to her children as Italy.

In a paper read before the Statistical Society the methods of determining the rates of mortality were described, and I collected information as to the treatment and management of children in Scotland, Norway, Sweden, France, and Austria. The subject was taken up in England by the Obstetrical Society, who published an able report, based on returns, on the birth and treatment of English children. I have not yet received papers from Russia or Italy.

The mortality of infants evidently depends, to some extent, on the midwifery of a country; on the way the children are fed by the mothers; on the water; and on the cleanliness observed, as well as the other sanitary conditions.

Age 5–10. Our young travellers now enter on their *sixth year* of life. They have left great numbers on the way. And nearly every one of the 736,818 survivors has been attacked by one disease or another; some by several diseases in succession. There is one fact in their favour: the majority of the zymotic diseases rarely recur. Each renders the body insusceptible of injury from diseases of its own kind, though not from other diseases. Medicine is still without any accurate determination of the numbers attacked to every death, but it is evident from the deaths that some hundreds of thousands of the survivors have had whooping-cough, measles, scarlet fever. Taking advantage of the non-recurrent law, Jenner, by his immortal discovery, substituted small-pox modified and mild for natural small-pox; and it is probable that the greater part of all the children at the age of five are vaccinated, or have had small-pox.

So the total deaths in the five years following are 34,309; 8743 of them from scarlatina, the principal plague of this age, 1364 from diphtheria, 4036 from fever: more than half of the deaths in this young age, are from miasmatic disease, in all 19,256. The brain and lung diseases levy also a certain tribute.

Age 10–15. But 702,509 survive and enter on this age, which culminates in puberty; and 684,563 pass through it into the next at the age of

29 "E-532 Documents"

15; for the deaths are fewer than at any other age. They amount to 17,946, of which 1901 are by scarlatina, 2842 by fever, 3526 by phthisis, these last two diseases already standing as the most deadly; in this period the change in girls is greater than the change in boys, and rather more of them die.

Age 15–20. Now the mortality increases especially among women, of whom 5263 die of consumption (phthisis), and 244 of childbirth, for at this age a few young girls marry with some risk to their lives. The tight ligatures that are so often and so unwisely placed round the waist interfere with respiration; and may, with their in-door-life, favour the development of phthisis. The deaths of males by consumption are 3811, by fever 1368: the deaths from both diseases being fewer than the deaths of females from the same causes. The violent deaths of 1387 males, against 193 females, go a long way towards redressing the inequality.

Melancholy suicide appears now among the causes of death; indeed 14 such deaths appear before the age of 15, but the numbers in this age amount to 94, of whom 46 are males, 48 are females. Insanity looms on the horizon, and there is an excess of fatal brain affections over affections either of the heart or lungs.

Age 20–25. At this age large numbers marry. The deaths are 28,705, of which nearly half, or no less than 13,785, are by phthisis. Fever is associated with it, as the great prevailing zymotic disease; the reign of the other zymoses of the young is almost over. The brain, heart, and lung begin again to suffer, and of their diseases more die. 1100 women die in childbirth.

If it is the age of love it is also the age of war, of dangerous work, and of crime: the violent deaths, exclusive of suicides, are 1677, without reckoning any death in foreign war.

Age 25–35. Of the million, 634,045 attain the age of 25, and 571,993 live to the age of 35. The period extends over double the time hitherto handled. It is the athletic, the poetic age: it is the prime of life: two thirds of the women are married; and now at its close is the mean of the period (33–34) when husbands become fathers, wives become mothers, the new generation is put forth. The deaths are separations; they leave widows and fatherless children behind. Of the 62,052 that die, 30,592 are men, 31,460 women; 2992 of the men, and only 309 of the women die by violence, suicide excepted; but 2901 women die in childbirth.

Consumption is the most fatal disease of the age; it is the cause of 27,134 deaths; women suffering more than men. Fever is fatal to fewer lives than it was earlier; but it is by far the most fatal of the zymotic diseases, and slays its 4197.

The local diseases of lungs, heart, and brain grow intenser in this period. We may now look back to the fate of a generation exposed to unfavourable conditions; such, for instance, as prevailed in the Liverpool District. There, of a million born, less than half, only 434,497, live to the age of 25; then 74,153 die in the ten years, leaving 360,344 alive at the age of 35. No less than 10,657 die of fever, 333 of suicide, 4850 of other violent deaths. The local diseases are exceedingly fatal; 1938 mothers die in childbirth.

In happier sanitary conditions 727,552 live to the age of 25, and 667,940 survive the age of 35. Only 3116 die of fever, 396 of suicide, and 2819 of other violent deaths.

Age 35-45. The losses are of 69,078 lives; 35,142 men, 33,936 women. The athletic age is now over; but the combined faculties of muscular and nervous energy are at their height. Women have borne more than half of their children, now they bear the rest. It is the age of fathers and mothers; criminality declines. Many of the structures now give way. Phthisis still predominates; fever snatches still its many victims; and the brain, heart, lungs, and bowels become more and more the seats of destructive disease. 564 persons commit suicide; 3280 die violent deaths, 2907 of them men, and 373 women; 2516 mothers die in childbirth.

While the deaths by fever are 3777 out of 571,993 attaining this age in England; 14,322 people die of it in the Liverpool District out of 360,344. The lung diseases in the two sets of conditions are fatal to 7452 and 13,967 lives.

In the healthy districts the deaths by fever are 2702, by diseases of the lungs 5261.

Age 45-55. This age is the middle arch of life: *nel mezzo del cammin di nostra vita,* for the million are reduced to half a million lives, a few months after the age of 45. They have produced the succeeding generation. The age of fertility is now nearly over in women; but a few lingerers bear children, and in the act 160 die. The deaths by all causes are 81,800; by fever 3749; diarrhoea, dysentery, and cholera 1944; by phthisis 16,468; by lung diseases 13,203, heart diseases and dropsy 10,041, brain diseases 9313, bowel and liver diseases 7917. The centres of life are sources of death. At this age, in their wretchedness, and in their weakness, 599 men, 204 women, in all 803, appeal rashly to the 'arbitrator of despairs, just death.' 2876 men, 478 women, in all 3354 persons, die violent deaths.

29*

Cancer, a formidable and dread disease that began to be fatal before, now destroys 4583 lives, 1140 men, 3443 women.

In unfavourable sanitary conditions out of a million lives in Liverpool, only 275,193 attain the age of 45, and 90,969 die in the following ten years; 12,504 of fever, 13,274 of phthisis, 24,417 of pulmonary diseases, 420 of suicide, and 4314 of other violent deaths.

In the Healthy districts of the country 606,019 attain the age of 45, and 71,938 die in the ten years following; 2306 by fever, 13,745 by phthisis, 10,012 by brain diseases, 10,451 by heart diseases and dropsy, 8234 by pulmonary diseases, 1022 by suicide, and 3030 by other violent deaths.

Age 55 and upwards. To the age of 55, near the middle of the possible lifetime of humanity in its present state, 421,115 attain; and from this point of time it is possible to look ahead, and discover the particular rocks, foes, collisions, tempests to be encountered, to be dreaded, or to be weathered by the fleet on its way to the utmost butt of existence, 'the very seamark of its journey's end.'

One thing to remark is, that the rate, the degree of danger, which has hitherto increased slowly, now increases at so much faster a pace, that although the number of lives grows less, the number of deaths increases in every one of the next 20 years, and is afterwards sustained for 10 years longer, until at last in the distance the living all sink into the elements from which they came.

Few will die of the non-recurrent zymotic diseases; some 29,803 in all will die of fever, diarrhoea, cholera, rheumatism, and other zymotic diseases; cancer will carry off almost as many as phthisis; the greatest mortality, however, will be experienced from diseases of the lungs; then will follow in their wake of destruction brain affections, apoplexies, paralysies; heart and artery diseases, often the remote cause of the other maladies, bowel and liver diseases, kidney diseases; many will still die violent deaths as they are less able to resist injuries than younger lives. Gout and intemperance will reap their later harvest; so will mortification, atrophy, debility, and the infirmities of old age: then comes the end.

Of 100 women living of the age of 55 and upwards, it is worthy of note that 11 are spinsters, 43 widows, and 46 wives; of 100 men 9 are bachelors, 24 widowers, 67 husbands. We now pass to the particular decenniads of life.

Age 55–65. The number of males and females surviving becomes equal at the age of 53, but at and after 55 the women exceed the men in number, as their mortality-rate is lower ever after the age of 39.

While 421,115 of both sexes enter this stage of life, 309,029 live on to the next; 112,086 die; only 9795 of fever, diarrhoea, dysentery, cholera, rheumatism, and other zymotic diseases. Cancer kills 5998 persons, 4035 of these women; consumption 10,445: the diseases most to be dreaded, and guarded against, especially by men, are affections of the lungs, and heart, of which 23,659 and 17,081 persons die: diseases of the brain are fatal to 15,678, of the stomach, intestines, and liver are fatal to 11,400. The vigour of life is somewhat subsiding; family cares perhaps accumulate, ambition is disappointed, and the mind sometimes gives way organically: the tendency to suicide is greatest at this age, and the greatest number of lives, 826, come to that melancholy close in this period. But 3155 are killed by violent deaths of various kinds; 2560 men, 595 women.

Age 65–75. 309,029 enter this age, and 161,124 leave it alive: 67 years is near the mean date at which their children give birth to their grand-children; the third generation. The 147,905 dying in this period succumb to the same classes of disease as were fatal in the previous decenniad; and still more succumb to lesions of the brain and heart, and lungs: the kidneys give way, but are never so fatal as affections of the higher organs. 11,256 of the deaths are from fever, diarrhoea, cholera, rheumatism, and other diseases of the miasmatic order; 9789 from five constitutional diseases; 92,391 from diseases of brain, heart, lungs, and other local organs; 3064 from violence, and there remain 31,405 referable in great part to a new head in the developmental class, *old age.*

The year of age 72 is that in which the greatest number of *men* die; and which may have led the psalmist to say, the days of the years of our life are three score years and ten; but these are 'days passed away in thy wrath,' in violation of the divine laws, and therefore are not necessarily the limit of healthy existence where the laws of life are observed.

Age 75–85. The numbers that enter this decenniad are 161,124, and the numbers that leave it alive are 38,565. More than half the numbers living have been married and are widowed. The 122,559 that die of recognised diseases at this age die chiefly of lung, brain, heart, and other local diseases; against such dangers they have to guard themselves. The number of such deaths is 51,838; then some 7229 persons die of miasmatic diseases; 131 of suicide, 1691 of violence. The cold weather is their great foe. But there remain 58,905 dying chiefly of atrophy, debility, and *old age.*

Age 85—to the end. The 38,565 aged pilgrims are no longer what they were; their strength is fading away, and they succumb to slight injuries, to cold, heat, want, or attacks which in their early years would have been shaken off: only 2153 live to the age of 95; and 223 to 100; finally by this Table at the age of 108 one solitary life dies.

*　　*　　*

Life has a pecuniary value. In its production and education a certain amount of capital is sunk for a longer or shorter time; and that capital, with its interest, as a general rule, reappears in the wages of the labourer, the pay of the officer, and the income of the professional man. At first it is all expenditure, and a certain necessary expenditure, goes on to the end to keep life in being, even when its economic results are negative.

The value of any class of lives is determined by valuing first at birth, or at any age, the cost of future maintenance; and then the value of the future earnings. Thus proceeding, I found the value of a Norfolk agricultural labourer to be 246*l*, at the age of 25: the child is by this method worth only 5*l* at birth, 56*l* at the age of 5; 117*l* at the age of 10; the youth 191*l* at the age of 15; the young man 234*l* at the age of 20; the man 246*l* at the age of 25, and 241*l* at the age of 30, when the value goes on declining to 138*l* at the age of 55, and only 1*l* at the age of 70; the cost of maintenance afterwards exceeding the earnings, the value becomes negative; at 80 the value of the cost of maintenance exceeds the value of the earnings by 41*l**. These values may be compared with the former cost of slaves in Rome, in the United States, and in the West Indies.

The amount of capital sunk in the education of professional men is not only greater, but it is probably at greater risk, and it has to remain longer under investment before it is returned. The maximum value of such a man is attained later in life, probably 40; and in the highest orders of the church, law, and politics, where experience and great weight of character are requisite, the life still increases in value at higher ages.

---

* See paper on Income Tax, by W. Farr, in Journal of Statistical Society, 1853. The return of wages was procured and carefully compiled by Sir John Kay Shuttleworth. It was from the best class of labourers in Norfolk.— Journal, vol. xvi. p. 43. Wages have since risen, and so has the cost of subsistence.

7 'We are all socialists now', 1890

(The Prime Minister, Lord Salisbury, in Lord's Debates, 19 May 1890)

Undoubtedly we have come upon an age of the world when the action of industrial causes, the great accumulation of population and many other social and economic influences have produced great centres of misery, and have added terribly to the catalogue of the evils to which flesh is heir. It is our duty to do all we can to find the remedies for those evils, and even if we are called Socialists in attempting to do it, we shall be reconciled if we can find those remedies, knowing that we are undertaking no new principle, that we are striking out no new path, but are pursuing the long and healthy tradition of English legislation.

8 The idea of town planning, 1908

(John S. Nettlefold, *Practical housing* (Letchworth: Garden City Press 1908), pp. 46–7)

Most town dwellers have to earn their living in a factory or an office, where light and air is comparatively restricted, and they cannot lead such a life with the greatest advantage to themselves and to the community, unless they have means within easy reach for rest and recreation in fresh air, as well as for work.

The most insular of Englishmen will scarcely attempt to maintain that opportunities of this nature are in any sense adequately provided in our towns. Millions of English town children have no playground within practical reach except the streets. The young men find it extremely difficult to obtain suitable cricket and football fields. Miss Elliston's expression, 'Ain't nowheres to play', exactly describes the circumstances of an overwhelming proportion of our rising generation, on whom depends England's future strength and prosperity.

Each year makes it harder for men to get allotments on which they can not only get rational enjoyment, but also materially increase the family food supply. The women have no place to go out to where they can enjoy an odd hour, and often find no better choice than the front-door step, or the nearest public-house.

We cannot by legislation make people healthy and happy, but we can give our town dwellers fewer temptations to irrational excitement, and more opportunities for beneficial enjoyment than they have at present.

We can, if we will, let light and air into our towns; we can, if we will, make the most and not the least of the sunshine.

Far from assisting rational town development, modern means of communication are at present only spoiling the beautiful and life-giving country districts, whereas they might be used to the great advantage of the whole nation, if only the rapid development and extension of our towns were carried out on a coherent and cohesive plan.

This unhappy state of affairs has drawn attention to our towns, and many clear-sighted cautious men are thinking that the remedy lies in giving Local Authorities comprehensive control over the development of the districts under their administration, with power to encourage and direct the numerous private agencies engaged in house-building rather than in encouraging them to undertake municipal house-building schemes, which can at the best only assist the very small proportion of the population directly provided for.

It has been already shown that existing legislation has resulted in a few good houses and a great many bad ones. It has discouraged instead of encouraged the provision of open spaces and playgrounds for the people, as part of the ordinary business of estate-development and house-building. No housing enterprise is generally useful unless it is carried out on sound business lines, which all and sundry can copy if they will.

In other countries the importance of town development has long been recognised, and some of their cleverest men have for many years past been engaged on the solution of the hundred and one difficult and complicated problems connected with such extension. Here in England, we have only just begun to realise the necessity for attending to the question at all, and yet there is none of greater national importance.

It has been thought by some that 'Back to the Land' would solve all our difficulties, and with the underlying idea of this popular cry I confess considerable agreement; but as practical men we must realise the conveniences for production and distribution afforded by our towns, the concentration of population caused by the introduction of machinery, and the fact that man is a gregarious animal. We must seek some way by which to bring the country to the town, and the town to the country.

It is with this idea in their minds that those most intimately connected with the practical work of housing reform in England, are so earnestly and strenuously advocating the policy of Town-planning.

9 Unemployment insurance, 1920

(*Parliamentary Debates, House of Commons*, 5th series, CXXV (1920), cols. 1739–44, 25 February 1920)

(*Sir R. Horne*) The problem of unemployment is one of the most difficult questions which confront the country, and I think I may also say one of the most important. Unemployment forms a tragedy in the life of the workman, and at all times it fills his mind with anxiety. Those who have had experience of distress can furnish us with many bitter recollections. The fear of unemployment is a cause of dispeace and disquietude. If we could once get rid of the dread of unemployment which affects the minds of the working men of this country I am sure we should be able to create a new spirit of harmony and happiness. The fear of unemployment is a great impediment to the country's prosperity. There are masses of people in the country to-day who believe that the more they produce the less will be wanted, and conversely that the less they do the more employment will be afforded. That belief, as most of us know, is erroneous; but so long as the fear of unemployment exists you will never succeed in completely eradicating it, and nothing would tend so much to increase output in the country, which is so urgently needed, as the creation of an assurance in the minds of the workpeople of this country that they would be kept free and would not drop into the lurch when the day of unemployment comes.

There are certain people who believe it is possible to devise a system in which there would be no unemployment. I wish I could hold that view, but I confess I cannot share their optimism. I cannot forget that a bad season, a blight in the crop of one of our great raw materials, may cause dislocation in the trade of any country which no provision you can make beforehand can cope with. Coming nearer home, even a strike in one part of the country may throw out of employment the people who depend upon the raw material produced by those who are on strike, and we have recently had an example of that fact. Looking at the whole matter in a prudent and tactful light, the only thing you can do is to anticipate unemployment and make provision against it. What provision can we make? Insurance is one of the best known expedients, and it is upon the road of insurance that this country has already progressed.

I hope the House will allow me for a moment or two to review the steps which have been taken, and the considerations which have been taken into account with regard to this great matter. Prior to the year

1911 there was no compulsory insurance in this country. You had a certain amount of voluntary insurance which had been created by the trade unions, but that only covers a very small portion of the working people of the country, and in many cases was given only upon special occasions like the breakdown of machinery or some other such dislocation in the work of a factory. In the early years of this century people who had been thinking about these matters had given great attention to the possibilities of State assurance. There were three general considerations which I think had been arrived at before the Act of 1911 was introduced. These three considerations were as follows: It appeared to people who had studied the problem, in the first place, that you must have compulsory insurance, and if you do not have it compulsory, the result would be that only a great majority of the bad risks of the country would come in your scheme and that would ruin any scheme. In the second place, it appeared necessary that benefits should be given in proportion or at least in some degree of proportion to the amount of contribution. It is perfectly obvious, if you do not do that, that you are taking the risk of the character of the individual workman, and that is a risk which in the question of unemployment no scheme which could be set up could really work against. In the third place, it seemed necessary to set up a system of Employment Exchanges. That was obviously necessary for the reason that you had to have a system by which you could check the workman who said that employment was not available. Accordingly, all thinking people who had been working on the subject had arrived at these three more or less definite views before 1911 as the foundation of any scheme. The 1911 Act was passed, and it was to some extent experimental. It dealt only with a selected number of the industries of the country. They were the building, the shipbuilding, and the engineering industries, taking these as large generic terms. There were good reasons for taking these industries to begin with. One was that they were exposed to some of the worst risks of unemployment. Secondly, because they had had that experience in the past they had adopted a certain amount of voluntary insurance in their Trade Unions which afforded data upon which you could build your scheme. It came into force in July, 1912, and the benefit began as from January, 1913. Each man had to pay a weekly contribution of $2\frac{1}{2}d$, the employer also paid a contribution of $2\frac{1}{2}d$ per week, and the State added one-third of the joint contributions of the employer and workman. The benefit which was given was $7s$ per week, and the same rate was applied to women as was applied to men. The rules which were laid down for working the scheme were, in general, based upon the three considerations which I have set forth. A man, in order to be qualified to receive the Unemployment Benefit, had to show that he was unemployed and that he had paid a certain number of contributions. Further, he was disqualified if he had left his work on his own accord, if he was on strike, or if he had refused suitable employment when offered him by the Employment Exchanges. It was not left entirely to officials to determine whether the refusal of Unemployment Benefit was justified or not. He was enabled to appeal to a Court of Referees which was set up under the statute, and difficult questions on which parties disagreed were determined by an umpire as a final authority. That scheme of insurance covered 2,500,000 employees of whom roughly 9,000 were women. . . .

. . . During the War a very large number of people came into industry, particularly women, who had never been in the occupations which they then adopted at any previous stage, and in considering the problems of reconstruction it was borne in upon everybody that there would be a large amount of unemployment immediately the War ceased, for which some provision ought to be made. Accordingly, in 1916, the unemployment insurance scheme was extended to a large number of other trades, particularly to munition workers and people working in the metal and chemical industries. That brought up the number of people insured to 4,000,000, and that was the figure at which unemployment insurance stood at the date of the Armistice. The number has slightly decreased since then, owing to the fact that a considerable number of people have gone right out of manual labour since then. Many women have given up the work that they previously did, and there has been a considerable shift in the working population. The result is that at the present time, giving the figures roughly, 3,750,000 are insured under the 1911 and 1916 Acts, of whom 1,000,000 are women. . . .

We now come down to the proposals which I am laying before the House. It has become obvious to everybody, I think, that insurance against unemployment must be much more widely extended than it has been in the past. The idea was to insure everybody against unemployment. As I have stated, there were only 3,750,000 out of all the workpeople of this country under the two previous insurance schemes. The proposals that we now make will bring the number insured up to about 12,000,000 of workpeople. That is making a very considerable advance upon what has been previously done. There are excepted from the scheme people engaged in agriculture and in domestic service.

Of course, the reason for these exemptions really is that it is anticipated that protests would be made against the inclusion of either of these classes, because unemployment in them has always been at a much lower figure than in the other industries of the country, and in making this advance we do not desire to raise any more complications than we can help. It seems to me that agricultural workers and domestic servants would be anxious to avoid the contributions in view of their greater immunity from the risks against which we are trying to guard. But we have provided in the Bill that we may by Order extend the benefits to any industry which desires them and which Parliament may regard as being entitled to be included. Accordingly, the machinery is there at any time to include people engaged in agriculture and domestic service if that is thought wise and proper.

10 Beveridge's plan for social security, 1942

(Sir William H. Beveridge, Social insurance and allied services, *B.P.P.*, 1942–3, VI, pp. 120–2)

300 *Scope of Social Security*: The term 'social security' is used here to denote the securing of an income to take the place of earnings when they are interrupted by unemployment, sickness or accident, to provide for retirement through age, to provide against loss of support by the death of another person, and to meet exceptional expenditures, such as those connected with birth, death and marriage. Primarily social security means security of income up to a minimum, but the provision of an income should be associated with treatment designed to bring the interruption of earnings to an end as soon as possible.

301 *Three Assumptions*: No satisfactory scheme of social security can be devised except on the following assumptions:

(A) Children's allowances for children up to the age of 15 or if in full-time education up to the age of 16;
(B) Comprehensive health and re-habilitation services for prevention and cure of disease and restoration of capacity for work, available to all members of the community;
(C) Maintenance of employment, that is to say avoidance of mass unemployment.

The grounds for making these three assumptions, the methods of satisfying them and their relation to the social security scheme are discussed in Part VI. Children's allowances will be added to all the insurance benefits and pensions described below in paras. 320-349.

302 *Three Methods of Security:* On these three assumptions, a Plan for Social Security is outlined below, combining three distinct methods: social insurance for basic needs; national assistance for special cases; voluntary insurance for additions to the basic provision. Social insurance means the providing of cash payments conditional upon compulsory contributions previously made by, or on behalf of, the insured persons, irrespective of the resources of the individual at the time of the claim. Social insurance is much the most important of the three methods and is proposed here in a form as comprehensive as possible. But while social insurance can, and should, be the main instrument for guaranteeing income security, it cannot be the only one. It needs to be supplemented both by national assistance and by voluntary insurance. National assistance means the giving of cash payments conditional upon proved need at the time of the claim, irrespective of previous contributions but adjusted by consideration of individual circumstances and paid from the national exchequer. Assistance is an indispensable supplement to social insurance, however the scope of the latter may be widened. In addition to both of these there is place for voluntary insurance. Social insurance and national assistance organised by the State are designed to guarantee, on condition of service, a basic income for subsistence. The actual incomes and by consequence the normal standards of expenditure of different sections of the population differ greatly. Making provision for these higher standards is primarily the function of the individual, that is to say; it is a matter for free choice and voluntary insurance. But the State should make sure that its measures leave room and encouragement for such voluntary insurance. The social insurance scheme is the greater part of the Plan for Social Security and its description occupies most of this Part of the Report. But the plan includes national assistance and voluntary insurance as well.

303 *Six Principles of Social Insurance:* The social insurance scheme set out below as the chief method of social security embodies six fundamental principles:

Flat rate of subsistence benefit
Flat rate of contribution
Unification of administrative responsibility
Adequacy of benefit
Comprehensiveness
Classification

304 *Flat Rate of Subsistence Benefit:* The first fundamental principle of the social insurance scheme is provision of a flat rate of insurance benefit, irrespective or the amount of the earnings which have been interrupted by unemployment or disability or ended by retirement; exception is made only where prolonged disability has resulted from an industrial accident or disease. This principle follows from the recognition of the place and importance of voluntary insurance in social security and distinguishes the scheme proposed for Britain from the security schemes of Germany, the Soviet Union, the United States and most other countries with the exception of New Zealand. The flat rate is the same for all the principal forms of cessation of earning—unemployment, disability, retirement; for maternity and for widowhood there is a temporary benefit at a higher rate.

305 *Flat Rate of Contribution:* The second fundamental principle of the scheme is that the compulsory contribution required of each insured person or his employer is at a flat rate, irrespective of his means. All insured persons, rich or poor, will pay the same contributions for the same security; those with larger means will pay more only to the extent that as tax-payers they pay more to the National Exchequer and so to the State share of the Social Insurance Fund. This feature distinguishes the scheme proposed for Britain from the scheme recently established in New Zealand under which the contributions are graduated by income, and are in effect an income-tax assigned to a particular service. Subject moreover to one exception, the contribution will be the same irrespective of the assumed degree of risk affecting particular individuals or forms of employment. The exception is the raising of a proportion of the special cost of benefits and pensions for industrial disability in occupations of high risk by a levy on employers proportionate to risk and pay-roll (paras. 86–90 and 360).

306 *Unification of Administrative Responsibility:* The third fundamental principle is unification of administrative responsibility in the interests of efficiency and economy. For each insured person there will be a single weekly contribution, in respect of all his benefits. There will be in each locality a Security Office able to deal with claims of every kind and all sides of security. The methods of paying different kinds of cash benefit will be different and will take account of the circumstances of insured persons, providing for payment at the home or elsewhere, as is necessary. All contributions will be paid into a single Social Insurance Fund and all benefits and other insurance payments will be paid from that fund.

307 *Adequacy of Benefit:* The fourth fundamental principle is adequacy of benefit in amount and in time. The flat rate of benefit proposed is intended in itself to be sufficient without further resources to provide the minimum income needed for subsistence in all normal cases. It gives room and a basis for additional voluntary provision, but does not assume that in any case. The benefits are adequate also in time, that is to say except for contingencies of a temporary nature, they will continue indefinitely without means test, so long as the need continues, though subject to any change of conditions and treatment required by prolongation of the interruption in earning and occupation.

308 *Comprehensiveness:* The fifth fundamental principle is that social insurance should be comprehensive, in respect both of the persons covered and of their needs. It should not leave either to national assistance or to voluntary insurance any risk so general or so uniform that social insurance can be justified. For national assistance involves a means test which may discourage voluntary insurance or personal saving. And voluntary insurance can never be sure of covering the ground. For any need moreover which, like direct funeral expenses, is so general and so uniform as to be a fit subject for insurance by compulsion, social insurance is much cheaper to administer than voluntary insurance.

309 *Classification:* The sixth fundamental principle is that social insurance, while unified and comprehensive, must take account of the different ways of life of different sections of the community; of those dependent on earnings by employment under contract of service, of those earning in other ways, of those rendering vital unpaid service as housewives, of those not yet of age to earn and of those past earning. The term 'classification' is used here to denote adjustment of insurance to the differing circumstances of each of these classes and to many varieties of need and circumstance within each insurance class. But the insurance classes are not economic or social classes in the ordinary sense; the insurance scheme is one for all citizens irrespective of their means.

11—And its fulfilment?

(National Insurance Act, 1946, 9 and 10 Geo. VI, c. 67)

An Act to establish an extended system of national insurance providing pecuniary payments by way of unemployment benefit, sickness benefit, maternity benefit, retirement pension, widows' benefit, guardian's allowance and death grant, to repeal or amend the existing enactments relating to unemployment insurance, national health insurance, widows', orphans' and old age contributory pensions and non-contributory old age pensions, to provide for the making of payments towards the cost of a national health service, and for purposes connected with the matters aforesaid.

(1st August 1946)

Be it enacted by the King's most Excellent Majesty, by and with the advice and consent of the Lords Spiritual and Temporal, and Commons, in this present Parliament assembled, and by the authority of the same, as follows:

PART I

Insured Persons and Contributions

1—(I) Subject to the provisions of this Act, every person who on or after the appointed day, being over school leaving age and under pensionable age, is in Great Britain, and fulfils such conditions as may be prescribed as to residence in Great Britain, shall become insured under this Act and thereafter continue throughout his life to be so insured.

(2) For the purposes of this Act, insured persons shall be divided into the following three classes:

(*a*) employed persons, that is to say persons gainfully occupied in employment in Great Britain, being employment under a contract of service;

(*b*) self-employed persons, that is to say persons gainfully occupied in employment in Great Britain who are not employed persons;

(*c*) non-employed persons, that is to say persons who are not employed or self-employed persons.

*   *   *

10—(I) Benefit shall be of the following descriptions:

(*a*) unemployment benefit;

(*b*) sickness benefit;

(*c*) maternity benefit, which shall include maternity grant, attendance allowance and maternity allowance;

(*d*) widow's benefit, which shall include widow's allowance, widowed mother's allowance and widow's pension;

(*e*) guardian's allowance;

(*f*) retirement pension;

(*g*) death grant.

12 New towns and the housing boom since 1945

(Report of the Ministry of Housing and Local Government, 1963, *B.P.P.*, 1963–64, XV, pp. 13, 40)

Since the end of the [1939–45] war, more than $4\frac{1}{4}$ million new houses [have] been provided in Great Britain, over three million of them in the previous ten years [i.e. since 1953]. Since the slum clearance campaign was resumed in 1956, more than 480,000 unfit houses [have] been cleared and the families living in them rehoused.

*   *   *

The growth of the 12 first-generation new towns[†] in England and Wales, designated between 1946 and 1950, to the level of population at which their corporations are at present expected to stop building is now about three-quarters complete. Their combined population is approaching half a million. Two corporations (Crawley and Hemel Hempstead) completed their task in 1962 and handed over to the Commission for New Towns. The remaining new towns are expected to reach their current population targets at various dates up to 1970. (Several of the towns may, however, be expanded beyond the size for which they are at present planned.)

About 100,000 dwellings have so far been completed in the new towns. Some 93% of these have been built by the corporations for letting. Of the houses built for sale, about five in every seven have been built by private enterprise.

[†] Basildon, Bracknell, Harlow, Hatfield, Stevenage, Welwyn Garden City, Crawley, Hemel Hempstead, Aycliffe, Corby, Cwmbran, Peterlee.

9

The consumer

The standard of living does not necessarily rise when output grows. If it did, the task of the economic historian would be both easier and duller. Growth of output of cotton textiles in the early nineteenth century appears to have benefited consumers—abroad as well as at home—as much as the producers in the cotton industry. Rapid population growth or as in recent years an increased ratio of dependants to workers may offset the benefits of technical change. A high rate of savings could have a similar effect. Until the repeal of the Corn Laws in 1846 and the import of basic foods from overseas in the later nineteenth century a harvest failure might well lead to severe hardship for the poor and to bread riots (*A1, 3*). In times past even more than today standards of living could vary substantially from one region to another. For all these reasons it is difficult to give a coherent account of changes in the standard of living. What is indisputable is that in 1900 manual labourers and their families fed far from well and had little money to spare (except for drink) (*A6, 8*; *B12*). By the 1930s diet was more filling, but not necessarily adequate for full health (*A10*). Patterns of expenditure since the Second World War show how far average earnings have risen above the bare needs of subsistence (*A9*).

Even more elusive than the standard of living is the quality of life. Measurement here is virtually impossible, and almost every judgment depends on personal taste and point of view. Following William Morris and D. H. Lawrence students of English literature have taken a romantic view of England before the Industrial Revolution. That period is largely outside the scope of this volume, but there are documents here that look back to an older and not necessarily better tradition that that of industrialism (*B5, 7*). Within the period covered by this volume tendencies hopeful and otherwise can be detected. Most people nowadays would agree that education may make for a fuller life (*B8, 15*). Most would regret the decay of craftmanship (*B11*), though they would not know of the drudgery that jobs commonly entailed before the use of machinery. The decline of religion and the rural exodus are both undeniable, but there would be less agreement about their undesirability (*B9, 13*). The motor car is perhaps the most ambiguous of modern developments (*B18*). Desired by nearly all, it is both a willing slave and a dangerous and tyrannical master.

A. THE STANDARD OF LIVING

1 A Cornish bread riot, 1773

(*Calendar of Home Office Papers, 1773–1775*, pp. 8–9. This letter was written by a gentleman in Bodmin.)

...We had the devil and all of a riot at Padstow. Some of the people have run to too great lengths in exporting of corn, it being a great corn country. Seven or eight hundred tinners went thither, who first offered the cornfactors seventeen shillings for twenty-four gallons of wheat; but being told they should have none, they immediately broke open the cellar doors, and took away all in the place without money or price. About sixteen or eighteen soldiers were called out to stop their progress, but the Cornishmen rushed forward and wrested the firelocks out of the soldiers' hands: from thence they went to Wadebridge, where they found a great deal of corn cellared for exportation, which they also took and carried away. ... We think 'tis but the beginning of a general insurrection, because as soon as the corn which they have taken away is expended, they will assemble in greater numbers armed, for 'tis an old saying 'The belly has no ears'.

2 Prosperous Lancashire, 1783

(Alfred P. Wadsworth and Julia de Lacy Mann, *The cotton trade and industrial Lancashire, 1600–1780* (Manchester University Press, 1931), pp. 390–1, quoting a Lancashire magistrate writing in 1787.)

The cotton or fustian manufacture is diffused throughout this township, and in the hardest times the poor have little reason to complain for want of labour; but few of them lay up anything against times of distress. Their constant resort to the neighbouring alehouses will but too well account for this. Indeed even here the increase of luxury hath been rapid. In my memory the labourer was well contented with leavened oaten bread, by the inhabitants called jannock, and but little tea was drunk; whereas now the finest wheaten bread is very generally

30*

consumed by the meanest of the poor, and the tea kettle is as necessary a utensil as the spinning wheel in almost every cottage.

In the month of July, 1783, when oatmeal was sold in Bolton market from 41s 9d to two guineas a load, and wheat from 38s to £2 3s 6d, there was so little appearance of want in this township that one evening I met a very large procession of young men and women with fiddles, garlands, and other ostentation of rural finery, dancing morris dances in the highway merely to celebrate an idle anniversary, or, what they had been pleased to call for a year or two, a fair at a paltry thatched alehouse upon the neighbouring common.

3 Hardship caused by high food prices, 1797

('Commerce' in *Encyclopaedia Britannica* (3rd ed. Edinburgh, 1797), V, p. 213)

... [I]f a computation be made of the hands employed in providing subsistence, and of those who are severally employed in supplying every other want, their numbers will be found nearly to balance one another in the most luxurious countries. From this we may conclude, that the article of food, among the lower classes, must bear a very high proportion to all the other articles of their consumption; and therefore a diminution upon the price of subsistence, must be of infinite consequence to manufacturers who are obliged to buy it. From this consideration, let us judge of the consequence of such augmentations upon the price of grain as are familiar to us; 30 or 40 per cent seems nothing. Now this augmentation operates upon two-thirds, at least, of the whole expense of a labouring man: let anyone who lives in tolerable affluence make the application of this to himself, and examine how he would manage his affairs, if, by accidents of rains or winds, his expenses were to rise 30 per cent without a possibility of restraining them; for this is unfortunately the case with all the lower classes.

4 An early estimate of the basic needs of labour, 1832

(William G. Rimmer, 'Working men's cottages in Leeds, 1770–1840', *Thoresby Miscellany*, *XIII* (Thoresby Society Transactions, XLVI, 1963), p. 199. The author of this estimate was Humphrey Boyle of Leeds.)

Least possible sum per week for which a man, his wife, and three children can obtain a sufficiency of food, clothing & other necessaries —Feby. 12th, 1832.

	s	d		s	d
Rent 2/-, fuel 9d, candle 3d	3	0	Brou[gh]t up	14	6½
Soap 3d, soda 1d, blue & starch 1½d		5½	Vegetables 1d per day		7
Sand, black lead, bees wax &c		2	Salt, pepper, mustard, vinegar		2
Whitewashing a cottage twice a year		½	7 pts beer 1½d		10½
1½st flour for bread—2/6d	3	9	Water		1
¼st flour for puddings—2/8d st		8	Schooling for 2 children		6
Eggs 2d, yeast 1½d		3½	Reading		2
1½ pints milk per day at 1¼d		11	Wear & tear in beds, bedding, brushes, pots, pans & other household furniture		6
¼ stone oatmeal 2/2d		6½	Clothing: husband 1/2, wife 8d	1	10
1 lb treacle 3½d, 1½ lb sugar at 7d lb	1	2	each child 4d	1	0
1½ oz tea at 5d, 2 oz coffee 1½d		10½			
5 lb meat 6d	2	6			
	14	6½		£1 0 3	

Besides the sum required for the fund which it is agreed every workman [ought] to lay in store for sickness and old age, I have set nothing down for butter, not being certain whether it is essential to health, although it is to be found in almost every cottage where the weekly income is not more than half the amount I have stated as necessary for the proper support of a family: tobacco, although it is in very general use, I have omitted for the same reason; neither have I reckoned anything for religious instruction, which is thought by great numbers of the people as necessary to their happiness as is their daily bread: something, therefore, ought to be allowed for it.

The above is not made out from my own knowledge of housekeeping only; I have elicited from the most intelligent and economical of my acquaintances their opinion upon the most weighty items of expenditure, which, if correct, would have made the amount rather more than is here set down. If, upon the most strict enquiry, no material alteration can be made in the detailed estimate of the necessary weekly expenditure of five persons, I conceive that a case will be made out that the average earnings of workmen are not sufficient for the proper support of their families; and will prove at the same time that if greater economy was practised, if less was spent at the public house, there would be a much greater degree of comfort in the workman's cottage than is to be met with at present.

H. Boyle

5 Social reform—and its limitations, 1872

(Thomas Brassey, *On Work and wages* (3rd ed. London, 1872), pp. 280–2)

... Subsequently to 1833 the Factories Acts, the Ten Hours Act, the Mines and Collieries Acts, the Acts relating to Merchant Seamen; the establishment of Loan Societies, the Post office Savings Banks, the Friendly and Benefit Building Societies; the creation of a National System of Education, the Penny Postage, the adoption of a new and more liberal fiscal policy; the facilities given for establishing public libraries and museums; the remission of the paper-duties and the creation of a cheap press; the enlargement of the franchise, which has given to the working classes an overwhelming share of political power; and last, and perhaps the greatest of these reforms, the extension of educational facilities to every child; testify to the generous spirit of our recent legislation in all that relates to the welfare of the industrial classes.

The importance of social reforms, and of securing the material well-being of the masses of our population, is now universally recognised. I confess my doubts as to the efficacy of legislation in such matters. It must be remembered that all national expenditure for the benefit of the working classes which is not reproductive, must be defrayed by additional taxes. Let the transfer of land be by all means facilitated, let railway communication between the centre of a great city and its suburbs be made as cheap as possible, let emigration be assisted by loans, if security can be taken for the repayment of such advances; but, granted that something may be done by these various means, I hesitate to admit that the State can be the chief instrument for elevating still higher the moral condition of the people. The work is too vast for any Government to undertake. It can only be accomplished by the self-help and self-sacrifice of the whole nation. And when all shall have done their duty in their several stations, the pressure of unforeseen calamity upon some unhappy individuals and the incapacity of others will leave a mass of suffering to our compassionate care, which it will tax our best energies to relieve. The poor we shall always have with us; and the great peers, the landowners, and the men who have become rich in commerce, must show themselves active in their sympathies for all just demands, benevolent and kindly in the presence of distress.

6 The poverty line, 1901

(B. S. Rowntree, *Poverty: a study of town life* (Macmillan, 1901), pp. 132–4)

... Allowing for broken time, the average wage for a labourer in York is from 18s to 21s; whereas, according to the figures given earlier in this chapter, the minimum expenditure necessary to maintain in a state of physical efficiency a family of two adults and three children is 21s 8d,[1] or, if there are four children, the sum required would be 26s.

It is thus seen that *the wages paid for unskilled labour in York are insufficient to provide food, shelter, and clothing adequate to maintain a family of moderate size in a state of bare physical efficiency.* It will be remembered that the above estimates of necessary minimum expenditure are based upon the assumption that the diet is even less generous than that allowed to able-bodied paupers in the York Workhouse, and that *no allowance is made for any expenditure other than that absolutely required for the maintenance of merely physical efficiency.*

And let us clearly understand what merely physical efficiency means. A family living upon the scale allowed for in this estimate must never spend a penny on railway fare or omnibus. They must never go into the country unless they walk. They must never purchase a halfpenny newspaper or spend a penny to buy a ticket for a popular concert. They must write no letters to absent children, for they cannot afford to pay the postage. They must never contribute anything to their church or chapel, or give any help to a neighbour which costs them money. They cannot save, nor can they join sick club or Trade Union, because they cannot pay the necessary subscriptions. The children must have no pocket money for dolls, marbles, or sweets. The father must smoke no tobacco, and must drink no beer. The mother must never buy any pretty clothes for herself or for her children, the character of the family wardrobe as for the family diet being governed by the regulation, 'Nothing must be bought but that which is absolutely

[1] This estimate is arrived at thus:	s	d
Food—two adults at 3s.	6	0
three children at 2s 3d	6	9
Rent—say	4	0
Clothes—two adults at 6d	1	0
three children at 5d	1	3
Fuel	1	10
All else—five persons at 2d	0	10
Total	21	8

necessary for the maintenance of physical health, and what is bought must be of the plainest and most economical description. Should a child fall ill, it must be attended by the parish doctor; should it die, it must be buried by the parish. Finally, the wage-earner must never be absent from his work for a single day.

If any of these conditions are broken, the extra expenditure involved is met, *and can only be met*, by limiting the diet; or, in other words, by sacrificing physical efficiency.

7 England and America compared, 1902

(*Mosely Industrial Commission to the United States of America, Oct.–Dec. 1902, Reports of the delegates* (London, 1903), pp. 54–5, 74–5. Report by Mr George N. Barnes of the Amalgamated Society of Engineers.)

But, before proceeding, it may be as well to state briefly the main conditions obtaining in Great Britain, so far as the engineering industry is concerned. Educational facilities here consist of the Board schools supplemented by a few scholarships, and technical schools, some of the latter being subsidised by public bodies, but all of which are, I believe, ineffective, as far as the working classes are concerned, because of the absence of preparatory schools. The training of the workman is confined mainly to the workshop, and workshop apprenticeship lasts five years, terminating at 21 years of age. The normal working week ranges from 48 to 54 hours. The standard minimum rate of wages at places here, corresponding to those visited in America, are as follows: —On the Clyde, Tyne, Wear, Tees, and the larger towns of Lancashire and Yorkshire 8*d* per hour, in Sheffield, London, Belfast, and the South Wales large towns it is $8\frac{3}{4}d$ in those shops working nine hours per day, and $9\frac{3}{4}d$ in those working eight hours. The average rate, however, for competent men is from 2 to 3 per cent higher. Overtime is paid for at about 11*d* per hour throughout, and the earnings of the pieceworkers amount to about the same. The pay of the operative engineers, otherwise the artificers, in the Royal Navy, begins at 5*s* 6*d* per day, and reaches its maximum at 7*s* 6*d*, except for the very small number who attain to warrant office rank, after many years' service, and who may then get 10*s* 6*d* per day.

Cost of Living—Rent ranges from 7*s* to 12*s* per week for from three to six rooms. Bread is 5*d* per 4lb loaf throughout. Coal varies considerably in price, according to distance from a coalfield, but may be said to average £1 1*s* per ton.

Saving.—Operative engineers in England save but little except for special purposes, and except those who, through building societies or trade unions, are part proprietors of their dwelling-houses, and these may number two per cent to three per cent of the total.

Just one more observation before proceeding. The wages quoted here following will not have reference to specialists or handy men, unless so stated. These latter are paid about 20 per cent lower than the mechanic in both countries, but in America they are far more numerous than in Great Britain, excepting in the Government workshops. The figures will have reference to the mechanic, meaning by that term the man who has served an apprenticeship, or who has acquired a knowledge and aptitude in the use of the tools of the trade, which enables him to earn mechanics' wages. Such men in this country are known as 'engineers,' and in America as 'machinists'.

※   ※   ※

How far is greater output in American factories due to—

(*a*) Longer hours of work?

(*b*) Greater speed at which the machinery is run?

Greater output is, I believe, almost entirely due to more machinery compared with hand work. To some extent a greater speed is also run because of the softer material, and the use of the best steel for cutting. Longer hours may also contribute somewhat, but not to the extent of the difference between those worked as compared with the number worked in Great Britain.

Are there any points in American practice which should, in your opinion, be imitated in English factories?

Yes. I would suggest (*a*) a freer use of machinery and latest appliances. (*b*) A more ready recognition and liberal reward on the part of employers of exceptional ability or initative of the workman. This might be carried out, either through bonus schemes or otherwise, as long as the standard time rate of wages was paid to all, and increased pay for increased effort was secured. (*c*) The standardisation of work where possible. In this respect I find America ahead of us. I might just cite railway stock wheels and axles as an instance. These have been reduced to a standard size and weight, and are supplied to the railway companies and other builders at extraordinarily cheap rates. One firm does an enormous business in wheels which are chilled to a certain depth, but guaranteed to run for so many thousand miles before the chilled part is worn through; and these are supplied to railway companies and builders all over America. (*d*) The re-modelling of the Patent Law, and re-organisation of the Patent Office, so as to secure a simple registration and Government search, and therefore reasonable guarantee of novelty to the inventor, and the reduction of fees charged to him. This is very important, and a matter upon which we cannot too promptly imitate our American cousins, who claim, I believe with a large amount of justification, that the American patents have given employment to enormous numbers of workpeople, and have led to the adoption of labour-saving and productive machinery. (*e*) The provision of secondary or continuation schools.

General Condition of Workers Outside the Factory

(*a*) Are the American workers better fed than the English?

(*b*) How does the price of food in America compare with that in England?

(*a*) I believe that the American, speaking generally, and of course not applying it to the poorest classes, eats more, and in more variety, than is good for him, and that the British mechanic with, and to some extent because of, his more simple diet is more physically efficient than the American mechanic. (*b*) The price of food in America, as compared with the price in Great Britain, is higher to the extent of, I should say, 25 to 30 per cent.

(*a*) Are the American workers better clothed than the English?

(*b*) How does the price of clothes in America compare with that in England?

There is but little difference in the clothing of the American as compared with the British mechanics, excepting that the Americans doff their working clothes, in many cases, before leaving the workshops. So far as clothing outside the workshop is concerned, I think the advantage is in favour of the American, from the point of view of cut. I do not want to entrench upon the report of the Tailors' delegate, but I may say that it seemed to me that tailoring was better done in America. The price of slop clothing is, I think, much about the same as here, but the price of good stuff—which is generally imported from England, and bears heavy import duty—well got up, is nearly double the English price.

(*a*) Are the American workers better housed than the English?

(*b*) How does rent in America compare with rent in England?

(*c*) Do more workers, relatively, own the houses they live in, than is the case in England? If yes, to what circumstances do you attribute this?

(*a*) I had but little opportunity of judging as to quality of housing, but having been in about a dozen houses, I may say that it appeared to me that the American house is, as a rule, larger than that of English workmen. I should not say, from the point of view of comfort, however, that it was any better. (*b*) Rent is about 30 to 40 per cent higher, taking the country through on both sides. (*c*) Yes. Far more workmen own houses than here. I should say at least twice as many. I am inclined to attribute this to the habits formed in early days of wooden houses. In the primitive settlements houses would be cheap, and the custom of house-owning would be general, and I rather think that this has become more or less stereotyped. I noted that ownership of houses was mostly prevalent in small towns, where early ideas are probably more tenaciously clung to, and where the houses are still mostly of wood, and therefore cheap.

How does the average wage in your trade in America, *expressed in money*, compare with the average wage in England?

The base rate of wages in America, as compared with Great Britain, taking not only the rate per normal week, which is higher, but the rate for overtime and night-work which is relatively lower, is, I should say, about 35 to 45 per cent higher for operative engineers, but there is a greater range of pay, and I should say that the maximum wage is probably 70 per cent higher. I am quoting present rates which, in America, have been increased considerably during the last three or four years.

How does the *value* of the American wage compare with that of the English, *cost of living being taken into account*?

The value of the American wage of the operative engineer, compared with that paid in Great Britain to the same, is, I should say, about 15 or 20 per cent higher on the base line, and proportionately more on the maximum.

Can the careful, sober, steady man, whilst keeping himself efficient, save more in America than in England?

He can, I believe, to the extent shown above, but as to efficiency I would qualify the answer by saying that such efficiency will probably last for a shorter time than in this country.

If yes, does he *in fact* save more, or not?

I should say he does.

Does gambling on horse racing, &c., enter as largely into the life of the American as of the English working man?

I do not think so.

8 The cost of living, 1904

(Board of Trade, *British and foreign trade and industrial conditions* [Second fiscal blue book, Cd. 2337, 1904], *B.P.P.*, 1905, LXXXIV, pp. 5, 32. These figures form the basis for the official cost-of-living index, 1914–56)

The following table shows the general results of the inquiry. It gives the average for the United Kingdom of the incomes and expenditures of the workmen's families to which the returns relate and the quantities consumed by them of the various articles of food in one week in summer in 1904. All children living at home, irrespective of age, have been included, but returns in which lodgers appeared have been excluded. N.B.—It should be understood that intoxicating drinks and tobacco are not included in the returns.

Average Weekly Cost and Quantity of Certain Articles of Food Consumed by Urban Workmen's Families in 1904

Limits of Weekly Income	*Under 25s*		*25s and under 30s*		*30s and under 35s*		*35s and under 40s*		*40s and above*		*All Incomes*	
Number of returns	261		289		416		382		596		1,944	
	s	d	s	d	s	d	s	d	s	d	s	d
Average weekly family income	21	4½	26	11¾	31	11¼	36	6¼	52	0½	36	10
Average number of children living at home	3·1		3·3		3·2		3·4		4·4		3·6	

Cost

	s	d	s	d	s	d	s	d	s	d	s	d
Bread and flour	3	0½	3	3¾	3	3½	3	4¼	4	3¾	3	7
Meat (bought by weight)	2	8	3	4¾	4	3½	4	5½	5	10½	4	5½
Other meat* (including fish)	0	7½	0	8¾	0	10	1	0	1	4	0	11¾
Bacon	0	6¾	0	9	0	10¼	0	11½	1	3¾	0	11½

* e.g., sheep's heads, tripe, heart, live r, pig's fry, tinned meats, rabbits.

Eggs	0	$5\frac{3}{4}$	0	$8\frac{1}{2}$	0	11	1	0	1	$4\frac{3}{4}$	1	0
Fresh milk	0	8	0	$11\frac{3}{4}$	1	$3\frac{1}{4}$	1	$4\frac{1}{4}$	1	$7\frac{3}{4}$	1	$3\frac{1}{4}$
Cheese	0	$4\frac{3}{4}$	0	$5\frac{1}{2}$	0	6	0	6	0	8	0	$6\frac{1}{2}$
Butter	1	2	1	7	1	$10\frac{1}{4}$	2	0	3	$0\frac{1}{2}$	2	$1\frac{1}{2}$
Potatoes	0	$8\frac{3}{4}$	0	$9\frac{3}{4}$	0	$10\frac{1}{4}$	0	$10\frac{1}{4}$	1	$1\frac{3}{4}$	0	11
Vegetables and fruit	0	$4\frac{3}{4}$	0	7	0	10	0	$11\frac{3}{4}$	1	$3\frac{3}{4}$	0	11
Currants and raisins	0	$1\frac{1}{2}$	0	$1\frac{3}{4}$	0	$2\frac{1}{4}$	0	3	0	$3\frac{3}{4}$	0	$2\frac{3}{4}$
Rice, tapioca, and oatmeal	0	$4\frac{1}{2}$	0	5	0	6	0	$5\frac{3}{4}$	0	7	0	6
Tea	0	$9\frac{1}{4}$	0	$11\frac{1}{4}$	1	$0\frac{3}{4}$	1	$1\frac{1}{4}$	1	5	1	$1\frac{1}{2}$
Coffee and cocoa	0	2	0	$3\frac{1}{4}$	0	$3\frac{1}{2}$	0	$4\frac{1}{4}$	0	$5\frac{1}{2}$	0	$3\frac{3}{4}$
Sugar	0	8	0	10	0	$10\frac{3}{4}$	0	$11\frac{1}{4}$	1	3	0	$11\frac{3}{4}$
Jam, marmalade, treacle and syrup	0	$4\frac{1}{4}$	0	$5\frac{1}{4}$	0	6	0	$6\frac{1}{2}$	0	$8\frac{3}{4}$	0	$6\frac{1}{2}$
Pickles and condiments	0	2	0	$2\frac{1}{4}$	0	$3\frac{1}{4}$	0	$3\frac{1}{2}$	0	$4\frac{1}{4}$	0	$3\frac{1}{4}$
Other items	1	$0\frac{1}{2}$	1	$3\frac{3}{4}$	1	$6\frac{1}{2}$	1	$10\frac{1}{2}$	2	$6\frac{1}{4}$	1	$9\frac{1}{2}$
Total expenditure on food	14	$4\frac{3}{4}$	17	$10\frac{1}{4}$	20	$9\frac{1}{4}$	22	$3\frac{1}{2}$	29	8	22	6

Quantities

	lbs	*lbs*	*lbs*	*lbs*	*lbs*	*lbs*
Bread and flour	28·44	29·97	29·44	29·99	37·76	32·04
Meat (bought by weight)	4·44	5·33	6·26	6·43	8·19	6·50
Bacon	0·94	1·11	1·19	1·38	1·82	1·38
	pts.	*pts.*	*pts.*	*pts.*	*pts.*	*pts.*
Fresh milk	5·54	7·72	9·85	10·34	12·63	9·91
	lbs.	*lbs.*	*lbs.*	*lbs.*	*lbs.*	*lbs.*
Cheese	0·67	0·70	0·79	0·77	1·02	0·83
Butter	1·10	1·50	1·69	1·89	2·78	1·96
Potatoes	14·05	15·84	16·11	15·87	19·93	16·92
Currants and raisins	0·42	0·50	0·62	0·80	0·91	0·70
Rice, tapioca, and oatmeal	2·54	2·64	2·93	2·55	3·38	2·95
Tea	0·48	0·55	0·57	0·59	0·72	0·60
Coffee and cocoa	0·15	0·18	0·20	0·23	0·29	0·22
Sugar	3·87	4·62	4·79	5·21	6·70	5·31

✳ ✳ ✳

In order to combine the above results so as to show the estimated change in the total cost of living to the working classes so far as indicated by changes in these four principal items, we have to attribute to each item its appropriate 'weight' or relative degree of importance.

After consideration of the available statistics of working-class expenditure, the following proportionate 'weights' have been adopted:

Food	7
Rent	2
Clothing	2
Fuel and light	1
Total expenditure on the above objects	12

It is to be understood that these weights are only approximate, and are not intended to represent exactly the proportional expenditure of the working classes on each of the above items.

If weight be allowed in the above proportions to the four series of figures in Table A, the following final 'Index Numbers' are obtained:

B—*Statement showing Estimated Changes in Cost of Living of the Working Classes, based on Cost of Food, Rent, Clothing, Fuel, and Light, in a series of averages for quinquennial periods. (Cost in the year 1900 = 100.)*

Period	Index Number of Cost of Living
Average of quinquennial period	
of which middle year is 1880	120·5
of which middle year is 1885	108·2
of which middle year is 1890	100·9
of which middle year is 1895	95·5
of which middle year is 1900	99·7

These figures show that in the twenty-year period 1880–1900 the cost of living of the working classes, as shown by the price they have to pay for the four principal items of necessary expenditure, has fallen, so that roughly 100 shillings in 1900 would do the work of 120 at the beginning of the period. In the first fifteen years of the period the fall was somewhat greater, but since that time there has been a rise of about 4 per cent. Much the greatest decline has been in the cost of food, which has fallen about twice as much proportionately as the total cost of living. The price of clothing has also declined, but at a much slower rate, while the cost of fuel and light has somewhat increased, and rents have risen considerably.

9 ... And fifty years after

(Cost of living advisory committee, Report on proposals for a new index of retail prices, *B.P.P.*, 1955–56, XIII, Appendix pp. 17–24. The pattern of consumption here reflected is that of 1953–4. Households where the head of the household was earning £20 a week or more are excluded from the index. So too are households whose income was mainly derived from old age pensions or national assistance.)

List of groups and sections into which the new index is to be divided.

Group and section	Weight
I Food [of which]	
1 Bread	19
2 Flour	4
6 Beef	24
7 Mutton and lamb	14
26 Sweets and chocolate	26
TOTAL FOOD	350
II Alcoholic drink	71
III Tobacco	80
IV Housing	87
V Fuel and light	55
VI Durable household goods	66
VII Clothing and footwear	106
VIII Transport and vehicles	
65 Petrol and oil	8
69 Rail transport	8
TOTAL TRANSPORT AND VEHICLES	68
IX Miscellaneous goods	59
X Services	58
TOTAL, ALL ITEMS	1000

10 Calories are not enough, 1936

(J. B. Orr, later Lord Boyd-Orr, *Food, health and income: report on a survey of adequacy of diet in relation to income* (Macmillan, 1936), pp. 49–50)

The food position of the country has been investigated to show the average consumption of the main foodstuffs at different income levels. The standard of food requirements and the standard of health adopted are not the present average but the optimum, i.e., the physiological standard, which, though ideal, is attainable in practice with a national food supply sufficient to provide a diet adequate for health for any member of the community. The main findings may be summarized as follows:

I Of an estimated national income of £3,750 millions, about £1,075 millions are spent on food. This is equivalent to 9s per head per week.

II The consumption of bread and potatoes is practically uniform throughout the different income level groups. Consumption of milk, eggs, fruit, vegetables, meat and fish rises with income. Thus, in the poorest group the average consumption of milk, including tinned milk, is equivalent to 1·8 pints per head per week; in the wealthiest group 5·5 pints. The poorest group consume 1·5 eggs per head per week; the wealthiest 4·5. The poorest spend 2·4d on fruit; the wealthiest 1s 8d.

III An examination of the composition of the diets of the different groups shows that the degree of adequacy for health increases as income rises. The average diet of the poorest group, comprising 4½ million people, is, by the standard adopted, deficient in every constituent examined. The second group, comprising 9 million people, is adequate in protein, fat and carbohydrates, but deficient in all the vitamins and minerals considered. The third group, comprising another 9 million, is deficient in several of the important vitamins and minerals. Complete adequacy is almost reached in group IV, and in the still wealthier groups the diet has a surplus of all constituents considered.

IV A review of the state of health of the people of the different groups suggests that, as income increases, disease and death-rate decrease, children grow more quickly, adult stature is greater and general health and physique improve.

V The results of tests on children show that improvement of the diet in the lower groups is accompanied by improvement in health and increased rate of growth, which approximates to that of children in the higher income groups.

VI To make the diet of the poorer groups the same as that of the first group whose diet is adequate for full health, i.e. group IV, would involve increases in consumption of a number of the more expensive foodstuffs, viz., milk, eggs, butter, fruit, vegetables and meat, varying from 12 to 25 per cent.

If these findings be accepted as sufficiently accurate to form a working hypothesis, they raise important economic and political problems. Consideration of these is outwith the scope of the investigation. It may be pointed out here, however, that one of the main difficulties in dealing with these problems is that they are not within the sphere of any single Department of State. This new knowledge of nutrition, which shows that there can be an enormous improvement in the health and physique of the nation, coming at the same time as the greatly increased powers of producing food, has created an entirely new situation which demands economic statesmanship. The prominence given to this new social problem at the last Assembly of the League of Nations shows that it is occupying the attention of all civilized countries. It is gratifying that the lead in this movement was taken by the British Empire.

B. THE QUALITY OF LIFE

1 Arthur Young's criticism of enclosure, 1801

Arthur Young, *An inquiry into the propriety of applying wastes to the better maintenance and support of the poor; with instances of the great effects which have attended their acquisition of property in keeping them from the parish* (Bury St. Edmunds, 1801), pp. 13, 42)

Go to an alehouse kitchen of an old enclosed country, and there you will see the origin of poverty and poor rates. For whom are they to be sober? For whom are they to save? (Such are their questions.) For the parish? If I am diligent, shall I have leave to build a cottage? If I am sober, shall I have land for a cow? If I am frugal, shall I have half an acre of potatoes? You offer no motives; you have nothing but a parish officer and a workhouse! Bring me another pot.

*　　　*　　　*

Objection VIII Wastes are as much property as my house. Will a farmer give up his right of commonage?

I will not dispute their meaning; but the poor look to facts, not meanings: and the fact is, that by nineteen enclosure bills in twenty they are injured, in some grossly injured. It may be said that commissioners are sworn to do justice. What is that to the people who suffer? It must be generally known that they suffer in their own opinions, and yet enclosures go on by commissioners, who dissipate the poor people's cows wherever they come, as well those kept legally as those which are not. What is it to the poor man to be told that the Houses of Parliament are extremely tender of property, while the father of the family is forced to sell his cow and his land because the one is not competent to the other; and being deprived of the only motive to industry, squanders the money, contracts bad habits, enlists for a soldier, and leaves the wife and children to the parish? If enclosures were beneficial to the poor, rates would not rise as in other parishes after an act to enclose. The poor in these parishes may say, and with truth, *Parliament may be tender of property*; *all I know is, I had a cow, and act of Parliament has taken it from me*. And thousands may make this speech with truth.

2 Malthus defied, 1822

(William Cobbett, *Cottage economy, containing information relative to the brewing of beer, making of bread, keeping of cows, pigs, bees, ewes, goats, poultry* (London, 1822), pp. 8–9. The italics are Cobbett's own.)

If the labourer have his fair wages; if there be no false weights and measures, whether of money or of goods, by which he is defrauded; if the laws be equal in their effect upon all men, if he be called upon for no more than his due share of the expences necessary to support the government and defend the country, he has no reason to complain. If the largeness of his family demand extraordinary labour and care, these are due from him to it. He is the cause of the existence of that family; and, therefore, he is not, except in cases of accidental calamity, to throw upon others the burden of supporting it. Besides, 'little children are as arrows in the hands of the giant, and blessed is the man that hath his quiver full of them'. This is to say, children, if they bring their *cares*, bring also their *pleasures* and *solid advantages*. They become, very soon, so many assistants and props to the parents, who, when old age comes on, are amply repaid for all the toils and all the cares that children have occasioned in their infancy. To be without sure and safe friends in the world makes life not worth having; and whom can we be so sure of as of our children? Brothers and sisters are a mutual support. We see them, in almost every case, grow up into prosperity, when they act the part that the impulses of nature prescribe. When cordially united, a father and sons, or a family of brothers and sisters, may, in almost any state of life, set what is called misfortune at defiance.

These considerations are much more than enough to sweeten the toils and cares of parents, and to make them regard every additional child as an additional blessing. But, that children may be a blessing and not a curse, care must be taken of their *education*.

3 Urban poverty, 1837

(Journal of the Statistical Society of London, I (1838–39), pp. 34–6)

Report of an Enquiry, conducted from House to House, into the State of 176 Families in Miles Platting, within the borough of Manchester, in 1837

By James Heywood, Esq. Read before the Statistical Society of London, on the 16th April, 1838.

The following Report has been prepared from the results of an enquiry undertaken at the request of the author, in a district of Manchester, with which the visitor who conducted the enquiry was previously well acquainted. Miles Platting is inhabited, for the most part, by the families of operatives, who are dependent upon manual labour for their subsistence; and a large proportion of the heads of the families included within this enquiry are hand-loom weavers. Their occupations are extremely laborious, their earnings very moderate, and their time of labour, when in full work, often amount to 14 hours per day. During the year 1837 many of the hand-loom weavers in this district did not find half employment; others were unable to earn more than 6s or 7s per week, and the most experienced and industrious of the class, by working 14 hours per day, frequently obtained for the full amount of their earnings only 12s per week.

Where the hand-loom weavers are employed in weaving plain thin cotton goods, similar to those manufactured by power-looms, their earnings seldom amount to 9s per week, and the majority of this class of workman can only get 6s or 7s per week. In weaving narrow shawls, a new kind of work, put out in the winter of 1837, the weekly earnings of a good weaver were 9s per week, from which 2d in every shilling, or 1s 6d on the gross earnings, must frequently be deducted for the expense of winding, leaving only 7s 6d as the net weekly earnings. A first-rate workman, weaving quiltings, may earn 12s per week; but, in this case, the winding and other expenses often amount to 3d in 1s, or 3s on the gross earnings, and thus their net earnings are diminished to 9s per week. In like manner a silk-weaver, working at plain sarsnets, may earn 12s per week; but the expenses of winding and the use of the loom usually amount to 3d in every shilling, or to 3s on the whole weekly earnings, and thus the net earnings of the silk-weaver are diminished to 9s per week. Whenever the hand-loom weaver works at his own loom, and is assisted in winding by his wife or family, he may consider the whole of his earnings as profit; but in the case of journeymen weavers, who hire the looms on which they work, and who pay, in addition, for the expense of winding, the total amount of the earnings will be necessarily lessened by the deduction of these concurrent expenses. The various occupations of the heads of families included within this enquiry are enumerated in the following table:

102 Hand-loom weavers	148 Brought forward
5 Silk weavers, winders, and warpers	5 Labourers
	1 Carter
3 Small-ware weavers	2 Smiths
2 Power-loom weavers	9 Tradesmen (hatters, joiners, &c)
8 Warpers and workers in factories	
1 Jacquard-loom manufacturer	1 Schoolmaster
6 Dyers	1 Designer
2 Fustian-shearers	1 Drawer-in of threads for the loom
5 Warehousemen	
3 Bricklayers	6 Washerwomen
11 Colliers	1 No occupation
148	176

Among the heads of families visited the number of married men is remarkable; and the total population of the families included within the enquiry comprehends the following individuals:

Heads of Families,	147	Married men
Heads of Families,	9	Widowers
Heads of Families,	20	Widows
	176	
	147	Wives of the heads of families
	232	Children under 10 years of age
	239	Children above 10 years of age
	113	Male lodgers, often journeymen
	49	Female lodgers
	34	Children with the lodgers
Total Population	990	

Of the 176 heads of families, 137 are English, 37 Irish, 1 Scotch, and 1 Welch.

It appears that the majority of the children receiving education are instructed in Sunday-schools; and it is worthy of notice that many of the children attend the Sunday-school at a very early age in Miles Platting, owing to the unwillingness of their parents to allow the elder children to attend the Sunday-school unless they take the younger children with them. Of course the maintenance of silence and order in the Sunday-school is rendered more difficult by the presence of a large number of infant scholars in the same room with the older children; and the attention of the more advanced scholars must be diverted by the process of elementary instruction which is required for the infant children.

The total number of children in the families visited is 505, viz.—

    232 Children under 10 years of age
    239 Children above 10 years of age
     34 Children with the lodgers
    ———
    505
    ═══

Of this number —

     63 attend both day and Sunday-schools
    208 attend Sunday-schools only
      8 attend day-schools only
      9 attend infant-schools
    ———

Making a total of 288 children at school.

There are very few of the heads of the families, included within this enquiry, who have formed the habit of reading, or are capable of understanding or enjoying a book. Many are either too illiterate, or too deeply sunk in indifference, or in animal gratification, to be easily convinced of the importance of mental culture or religion. There are, however, others who may be regarded as sincerely religious characters.

In the following table, where the heads of families are really in connection with the Church of England, or with any other religious denomination, they are classified accordingly; but wherever they profess no particular attachment to any one Protestant sect, they are considered uncertain. Of 176 heads of families, it appears that—

    31 belong to the Church of England
    23 are Roman Catholics
    13 Methodists
     4 Unitarians
     3 Baptists
     1 Scotch Presbyterian
     1 Independent
     3 Deistical
    97 Uncertain
    ———
    176
    ═══

Of the 176 heads of families visited—

    35 attend public worship regularly
    45 occasionally
    36 seldom
    60 do not attend public worship
   —
  176
  ═══

130 profess to be able to read
 15 read imperfectly
 23 cannot read
  8 are not particularized as to their power of reading
  —
176
═══

  78 possess a Bible and a Testament
   2 possess a Bible
  35 possess a Testament
  97 possess a Prayer-book, or Hymn-book, or both, in addition sometimes to the Bible or Testament
  36 possess other works, chiefly religious
  37 possess no books.

4 London coffee houses, 1840

(Select committee on import duties, *B.P.P.*, 1840, V, qq. 2738–87)

Mr William Hare, Mr James Pamphilon, Mr J. B. Humphreys, Mr Thomas Letchford, and *Mr James Rogers, called in*; *and Examined.*

(*Mr Ewart*) You are all keepers of coffee-houses in London?—(*Mr Hare*) We are.

Will you state where you respectively carry on that business?—I keep the Colonial Coffee-house, 78, Lombard-street. (*Mr Pamphilon*) I keep a coffee-house at Nos. 3 and 4, Sherrard-street, Haymarket. (*Mr Humphreys*) I keep the Crown Coffee-house, 41, High Holborn. (*Mr Letchford*) Mine is the British Coffee-house, 37, High-street, Bloomsbury. (*Mr Rogers*) I keep the Angel Coffee-house, St. Clement's.

Is coffee alone consumed, or principally, in your houses?—(*Mr Hare*) Principally coffee and tea.

Have the number of coffee-houses in London increased much of late years?—(*Mr Humphreys*) Very materially; the annual increase has been nearly 100 per annum.

Do you recollect the time when there were scarcely any coffee-houses in London?—Yes; when I was comparatively a young man, there were not above 10 or 12 coffee-houses in London, about 25 years ago. There are now, I should think, from 1,600 to 1,800.

Can you give the Committee any idea of the increase in the last six years?—I should think the average of increase, in the last six years, has been about 100 per annum.

There has been an immense increase in the consumption of coffee in London, in consequence of the establishment of that great number of coffee-houses?—There has.

Has the charge for coffee to the consumer been reduced, in consequence of this competition?—Very materially. About 25 years ago there was scarcely a house in London where you could get any coffee under 6d a cup, or 3d a cup; there are now coffee-houses open at from 1d up to 3d. There are many houses where the charge is 1d, where they have 700 to 800 persons in a day. There is Mr Pamphilon, who charges $1\frac{1}{2}d$ a cup; and he has from 1,500 to 1,600 persons a day.

In what vicinity?—In the vicinity of the Haymarket, in Sherrard-street.

(To *Mr Pamphilon*) What are the class of persons that frequent your house?—We have all classes, from hackney-coachmen and porters to the most respectable classes. I have three rooms; and the more respectable classes are in the best rooms. We have a great many foreigners.

What is the number daily at the most frequented coffee-houses in London?—Mine is the largest in London.

What are the hours of opening and of closing coffee-houses?—The hours are from four o'clock to 11 at night; I open at half past five, and shut up at half-past 10 in the evening.

Do you take in a great number of newpapers and periodicals?—Yes, a very large supply of papers; I take in 43 London daily papers, and the reason the number is so large is, that I have five or six copies of some of them; I have eight copies of the Morning Chronicle.

(*Chairman*) Is that solely for the use of those that come in to take coffee?—Yes.

Then a man that pays $1\frac{1}{2}d$ for a cup of coffee has the opportunity of reading those papers?—He can read everything I have there.

How long is he allowed to remain?—An hour, or longer if he thinks proper. We take in seven country papers, six foreign papers, 24 magazines every month, four quarterly reviews, and 11 weekly periodicals, and any customer who comes in and has a cup of coffee, for which he pays $1\frac{1}{2}d$, he can read anything we have.

(*Mr Villiers*) Is it owing to the lowness of the price, that you have so many customers?—It is owing partly to the attraction of the newspapers and periodicals.

(*Mr Ewart*) Is it not owing, in part, to the change in the habits of society, that the people are more inclined to consume coffee, and sober beverages of that description, than they used to be?—Yes; they used to have nothing to go to but a public-house.

(*Mr Villiers*) You observe an increasing rate for coffee?—Yes. And that has increased with the reduction of price?—Yes.

You do not sell spirits at all?—Not at all, and we do not even cook a chop or a steak.

(*Chairman*) Have you found the desire to read the newspapers increase?—Yes.

What kind of people are they that chiefly frequent your house?—We have a few hackney-coachmen; we have post lads from Regent-street, and we have mechanics of all classes. In the best room we have occasionally foreign couriers and some gentlemen that live in the neighbourhood.

Are the majority of them what may be called artizans?—Yes.

(*Mr Ewart*) There are some coffee-houses entirely devoted to workmen, are there not?—Yes, where the price is rather lower, mechanics and labourers.

Are there many other coffee-houses in London which take in a great number of periodicals besides yours?—A great many. (*Mr Humphreys*) The amount I used to pay for newspapers, prior to the reduction of the duty, was 400*l* a year for newspapers, magazines, and the cost of binding the back numbers up for the use of my customers.

No inebriety can possibly occur in your establishment?—No; and I never heard an indecent expression, and I have never seen a drunken man in my house, with two exceptions, ever since I have been in Holborn.

You cannot sell any intoxicating beverage?—We are very seldom asked for it: we found that it would interfere with the habits of gentlemen that frequent our houses for the purpose of taking coffee.

Your interest is to consult the wishes of your general customers? —It is.

(*Mr Villiers*) It is the particular beverage that you sell which is the great attraction to the persons that come to your house?—Yes. I have upon the average 400 to 450 persons that frequent my house daily; they are mostly lawyers' clerks and commercial men; some of them are managing clerks, and there are many solicitors likewise, highly respectable gentlemen, who take coffee in the middle of the day in preference to a more stimulating drink. I have often asked myself the question where all that number of persons could possibly have got their refreshment prior to opening my house. There were taverns in the neighbourhood, but no coffee-house, nor anything that afforded any accommodation of the nature I now give them; and I found that a place of business like mine was so sought for by the public, that shortly after I opened it I was obliged to increase my premises in every way I could, and at the present moment, besides a great number of newspapers every day, I am compelled to take in the highest class of periodicals. For instance, we have eight or nine quarterly publications, averaging from 4s to 6s, and we are constantly asked for every new work that has come out. I find there is an increasing taste for a better class of reading. When I first went into business many of my customers were content with the lower priced periodicals; but I find, as time progresses, that the taste is improving, and they look out now for a better class of literature.

(*Mr Villiers*) Do you know whether there are any shops which the very lowest class of society resort to for coffee?—Mr Letchford keeps a coffeehouse in St Giles's.

(*Chairman* to *Mr Letchford*) How long have you kept that house?— Seven years. I established it myself.

What has been the increase in the number of persons frequenting it?—I have been there seven years, and I suppose now I have from 700 to 900 a day frequenting my house.

Are they chiefly Irish?—No; my house is in High-street, St Giles's, and the class who use my house are a hard-working class of people.

Working men?—Hard-working men.

(*Mr Villiers*) To whom the cost of refreshment would be an object? —Yes.

(*Chairman*) What do you charge per cup?—My front room is 1d, my back room $1\frac{1}{2}d$, and my up-stairs is 3d.

How many would each room hold?—My front shop will hold about 46.

Do you supply them with bread?—Yes.

At the ordinary price?—Yes.

(*Mr Villiers*) Which of the three rooms is the most numerously attended?—The hard-working class; the cheapest.

What newspapers do you take in?—Nine daily papers.

At what time of the day do the people come that pay 1d?—From 4 in the morning to 10 at night.

They are constantly coming?—Constantly coming.

Men and women?—No women.

Does a man come and get his breakfast?—Yes; he comes in the morning at 4 o'clock, and has a cup of coffee, a thin slice of bread and butter, and for that he pays 1½d; and then again at 8, for his breakfast, he has a cup of coffee, a penny loaf and a pennyworth of butter, which is 3d; and at dinner time, instead of going to a public-house, at 1 o'clock, he comes in again, and has his coffee and his bread and brings his own meat. I do not cook for any one.

(*Chairman*) Do you sell any beer?—No.

Do you sell any tea?—Yes.

What is the proportion of tea that you use to coffee?—(*Mr Humphreys*) About one-third.

What do you charge for a cup of tea?—The trade generally, where they sell coffee at 1½d a cup, charge 2d for tea.

5 Crime in the countryside, 1842

(Reproduced by kind permission from a poster belonging to Mr R. Nott of Topsham, Devon)

REWARDS

OFFERED BY THE

Woodbury Association,

FOR THE

protection of property

For every Burglary, Highway or Footpad Robbery or feloniously
assaulting with such an intent, ... £ 5 0 0
Stealing Horses, Cows, Oxen, Calves, Sheep or Lambs, or
feloniously Shearing Sheep 5 0 0
Setting Fire to any Dwelling House, Barn, or Out-house, Corn
Stack, Hay Rick, or the like, ... 5 0 0
Robbing Orchards or Gardens, Stealing any kind of Corn or
Grain, Hay or Straw, Poultry, Bees, Potatoes, Turnips, Peas or
like produce, Gates, Hurdles, Rails, Posts, or Implements of
Husbandry, Night Hunting, or Poaching by Night; wilfully or
maliciously injuring or maiming any Horse, Bullock, Sheep,

Pig or Cattle of any kind, ... 2 0 0
Not exceeding £2, nor less than 10s, at the discretion of the
    Committee.
Feloniously Milking Cows, cutting down Trees, Barking Timber,
    or Sticks likely to become Timber, or cutting Coppice or Living
    Wood of any kind, ... 1 0 0
Not exceeding £1 nor less than 10s, at the discretion of the Com-
    mittee.
But no claim to any of the above Rewards to be allowed to any
    Member of the Association, for information given by himself,
    in his own particular case of suffering.

6 The Rochdale pioneers, 1844

(Catherine Webb, ed. *Industrial co-operation: the story of a peaceful
revolution, being an account of the history, theory and practice of the
co-operative movement in Great Britain and Ireland* (Manchester:
Co-operative Union, 1904), pp. 68–9)

The objects of this Society are to form arrangements for the pecuniary benefit and improvement of the social and domestic condition
of its members, by raising a sufficient amount of capital, in shares of
one pound each, to bring into operation the following plans and
arrangements:
The establishment of a Store for the sale of provisions, clothing,
etc.
The building, purchasing, or erecting a number of houses, in which
those members desiring to assist each other in improving their domestic and social condition may reside. To commence the manufacture
of such articles as the Society may determine upon, for the employment of such members as may be without employment, or who may
be suffering in consequence of repeated reductions in their wages.
As a further benefit and security to the members of this Society,
the Society shall purchase or rent an estate or estates of land, which
shall be cultivated by the members who may be out of employment
or whose labour may be badly remunerated.
That, as soon as practicable, this Society shall proceed to arrange
the powers of production, distribution, education and government;
or, in other words, to establish a self-supporting home colony of united
interests, or assist other societies in establishing such colonies.
That, for the promotion of sobriety, a Temperance Hotel be opened
in one of the Society's houses as soon as convenient.

7 Miners at play

(Albert L. Lloyd, ed. *Come all ye bold miners: ballads and songs of the coalfield* (Lawrence & Wishart, 1952), pp. 36–7. The earliest known printing of this ballad was in 1865.)

The Cock-fight

Come all ye colliers far and near,
I'll tell of a cock-fight, when and where
Out on the moors I heard them say,
Between a black and our bonny grey.

First come in was the Oldham lads;
They come with all the money they had.
The reason why they all did say:
'The black's too big for the bonny grey.'

It's into the pub to take a sup,
The cock-fight it was soon made up.
For twenty pound these cocks will play,
The charcoal-black and the bonny grey.

The Oldham lads stood shoutin round:
'I'll lay ye a quid to half a crown,
If our black cock he gets fair play,
He'll make mincemeat of the bonny grey!'

So the cocks they at it, and the grey was tossed,
And the Oldham lads said: 'Bah, you've lost!'
Us collier lads we went right pale,
And wished we'd fought for a barrel of ale.

And the cocks they at it, one, two, three,
And the charcoal-black got struck in the eye.
They picked him up, but he would not play,
And the cock-fight went to our bonny grey.

With the silver breast and the silver wing,
He's fit to fight in front of the king.
Hip hooray, hooray, hooray!
Away we carried our bonny grey.

8 The aims of elementary education

(i) Mid-Victorian defeatism, 1861

(Royal Commission on popular education in England [Newcastle Commission] Report, *B.P.P.*, 1861, XXI, Part 1, p. 243. The words are those of an assistant commissioner, quoted approvingly in the report.)

...Even if it were possible, I doubt whether it would be desirable, with a view to the real interests of the peasant boy, to keep him at school till he was 14 or 15 years of age. But it is not possible. We must make up our minds to see the last of him, as far as the day school is concerned, at 10 or 11. We must frame our system of education upon this hypothesis; and I venture to maintain that it is quite possible to teach a child soundly and thoroughly, in a way that he shall not forget it, all that is necessary for him to possess in the shape of intellectual attainment, by the time that he is 10 years old. If he has been properly looked after in the lower classes, he shall be able to spell correctly the words that he will ordinarily have to use; he shall read a common narrative—the paragraph in the newspaper that he cares to read—with sufficient ease to be a pleasure to himself and to convey information to listeners; if gone to live at a distance from home, he shall write his mother a letter that shall be both legible and intelligible; he knows enough of ciphering to make out, or test the correctness of, a common shop bill; if he hears talk of foreign countries he has some notions as to the part of the habitable globe in which they lie; and underlying all, and not without its influence, I trust, upon his life and conversation, he has acquaintance enough with the Holy Scriptures to follow the allusions and the arguments of a plain Saxon sermon, and a sufficient recollection of the truths taught him in his catechism, to know what are the duties required of him towards his Maker and his fellow man. I have no brighter view of the future or the possibilities of an English elementary education, floating before my eyes than this. If I had ever dreamt more sanguine dreams before, what I have seen in the last six months would have effectually and for ever dissipated them.

In such inspection of schools as time and opportunity allowed me to make, I strictly limited myself to testing their efficiency in such vital points as these; never allowing myself to stray into the regions of English grammar, or English history, or physical science, unless I had previously found the ground under the children thoroughly firm, and fit to carry, without risk of settlements, a somewhat loftier and more decorated superstructure. Then it was but common justice to a conscientious teacher to take note of and show that one appreciated the higher mark at which he had aimed. Teachers look for such recognition, and cherish it as one of their best rewards.

(ii) High ideals of the Board of Education, 1905

(Board of Education, Suggestions for the consideration of teachers and others concerned in the work of public elementary schools, B.P.P., 1905, LX, pp. 14–15 The writer of this celebrated handbook was Sir Robert Morant.)

Examinations have been largely discredited by their mechanical use for the inspection of schools. The assessment of the Parliamentary Grant on the results of an examination, which became more and more formal as the numbers of children in the classes increased, led inevitably to cram. To put a pecuniary value on the success of a child in giving correct answers to questions ranging over a precise and limited field, was a sure way to spoil teaching, to weaken or destroy the interest of the pupil, and to misdirect the whole purpose of school life. Inspection of methods of teachings is now substituted for the assessment of a school by the answers of individual scholars to selected questions; but the Inspector will necessarily require to test from time to time the efficiency of these methods, and the knowledge which the pupils have obtained not merely as regards its quantity, but as regards its quality of clearness and thoroughness.

And from the teacher's point of view the test which he should apply to the value of his own teaching may, from time to time, be usefully supplemented by the test of an outside opinion, which has been formed by estimating and comparing the value of work of a similar character in the schools of a wide area often presenting a considerable variety of circumstances. The judgment formed by such an observer will clearly be useful to the teacher inasmuch as it is based upon experience ampler than that which he himself, in the nature of the case, has ordinarily been able to acquire. Experienced observers, however, sometimes find it impossible to judge the efficiency of teaching by such immediate effects upon a class as are visible to an onlooker, and may find it necessary to examine the scholars. Examination is educationally useful if carefully applied in the ordinary course of study; and an external examination, if conducted with due regard to the teaching which has been given, can be made reasonably free from the obvious evils of an examination to which all teachers alike are forced to adapt their teaching. This machinery is therefore at the disposal of the Inspector for such occasional use as may seem to him desirable. If, therefore, the Inspector thinks that he needs some further information than he can obtain by hearing a class taught or studying the records of examinations held by the teachers, it will be necessary for him in the last resort to have recourse to an examination of his own.

(e) The Teacher and his Work

The essential condition of good education is to be found in the right attitude of the teacher to his work. The greatest of human achievements, whether it be the attainment of an ideal of conduct, the mastery of the forces of nature, or the perfect expression in language of thought or fact, is the outgrowth of the individual desire to know and to do, which begins with the active curiosity of the child in face of the external world. Let the teacher realise this, and he will not fail to perceive that the value of any act of the teaching process lies not in the intrinsic utility of the subject taught, nor in the trained and skilful application of the process itself, but in the way in which it calls into play the natural activities of the children, and develops in them a sense of their powers, and of the added mastery of these which each succeeding use secures.

The teacher, therefore, must know the children and must sympathise with them, for it is of the essence of teaching that the mind of the teacher should touch the mind of the pupil. He will seek at each stage to adjust his mind to theirs, to draw upon their experience as a supplement to his own, and so take them as it were into partnership for the acquisition of knowledge. Every fact on which he concentrates the attention of the children should be exhibited not in isolation, but in relation to the past experience of the child; each lesson must be a renewal and an increase of that connected store of experience which becomes knowledge. Finally all the efforts of the teacher must be pervaded by a desire to impress upon the scholars, especially when they reach the highest class, the dignity of knowledge, the duty of each pupil to use his powers to the best advantage, and the truth that life is a serious as well as a pleasant thing.

The work of the public elementary school is the preparation of the scholar for life; character and the power of acquiring knowledge are valuable alike for the lower and for the higher purposes of life; and though the teacher can influence only a short period of the lives of the scholars, yet it is the period when human nature is most plastic, when good influence is most fruitful, and when teaching, if well bestowed, is most sure of permanent result.

9 Diminishing interest in religion, 1872

(Thomas Cooper, *The life of Thomas Cooper written by himself* (Hodder & Stoughton, 1872), pp. 392–4. Cooper, formerly a Chartist, was at this period of his life an itinerant preacher and lecturer.)

With 1870 I returned to my inquiry, and devoted January, February, March, and April again to Lancashire—renewing my work chiefly in the towns I had visited a year before, and entering a few new places. My sorrowful impressions were confirmed. In our old Chartist time, it is true, Lancashire working men were in rags by thousands; and many of them often lacked food. But their intelligence was demonstrated wherever you went. You would see them in groups discussing the great doctrine of political justice—that every grown-up, sane man ought to have a vote in the election of the men who were to make the laws by which he was to be governed; or they were in earnest dispute respecting the teachings of Socialism. *Now*, you will see no such groups in Lancashire. But you will hear well-dressed working men talking, as they walk with their hands in their pockets, of 'Co-ops' (Co-operative Stores), and their shares in them, or in building societies. And you will see others, like idiots, leading small greyhound dogs, covered with cloth, in a string! They are about to race, and they are betting money as they go! And yonder comes another clamorous dozen of men cursing and swearing and betting upon a few pigeons they are about to let fly! As for their betting on horses—like their masters!—it is a perfect madness.

Except in Manchester and Liverpool—where, of course, intelligence is to be found, if it be found anywhere in England,—I gathered no large audiences in Lancashire. Working men had ceased to think, and wanted to hear no thoughtful talk; at least, it was so with the greater number of them. To one who has striven hard, the greater part of his life, to instruct and elevate them, and who has suffered and borne imprisonment for them, all this was more painful than I care to tell.

10 American impressions of British life, 1883

(Andrew Carnegie, *An American four-in-hand in Britain* (Sampson Low, 1883), pp. 150–1, 323–4)

... We see the Black Country now, rows of little dingy houses beyond, with tall smoky chimneys vomiting smoke, mills and factories at every turn, coal pits and rolling mills and blast furnaces, the very bottomless pit itself; and such dirty, careworn children, hard-driven men, and squalid women. To think of the green lanes, the larks, the Arcadia we have just left. How can people be got to live such terrible lives as they seem condemned to here? Why do they not all run away to the green fields just beyond? Pretty rural Coventry suburbs in the morning and Birmingham at noon; the lights and shadows of human existence can rarely be brought into sharper contrast. If 'Better fifty years of Europe than a cycle of Cathay', surely better a year in Leamington than life's span in the Black Country! But do not let us forget that it is just Pittsburgh over again; nay, not even quite so bad, for that city bears the palm for dirt against the world. The fact is, however, that life in such places seems attractive to those born to rural life, and large smoky cities drain the country; but surely this may be safely attributed to necessity. With freedom to choose, one would think the rush would be the other way. The working classes in England do not work so hard or so unceasingly as do their fellows in America. They have ten holidays to the American's one. Neither does their climate entail such a strain upon men as ours does.

I remember after Vandy and I had gone round the world and were walking Pittsburgh streets, we decided that the Americans were the saddest-looking race we had seen. Life is so terribly earnest here. Ambition spurs us all on, from him who handles the spade to him who employs thousands. We know no rest. It is different in the older lands—men rest oftener and enjoy more of what life has to give. The young Republic has some things to teach the parent land, but the elder has an important lesson to teach the younger in this respect.

*　　*　　*

I do not think I have spoken of the announcements of amusements seen everywhere during the trip throughout the rural districts: band competitions, cricket matches, flower shows, wrestling matches, concerts, theatricals, holiday excursions, races, games, rowing matches, football contests, and sports of all kinds. We are surprised at their number, which gives incontestable evidence of the fact that the British people work far less and play far more than their American cousins do. No toilers, rich or poor, like the Americans! The band competitions are unknown here, but no doubt we shall soon follow so good an example and try them. The bands of a district meet and compete for prizes, which stirs up wholesome rivalry and leads to excellence. We saw eight gathered for competition in one little town which we

32　"E-532 Documents"

passed, and the interest excited by the meet was so great as to put the town *en fête*. I do not know any feature of British life which would strike an American more forcibly than these contests.

11 From craft and custom to machinery and contract, 1889

(George Sturt, *The wheelwright's shop* (Cambridge University Press, 1923), pp. 200–2)

What was to be done? How long I thought it over is more than I can at all tell now; but eventually—probably in 1889—I set up machinery: a gas-engine, with saws, lathe, drill and grindstone. And this device, if it saved the situation, was (as was long afterwards plain) the beginning of the end of the old style of business, though it did just bridge over the transition to the motor-trade of the present time.

I suppose it did save the situation. At any rate there was no need for dismissals, and after a year or two there was trade enough—of the more modern kind—to justify my engaging a foreman, whom I ultimately took into partnership. It proved a wise move from every point of view save the point of sentiment. The new head had experience and enterprise enough, without offending the men too, to develop the new commercial side—the manufacture of trade-vans and carts—when the old agricultural side of the business was dying out. Wood-work and iron-work were still on equal terms. Neither my partner nor myself realised at all that a new world (newer than ever America was to the Pilgrim Fathers) had begun even then to form all around us; we neither of us dreamt that the very iron age itself was passing away or that a time was actually near at hand when (as now) it would not be worth any young man's while to learn the ancient craft of the wheelwright or the mysteries of timber-drying. It might be that improved roads and plentiful building were changing the type of vehicles wheelwrights would have to build; but while horses remained horses and hill and valley were hill and valley, would not the old English provincial lore retain its value? We had no provocation to think otherwise, and yet:—

And yet, there in my old-fashioned shop the new machinery had almost forced its way in—the thin end of the wedge of scientific engineering. And from the first day the machines began running, the use of axes and adzes disappeared from the well-known place, the saws and saw-pit became obsolete. We forgot what chips were like. There, in that one little spot, the ancient provincial life of England was put into a back seat. It made a difference to me personally, little as I dreamt of such a thing. 'The Men', though still my friends, as I fancied, became machine 'hands'. Unintentionally, I had made them servants waiting upon gas combustion. No longer was the power of horses the only force they had to consider. Rather, they were under the power of molecular forces. But to this day the few survivors of them do not know it. They think 'Unrest' most wicked.

Yet it must be owned that the older conditions of 'rest' have in fact all but dropped out of modern industry. Of course wages are higher— many a workman to-day receives a larger income than I was ever able to get as 'profit' when I was an employer. But no higher wage, no income, will buy for men that satisfaction which of old—until machinery made drudges of them—streamed into their muscles all day long from close contact with iron, timber, clay, wind and wave, horse-strength.

12 The drink problem, 1898

(Joseph Rowntree and Arthur Sherwell, *The temperance problem and social reform* (Hodder & Stoughton, 1899), pp. 7–10, 364)

In 1898, for example, the total expenditure on alcoholic beverages in the United Kingdom amounted to £154,480,934,[2] a sum equal to—

(1) Nearly one-and-a-half times the amount of the national revenue, or
(2) All the rents of all the houses and farms in the United Kingdom. It exceeded by two-and-a-fifth millions (i.e. £2,199,211) the total expenditure on the same beverages in 1897.

The population of the United Kingdom in 1898, according to the official estimate, was 40,188,927 persons, so that the total amount expended on alcoholic liquors in that year represents an average expenditure of £3 16s 10½d for each man, woman, and child in the kingdom, or £19 4s 4½d for each family of five persons. These averages are, of course, purely arithmetical, the actual expenditure, both for individuals and families, varying, in the case of consumers, from sums relatively small to a large proportion of the entire personal or family

---

[2] These figures include the expenditure on British wines, cider, etc., which in 1898 amounted to £1,500,000. In estimating the total expenditure, the cost of British spirits has been taken at 20s per gallon; foreign and colonial spirits at 24s per gallon; beer at 54s per barrel; and wine at 18s per gallon. See 'The Annual Drink Bill', *Times*, February 24th, 1899.

income, while in the case of a considerable proportion of the population (i.e. the non-consumers) no expenditure at all was incurred.[1]
The seriousness of the figures becomes more fully apparent when
we consider them in their relation to the working classes. It is impossible, of course, to fix precisely the amount spent by the working
classes on alcoholic drinks, inasmuch as a considerable proportion
of the expenditure is made from indeterminate sources; i.e., either
(1) what is vaguely called 'pocket money', or (2) money that is inter-
cepted before it reaches the home. For this reason, so-called 'Family
Budgets' are on this point entirely unreliable and misleading. It has,
however, been authoritatively estimated[2] that of the total sum represented by the national drink bill, at least two-thirds are spent by the
working classes, who constitute, approximately, 75 per cent of the
population.[3] That is to say, of the £154,000,000 spent on drink in the
United Kingdom in 1898, more than £100,000,000 must, according to
this estimate, have been spent by 30,000,000 persons (representing
6,000,000 families)[4] belonging to the working classes. In other words,

---

[1] If we deduct the non-consuming classes (i.e., children under 15, and the
three millions of non-drinkers above the age of 15), the average annual
expenditure of the remaining twenty-three millions (i.e., the actual consumers)
in nearly doubled, amounting to £6 18s 11d per individual.

[2] Professor Leone Levi, *Journal of the Statistical Society*, March, 1872.
This estimate has recently been accepted in a statement prepared in the
interests of the Trade for the Royal Commission on Liquor Licensing Laws
(July, 1897), and subsequently published as a pamphlet by Messrs Peter Walker and Son, of Warrington, See *The Fallacies of Teetotalers*, p. 13.

[3] In 1882, a Special Committee of the British Association estimated the
proportion of the labouring classes at 70 per cent of the population. A similar
proportion, based, however, upon different figures, is given by Mulhall,
Dictionary of Statistics, p. 320; and by Professor Leone Levi, *Report of the
British Association* (1888), p. 361; and *Wages and Earnings of the Working
Classes* (1885), p. 2. For the purposes of the present discussion, however, it
will be safer to take the slightly higher estimate given above, which more than
allows for the relatively greater increase of the working classes in the intervening years as compared with the growth of the population as a whole.

[4] This estimate is based upon a proportion of five persons to a family. The
proportion in 1891, according to the official census returns, was 4·73. Mulhall
(*Dictionary of Statistics*) estimated the number of working-class families
in 1889 at 4,774,000, and Professor Leone Levi (1885) at 5,600,000. The latter
estimate (Professor Levi's) is however, based upon a proportion of 4·64
persons to a family. On the basis of five persons chosen above, the number
of working-class families in 1885, according to Professor Leone Levi, would
be 5,200,000—an estimate which, allowing for the growth of four millions
in the population of the United Kingdom since 1885, confirms the estimate
adopted above.

every working-class family spent on an average, in 1898, no less than
£16 13s 4d, or 6s 5d per week on alcoholic liquors, a sum which (assuming the average income of a working-class family to be thirty-five
shillings per week) is equal to more than one-sixth of the entire family
income.

*   *   *

The Need for Constructive as well as Controlling Reforms

But it is important here to urge—what is too often forgotten by
those interested in this question—that the adoption of the controlling system, while effective in withdrawing the traffic from private
hands, would not, by itself, adequately solve the problem. For what
is the problem? Men go to the public-house—young men especially—
quite as much for social intercourse, and for escape from their surroundings, as for drink. The love of drink—as the *Times* pointed out
twenty-five years ago—'is a symptom only: it is not the disease, and
we should be wrong to deal with it as if it were. A man drinks, not
only because his brute nature is strong and craves the stimulus, but
because he has no other interests, and must do something; or because
his home is uncomfortable and his life dull, and he needs some real
enjoyment; or because he is fond of company, and only wishes to
be like the rest.'

13 Flight from the land, 1902

(H. Rider Haggard, *Rural England, being an account of agricultural
and social researches carried out in the years 1901 & 1902* (new ed.
2 vols. Longmans, 1906), II, pp. 545–6.)

In face of these advantages, however, the rural labourer has never
been more discontented than he is at present. That, in his own degree,
he is doing the best of the three great classes connected with the land
does not appease him in the least. The diffusion of newspapers, the
system of Board school education, and the restless spirit of our age
have changed him, so that now-a-days it is his main ambition to
escape from the soil where he was bred and try his fortune in the cities.
This is not wonderful, for there are high wages, company, and amusement, with shorter hours of work. Moreover on the land he has no
prospects: a labourer he is, and in ninety-nine cases out of a hundred
a labourer he must remain. Lastly, in many instances, his cottage
accommodation is very bad; indeed I have found wretched and insufficient dwellings to be a great factor in the hastening of the rural

exodus; and he forgets that in the town it will probably be worse. So he goes, leaving behind him half-tilled fields and shrinking hamlets. Moreover, even of those young men who remain but few care to become masters of their work. Here is an instance of which I have just been told, in September, 1902. The Technical Committee of the Norfolk County Council allotted to Ditchingham and a group of three or four other parishes, £9 to be given in prizes at a ploughing competition. From the whole parish of Ditchingham with its population of about 1,100 but one man has entered—a servant of my own— and from the group of parishes, I am informed, *not a single lad is forthcoming*, although a sum of £3 was set aside to be given as prizes in the boys' ploughing class. The fact is, of course, that the youth of this, as of other districts, does not wish to learn to plough, even when bribed so to do with prizes, and that here, before long, ploughmen, or any skilled labourers, will, to all appearances, be scarce indeed.

14 A wasteful society, 1905

(L. G. Chiozza-Money, *Riches and Poverty* (Methuen, 1905), pp. 318–21)

If our national income had but increased at the same rate as our population since 1867 it would, in 1905, amount to but £1,200,000,000. As we have seen, it is now about £1,700,000,000. Yet the Error of Distribution remains so great that while the total population in 1867 amounted to 30,000,000, we have to-day a nation of 30,000,000 poor people in our rich country, and many millions of these are living under conditions of degrading poverty. Of those above the line of primary poverty, millions are tied down by the conditions of their labour to live in surroundings which preclude the proper enjoyment of life or the rearing of healthy children. The comparatively high wages of London are accompanied by rents high in proportion and frequently by waste of income and time upon travelling expenses. In so far as the manual labourers have been reduced in proportion to population it has been to swell the ranks of black-coated working men, clerks, agents, travellers, canvassers, and others, whose tenure of employment is precarious, whose earnings are very low, and whose labour as we have already noted is largely waste.

We have won through the horrors of the birth and establishment of the factory system at the cost of physical deterioration. We have purchased a great commerce at the price of crowding our population into the cities and of robbing millions of strength and beauty. We have given our people what we grimly call elementary education and robbed them of the elements of a natural life. All this has been done that a few of us may enjoy a superfluity of goods and services. Out of the travail of millions we have added to a landed gentry an aristocracy of wealth. These, striding over the bodies of the fallen, proclaim in accents of conviction the prosperity of their country. . . .

Blessed indeed are the Rich, for theirs is the governance of the realm, theirs is the Kingdom. Theirs is a power above the throne, for it has been a maxim of British politics that our government should be a poor government, and a poor government cannot contend in the direction of affairs with the imperium of wealth. This may be illustrated by our attempts to 'educate' the mass of the people. For a few brief years the government, with small funds raised with timorous hands, does a little to form the mind and character of the child. Even in these early years it consents that the future proud citizen of Empire shall be improperly fed and badly housed. These early moments passed, the mockery of 'education' ceases, and the child, taught by the State to read, to write, and to cipher, becomes a unit of industry. At this point begins the serious training of the citizen. Forthwith he is inducted into some more or less worthy employment, that employment, as we have seen, resulting from the great expenditure of the few and the poor expenditure of the many. Careers are thus chiefly shaped by the wealthy, for theirs is the greatest call. The demand for luxuries is too great; the demand for necessaries is too small; the unit of industry is fortunate, therefore, if he is inducted into useful service. The State washes its hands of his development. The educational sham over, the real education of life begins. So far as the State calls for privates of industry it is chiefly to make them soldiers, sailors, makers of guns, builders of battle-ships. The development of all things useful, of railways, of canals, of roads, of cities, of houses, is resigned to the blind call for commodities and the intelligence of individuals who, in search of private gain, seek, without regard to the national well-being, to profit by that blind call.

Yet the manner in which its people are employed matters everything to a nation. It is not sufficient to give the child a smattering of knowledge. We need to take a collective interest in the general education of our citizens, and that education is the result of expenditure. The consumer gives the order. Given a fairly equable distribution of income, the call will be as to the greater part for worthy things, as to the smaller part for luxuries. Given a grossly unequal distribution, and the call for luxuries will be so great as to divert a considerable part of the national labour into channels of waste and degradation. To keep a government poor is to keep it weak. The poor government may resolve to educate, but it will have no means to carry out its resolve; its teachers will be underpaid; its schools inefficient. The poor government may pass Housing Acts; it will but call for better houses that will not come when it does call for them. The poor government may piously resolve to create small holdings; there will be no means to carry out the pious resolve. The poor government may, at periodic intervals, look the question of Unemployment in the face; its legislation will but reflect its poverty, and be in its provisions an acknowledgment that the power to employ, the power to govern, is in other hands.

15 Adult education, 1914

(R. H. Tawney, 'An experiment in democratic education', *Political Quarterly*, No. 2 (May 1914), 62–84, reprinted in *The radical tradition: twelve essays on politics, education and literature*, ed. Rita Hinden (Allen & Unwin, 1964), pp. 80–1)

University Tutorial Classes are not, in short, an alternative to a university education, a *pis aller* for those who cannot 'go to a university'. Nor are they merely a preparation for study in a university. They are themselves university education, carried on, it is true, under difficulties, but still carried on in such a way as to make their promotion one among the most important functions of a university. If this is not yet fully recognized it is because one of the besetting sins of those in high places in England—it is not that of the working classes—is the bad utilitarianism which thinks that the object of education is not education, but some external result, such as professional success or industrial leadership. It is not in this spirit that a nation can be led to believe in the value of the things of the mind. In the matters of the intellect, as in matters of religion, 'High Heaven rejects the lore of nicely calculated less or more'. And it is, perhaps, not fanciful to say that the disinterested desire of knowledge for its own sake, the belief in the free exercise of reason without regard to material results and because reason is divine, a faith not yet characteristic of English life, but which it is the highest spiritual end of universities to develop, finds in the Tutorial Classes of the Workers' Educational Association as complete an expression as it does within the walls of some university cities. To these miners and weavers and engineers who pursue knowledge with the passion born of difficulties, knowledge can never be a means, but only an end; for what have they to gain from it save knowledge itself?

Historians tell us that decadent societies have been revivified through the irruption of new races. In England a new race of nearly 900,000 souls bursts upon us every year. They stand on the threshold with the world at their feet, like barbarians gazing upon the time-worn plains of an ancient civilization. If, instead of rejuvenating the world, they grind corn for the Philistines and doff bobbins for mill-owners, the responsibility is ours into whose hands the prodigality of Nature pours life itself, and who let it slip aimlessly through the fingers that close so greedily on material riches.

16 Public squalor without private affluence, 1933

(J. B. Priestley, *English journey, being a rambling but truthful account of what one man saw and heard and felt and thought during a journey through England during the autumn of the year 1933* (Heinemann and Gollancz, 1934), p. 302)

'Do you know what's the biggest town—either on the railway or on the direct road—between Newcastle and London? I'll bet you'll never guess. Well, this is it—Gateshead. You can catch a lot of people out with that. Gateshead's the biggest town between Newcastle and London. It's got more than a hundred and twenty-five thousand people in it, Gateshead has'.

And that, as he went on to admit, is about all Gateshead has in it. One hundred and twenty-five thousand people, but no real town. It has fewer public buildings of any importance than any town of its size in the country. If there is any town of like size on the continent of Europe that can show a similar lack of civic dignity and all the evidences of an urban civilisation, I should like to know its name and quality. No true civilisation could have produced such a town, which is nothing better than a huge dingy dormitory. I admit that it is only just across the river from Newcastle, which has some amenities. But Gateshead is an entirely separate borough. It has a member of Parliament, a mayor, nine aldermen, and twenty-seven councillors; and I hope they are all delighted with themselves. They used to build locomotives in Gateshead, very fine complicated powerful locomotives, but they never seem to have had time to build a town. A place like this belongs to the pioneer age of industrialism, and unfortunately the industry appears to be vanishing before the pioneers themselves have time to make themselves comfortable. It is a frontier camp of bricks and mortar, but no Golden West has been opened up by its activities. If anybody ever made money in Gateshead, they must have taken great care not to spend any of it in the town. And if nobody ever did make money in the town, what is it there for? It cannot be there for fun. Gateshead is not Somebody's Folly.

17 Contrasts: Britain depressed and Britain prosperous

(Tom [Thomas H.] Harrisson and others, *Britain revisited* (Gollancz 1961), pp. 29–30, 33, 207–8)

Touches of Colour

85 Davenport Street, one in a continuous row of bug-ridden houses where we lived for years, looked very much the same in the smoke-filtered summer sunlight (. . . . . .). But there was one vivid difference there at once: a handsome negro with a lively tie, leaning on the gate of the house next door, No. 87. And presently, out came another coloured gentleman, in bus conductor's uniform. None of us could remember seeing a coloured man in Worktown [Bolton] before. The nearest thing then was seeing a chimney-sweep—which was considered 'lucky'.

Closer inspection showed that there were many smaller differences in Davenport Street, though. Lots more bright colours, especially on the doors. Many of the old, standard, clumsy, brown, cracked wooden doors replaced with finer wood or ply, including Yale locks, letter slots and good quality chromium handles. The windows of many houses—subsequent counts gave over a third in some streets—have been decorated from the outside with do-it-yourself lead strips, sometimes in intricate patterns. Before the war, window leading was almost a monopoly of the better-off home where it was 'built-in'. Now it is becoming standard for all. . . .

Continentalia

Down at the bottom of Davenport Street, what had once been an ordinary corner grocer's now had smart new wooden slat-panelling over the front and a big notice:

CONTINENTAL AND DELICATESSEN SHOP

This reflected what we soon found to be one of the more important lesser changes in Worktown's outlook: a wider acceptance of the world beyond even Blackpool. Unthinkable in the thirties, successful now, were two Chinese, Spanish and Greek restaurants, an Indian one (often open after midnight). It used to be impossible to get anything except fish and chips after 7.30 p.m. The Nevada Skating Rink, first opened as such in 1909, cinema in the thirties, reverting to rink in 1955, currently boasts a 'Dutch Bar' for snacks, a 'Swiss Bar' for soft drink and sweets. Brand-new is the Ukrainian Society (with 100 members), the French Circle (founded 1950) and a thriving International Club (1956). Estimates give up to ten times as many Worktowners going abroad as pre-war. Cook's are currently opening large new premises beside the Town Hall. A travel agency in the main shopping street was window-plugging tours to Russia—July 1960.

The town's oldest established wine-merchants (we could seldom afford his good sherry in earlier days) told one of us:

'In Worktown people are *thinking* about wine nowadays. People are much more adventurous. We stock and sell *eleven* types of Vodka alone. The number of working class who travel abroad and get the taste is remarkable'. (11 August 1960)

Clothes Colours

A more conspicuous change is among women. Counts, for instance, made at the same place and time and on the same dates as in those earlier summers give:

	% of all women	
	1940	*1960*
Women wearing mainly:	%	%
Brown	21	2
Grey	19	5
Blue	16	24
Green	12	14

Perhaps the largest single change, however, in summer 1960, is to one woman in three wearing a main dress scheme of definite patterns, usually floral, away from simple colour. 16% had strongly floral designs, often associated with hoop skirts; 10% had checks—of which there were none in 1940 counts (in those days checks were almost confined to upper middle-class women in Worktown, few of whom came through this street on a Saturday afternoon).

These figures do not adequately reflect the gaiety of 1960 dress as compared with 1940, nor the wide range of intricate detail (and often of mixed-up colour), as compared with the comparative *drabness* of 1940. In 1940 over half, but in 1960 one in twenty only, of all women's dresses could be called monotone.

This feeling of gaiety in colour is all through the town, on street doors and repainted house fronts, in women's shoes, men's ties and funeral robes for the deceased. Most outsiders would still call Worktown 'drab'. In many parts of the sprawling town-land the eye can see no tree, no growing thing. But out of this drabness now bud vivid gestures and blossom splashes of wide gaiety.

Jobs— (ut of 'Insecurity'

There were 13,855 *registered* unemployed in Worktown, when the present writer first went to work there in 1936. Through 1937 the figure seldom fell below 10,000. There were under 1,000 men unemployed in Worktown in the summer of 1960.

In 1936–7 it is fair to say that the whole *atmosphere* breathed insecurity and dread of unemployment. In 1960 we seldom felt such winds of fear, either among old-timers or younger folk.

*   *   *

In the result, TV's growth temporarily affected people much as they expected. But there are numerous signs that these effects are now wearing or easing off, as the adaptable Briton makes new adjustments. Nor do these adjustments differ *essentially* from others required by the same technology which has made TV possible; the technology that has produced Washing Machines, Spin Dryers, Refrigerators, Hifi, cheap and common place now where before they were for the few and well-off, if at all. Much of the leisure which TV is taking up is thereby a new leisure, too.

Here we come to a vital question which is seldom asked by those who feel that TV has 'kept people indoors when they should be out', and all that. What *did* people do on a winter's day or a Sunday afternoon of rain before TV—especially if they hadn't much money? The short answer for millions of people then: 'Nothing much'. They 'messed about' or did odd things (e.g. read pulp mags and Westerns, argued, titivated each other), which can hardly be said to be 'worse than looking into polypact television'. This is a simplified abbreviation of the past: but no one who lived long in a working-class street in the thirties can really dispute the basic truth.

The simple fact is, that on a winter's evening *before* the days of television, in a less affluent society, the great majority of people spent their time half-listening to the sound radio (on Sunday largely commercial sound), and doing all sorts of odds and ends indoors. When they could, they also went out to the cinema, the greyhounds, all-in wrestling, the sea-side (though seldom into the sea), and so on—but not very frequently; and some of them never went out at all in foul weather, or on the Sabbath. There has been some movement away from going out to these things, though it is doubtful if this movement is necessarily permanent. On the other hand, many of the people who no longer are going so much to cinemas (for instance) are no longer doing so for two quite other reasons:

1. Because they are going and doing other, and often more expensive things—such as driving out in a car, which they now own; sailing a small boat on one of the reservoirs; fishing; running a motor bike into a city for something there; going on evening train excursions; taking meals out in a new growth of restaurants and cafés.
2. A lot of people were getting fed up with the standard of cinema films in any case. The same applies to other sorts of outing as well; but most sorts of choice at this level, in a modern society, require money.

18 The motor age, 1963

(*Traffic in towns: report of the steering group* [the Buchanan Report] (H.M.S.O., 1963), paras. 6–7)

In Britain the Motor Age is still at a comparatively early stage. We are approaching the crucial point when the ownership of private motor vehicles, instead of being the privilege of a minority, becomes the expectation of the majority. At present there are in Great Britain about 16·4 million families and 6·6 million cars (excluding buses and lorries, and 1·8 million motor cycles. When allowance has been made for cars that are not privately owned, and for families that own more than one car, it is not yet true to say that more than a bare majority of families own a motor vehicle—and perhaps not quite that.

There is no doubt that the desire to own a car is both widespread and intense. The number of people who genuinely do not desire to possess their own private means of transport must be very small, and we think it is safe to base estimates of the future on the assumption that nearly all the families who, at any time can afford to own a car (or who think they can) will in fact do so. The ownership of private cars is a direct function of real incomes. Not only, therefore, will the total number of cars increase as average real incomes rise, but the most rapid increase will come sooner rather than later, for the reason that there is a very large group of family incomes now just on the verge of the car-affording class.

10

Economic policy and taxation

Wise policy springs from knowing one's limitations. The first and last documents in this chapter, the one protesting at a disastrous war, the other offering to surrender economic sovereignty to the European Economic Community resemble each other in that both recognise limits to England's economic (and military) power. The intervening two centuries were the time when English statesmen had to worry least about the limits of their country's power. After Waterloo governments became modest in the extreme. Abdicating from responsibility for the control of foreign trade (*3–5*), lacking the will or the knowledge to protect factory children and town dwellers from the early consequences of industrialism, they contented themselves with raising as little taxation as was necessary to carry out the exiguous duties of public administration and of defence. It is true that by the end of the eighteen eighties government and particularly local government had acquired a long list of functions, many of them new in the nineteenth century. Nevertheless, England remained, considering its wealth and the complexity of its economic life, lightly governed and lightly taxed down to the death of Queen Victoria and beyond. The turning point was the budget of 1909. It deliberately set out to raise the level of taxation on large incomes not only to finance naval rearmament but also to provide money for social reform. Despite the furore that the budget aroused there can be little doubt that the country's system of taxation remained regressive after 1909 as before, i.e. the poor paid a larger proportion of their income in taxes than the rich. Local rates, beer and tobacco duties, taxes on tea and sugar, all took a larger slice of the income of the poor than of the rich, and more than offset the still modest rates of income tax that affected incomes of £160 a year and upwards. Both the world wars led to a permanently higher level of taxation in the succeeding years of peace, partly because of the increased burden of debt, partly because of the wider range of social services demanded by the electorate. But the tradition of low taxes died hard: there were determined attempts to reduce government expenditure after the First World War (*8*) and again in the economic depression of 1931. Similar but milder attempts have been made in the periodic balance of payments crises that have afflicted the economy since 1945 (*11*). Government has always been interested in promoting prosperity or what is now called economic growth. Policy to this end may have changed—from Peel's free trade budgets in the 1840s (*4*) to protection (*9*) or to direct management of the economy to produce a 'high and stable level of employment' (*10, 11*). Since 1945 international co-operation in the shape of the International Monetary Fund (I.M.F.) and the General Agreement on Trade and Tariffs (G.A.T.T.) has had the support of successive governments although there have been difficulties in reconciling economic growth with fixed exchange rates and with freer trade. The recent and eventually successful applications to join the 'Common Market' (*12*) are further evidence of governmental interest in the promotion of prosperity, this time by the painful method of exposure to foreign competition. The ends of economic policy do not change much; it is the means that give the subject its variety and interest.

1 The disasters of war, 1781

(*Addresses, remonstrances and petitions to the throne presented from the court of aldermen* (Corporation of the City of London, 1865), pp 60-2)

In a Meeting or Assembly of the Mayor, Aldermen, and Liverymen of the several Companies of the City of *London*, in Common Hall assembled, at the *Guildhall* of the said City, on Thursday, the Sixth day of December, 1781,

A Motion was made and Question put, That an humble Address, Remonstrance, and Petition be presented to His Majesty, from the Lord Mayor, Aldermen, and Livery of the City of *London*, in Common Hall assembled, on the present alarming situation of public affairs; which Address, Remonstrance, and Petition was agreed to, as follows:

To the King's Most Excellent Majesty

The humble Address, Remonstrance, and Petition of the Lord Mayor, Aldermen, and Livery of the City of *London*, in Common Hall assembled.

May it please your Majesty,

Impressed with an awful sense of the dangers which surround us;

feeling, for ourselves and our posterity, anxious for the glory of a country hitherto as much renowned for the virtues of justice and humanity as for the splendour of its arms, we approach your Throne with sentiments becoming citizens at so alarming an hour, and at the same time with that respect which is due to the monarch of a free people and a prince of the illustrious House of Brunswick, to which we feel ourselves, in a peculiar manner, attached by all the ties of gratitude and affection.

It is with inexpressible concern that we have heard your Majesty declare, in your speech to both Houses of Parliament, your intention of persevering in a system of measures which has proved so disastrous to this country.

Such a declaration calls for the voice of a free and injured people. We feel the respect due to Majesty; but in this critical and awful moment to flatter is to betray.

Your Majesty's Ministers have, by false assertions and fallacious suggestions, deluded your Majesty and the nation into the present unnatural and unfortunate war.

The consequences of this delusion have been that the trade of this country has suffered irreparable losses, and is threatened with final extinction.

The manufactures in many valuable branches are declining, and their supply of materials rendered precarious by the inferiority of your Majesty's fleet to that of the enemy in almost every part of the globe.

The landed property throughout the kingdom has been depreciated to the most alarming degree.

The property of your Majesty's subjects vested in the Public Funds has lost above one-third of its value.

Private credit has been almost wholly annihilated by the enormous interest given in the public loans, superior to that which is allowed by law in any private contract.

Such of our brethren in America as were deluded by the promises of your Majesty's Ministers and the proclamations of your generals, to join your Majesty's standard, have been surrendered by your Majesty's armies to the mercy of their victorious countrymen.

Your Majesty's fleets have lost their wonted superiority.

Your armies have been captured.

Your dominions have been lost.

And your Majesty's subjects have been loaded with a burthen of taxes which, even if our victories had been as splendid as our defeats have been disgraceful, if our accession of dominion had been as fortunate as the dismemberment of the empire has been cruel and disastrous, could not in itself be considered but as a great and grievous calamity.

We do, therefore, most humbly and earnestly implore your Majesty to take all these circumstances into your Royal consideration; and to compare the present situation of your dominions with that uncommon state of prosperity to which the wisdom of your Royal ancestors, the spirit and bravery of the British people, and the favour of Divine Providence—which attends upon principles of justice and humanity—had once raised this happy country, the pride and envy of all the civilised world.

We do beseech your Majesty no longer to continue in a delusion from which the nation has awakened; and that your Majesty will be graciously pleased to relinquish entirely and for ever the plan of reducing our brethren in America to obedience by force, a plan which the fatal experience of past losses has convinced us cannot be prosecuted without manifest and imminent danger to all your Majesty's remaining possessions in the Western world.

We wish to declare to your Majesty, to Europe, to America itself, our abhorrence of the continuation of this unnatural and unfortunate war, which can tend to no other purpose than that of alienating and rendering irrecoverable the confidence of our American brethren, with whom we hope to live upon the terms of intercourse and friendship so necessary to the commercial prosperity of this kingdom.

We do, therefore, further humbly implore your Majesty that your Majesty will be graciously pleased to dismiss from your presence and councils all the advisers, both public and secret, of the measures we lament, as a pledge to the world of your Majesty's fixed determination to abandon a system incompatible with the interests of your crown and the happiness of your people.

2 The income tax repealed, 1816

(*Parliamentary Debates*, first series, XXXIII (18 March 1816), cols. 421-2, 429, 435, 440-1, 450-1)

Motion for the Continuance of the Property Tax The House having resolved itself into a committee of ways and means, for raising the supply granted to his majesty, Mr Brogden in the chair,

The *Chancellor of the Exchequer* rose to submit his proposition for the continuance of the property tax. He expressed his hope that when the very important and serious nature of the subject was considered, the committee would apply itself to the question with that calmness and impartiality which its importance demanded. He was now to submit for the serious and deliberate consideration of parliament his proposition for the continuance of the property tax: and as the House had already sanctioned the estimates for a considerable naval and military establishment, it was an obvious consequence, that the necessary means must be allowed for the support of those establishments. It was said, that the general sense of the country was strongly adverse to the continuance of this tax; but he could not consider that opinion, in whatever degree it might exist, otherwise than as a prejudice, arising partly from feelings of very natural impatience, and partly from misapprehensions and misrepresentations, which, he was persuaded, would be removed when the subject came to be better considered. He had been told of the vast number of petitions which had come up against the measure. Still he thought it his duty to submit the question for the calm and deliberate consideration of parliament, and by its judgment he would abide. The petitions contained only the sentiments of a very small proportion of the people (*Hear, hear!*). It could not be contended that the signatures to those petitions, although very numerous, contained a very great proportion of the whole population of the country: nor could it be expected that many persons would carry disinterestedness and public spirit so far as to petition in favour of a heavy tax: though, in fact, there were on the table several petitions praying for its continuance with more or less modification. But, even if a much greater number had petitioned against the measure, he should still consider it his duty to submit it to the consideration of the House, and to give parliament an opportunity of judging of its propriety, by laying before them the grounds upon which he thought it fitting to make the proposition. The petitioners had only attended to the pressure upon themselves, which they were naturally anxious to remove, because they thought it no longer necessary. But he was persuaded that such would not have been their judgment, if they had had an opportunity of being fully acquainted with the whole matter, and of deliberating calmly and impartially upon the subject.

*   *   *

He was unwilling in the present stage to trouble the committee by entering into particulars, but he trusted he had proved that the present state of the public credit rendered it much more expedient to raise the supplies of the year by a tax than by a loan, and he thought it would be easy to prove that no other tax could raise the large sum required with so little pressure as the property tax. Among the many opinions to which he had listened, he had hardly ever heard the general equality and equity of the principle of the property tax denied—[*Hear, hear! from the opposition side*)—he meant compared with the circumstances of the individual and the weight of other taxes; those upon articles of consumption bore with a very unequal pressure, not being in proportion to the resources, but to the expenditure of the individual, which a thousand circumstances might affect.

*   *   *

Mr William Smith stated, that the consideration of this tax had long occupied his attention: he had, however, not found an opportunity of addressing the House before, and wished now to deliver his sentiments on the subject. Whether in war or in peace, he had on all occasions denied his assent to this tax: and he thought the objections against it so general, as to apply to all times and all situations. Those objections were not directed against the amount alone, but against the very principle of the tax. As far as the commercial world was concerned, the great evil of the tax consisted in that inquisitorial visiting which laid open to the world, with the most ruinous effect, the exact situation of every man's affairs, however he might wish for concealment. And another important objection, hitherto little dwelt on, was the unfair manner in which incomes arising from professions were rendered subject to this tax. As to commercial men, the disquiet they must experience at having their concerns laid open to the world would be very little alleviated by any of the expedients or modifications now suggested by the right hon. gentleman opposite. It was probable that a large proportion of the commercial interest must now be liable to heavy losses; and it followed that a trader must either pay five per cent on a supposed profit, or go to the commissioner and confess his loss: rather than do this, numbers would pay the tax, so that it would be a tax, not on income, but on loss.

*   *   *

Mr Baring having with some difficulty obtained a hearing, said he should not trespass long on the attention of the committee, as such anxiety prevailed for a coming to decision. In another state of the House he should have wished to go into details; but that seemed now quite impossible. The right hon., the treasurer of the navy (*Mr Rose*) was a general panegyrist of all taxes and tax-gatherers—to

33*

answer him was quite superfluous. When the hon. member for Bristol (*Mr Hart Davis*) rose, he had anticipated that he would find some good reasons for supporting the ministers and deserting his constituents. But from whence he had drawn the conclusion, that the cows of the people of Bristol would be raised by the operation of the tax from 10*l* to 11*l* each, it was out of the power of ingenuity to imagine. That hon. member had been lavish in his praise of the 'luminous' statement of the chancellor of the exchequer. If he (*Mr B.*) had thought that the statement of the chancellor of the exchquer had been any thing but a desperate plunge in default of all argument—and if it had made any impression on the committee, he, though he had not much opinion of his own ability, should have taken up some of the time of the House to refute it. But he should set the member for Bristol right as to one fact connected with that statement. The member for Bristol had supposed that the tax was to last only for two years. The chancellor of the exchequer had that night stated no such thing. Sometimes, indeed, the ministers held forth the promise of the tax being put an end to in two years, but when they felt a little stronger, no such promise was made. This promise was like the French constitution, which had been compared to an umbrella, which was held up by the king in bad weather only. It was now pretended that it was to be continued only for the payment of an uncertain sum, which seemed to be about 14 millions. Was it probable that the tax, pared down as it would be, with little left but its inquisitorial power and its inequality, would produce that sum in two years? It was improbable, as long as the ministers could show any tail of the war expense, that they would give it up. If they could get the tax once fixed in time of peace, they would gain their object, and not a member of that House would ever see an end of it. No one who had the least experience of the conduct of government, could fail to see, that instead of producing a redemption of the debt, it would produce wasteful establishments and extravagant expenditure. It could only be by ridding the country of the tax that ministers could be induced to practise any economy. A right hon. gentleman (*Mr Rose*) had compared the distress of the country now to that of 1797; and had argued, that as the distress was removed by imposing the tax then, to impose the tax now would afford the same relief. But between the distress of 1797 and that of the present moment there was no analogy. At that time there was a poor exchequer and a rich country. Now the case was the reverse—we had a rich exchequer and a poor country. The remedy now applied should be the reverse of that applied in

1797. Instead of imposing additional taxes to relieve the exchequer, the object should be to take off as much of the taxes as was possible to relieve the country.

*   *   *

Mr Wilberforce conceived it fair to conclude, that ministers were actuated by what they considered a strong sense of duty, in persevering in a measure so unpopular. Yet, after the concession of his noble friend, that a great feeling of expectation had been entertained by the country that the tax would expire with the war, he thought its renewal would be harsh under any circumstances. But, under what circumstances was it proposed, when it was rather expected that a bonus would be given to the people? His noble friend had given a tremendous view of the subject. If he understood him rightly, he had shown that at the end of two years the necessity for the continuance of the tax would be as strong as it was now said to be. He had therefore proved too much. He had proved that the country would never obtain the relief which it ought to receive, and that the only means of gaining it, was the curtailment of her expenditure. (*Repeated shouts of Hear, hear, hear, for several minutes.*) The question was, which should be relieved, the money market, or the people? When the people of England should know the amount thus imposed on them, they would think that the House had been carried away by speculations on general principles, and had not felt for but contemned those distresses, which were very great indeed, and were increased, as his noble friend himself had said, by the confident expectation of relief—an expectation which would have been sacrificed to the wish of assisting the money market. (*Here the shouts of Hear! hear! were so frequent, and the anxiety of the House to divide, so tumultuous, that little more became audible in the gallery, but by the thundering peal which attended the close of the hon. member's speech, it was understood that he had once more expressed his determination to vote against the tax.*)

The House then divided:

For the continuance of the Property Tax 201
Against it 238
    Majority against the continuance of the Property Tax 37

As soon as the numbers were announced in the House, a loud cheering took place, which continued for several minutes. Similar exultation was manifested by the crowd of strangers in the lobby and the avenues of the House.

3 An influential report, 1840

(Select committee on import duties, *B.P.P.*, 1840, V, pp. iii–vii)

The Evidence is of so valuable a character, that Your Committee could hardly do justice to it in detail, unless they were to proceed, step by step, to a complete analysis, which the advanced period of the Session will not allow them to do. They must, therefore, confine themselves to reporting the general impressions they have received, and submit the Evidence to the serious consideration of The House, persuaded that it cannot be attentively examined without producing a strong conviction that important changes are urgently required in our Custom-house legislation.

The Tariff of the United Kingdom presents neither congruity nor unity of purpose; no general principles seem to have been applied.

The Schedule to the Act 3 & 4 Will. 4, c. 56, for consolidating the Customs Duties, enumerates no fewer than 1,150 different rates of duty chargeable on imported articles, all other commodities paying duty as unenumerated; and very few of such rates appear to have been determined by any recognised standard; and it would be difficult for any person unacquainted with the details of the Tariff to estimate the probable amount of duty to which any given commodity would be found subjected. There are cases where the duties levied are simple and comprehensive; others, where they fall into details both vexatious and embarrassing.

The Tariff often aims at incompatible ends; the duties are sometimes meant to be both productive of revenue and for protective objects, which are frequently inconsistent with each other; hence they sometimes operate to the complete exclusion of foreign produce, and in so far no revenue can of course be received; and sometimes, when the duty is inordinately high, the amount of revenue becomes in consequence trifling. They do not make the receipt of revenue the main consideration, but allow that primary object of fiscal regulations to be thwarted by an attempt to protect a great variety of particular interests, at the expense of the revenue, and of the commercial intercourse with other countries.

Whilst the Tariff has been made subordinate to many small producing interests at home, by the sacrifice of Revenue in order to support these interests, the same principle of preference is largely applied, by the various discriminatory Duties, to the Produce of our Colonies, by which exclusive advantages are given to the Colonial Interests at the expense of the Mother Country. Your Committee would refer to the Evidence respecting the articles of Sugar and Coffee, as examples of the operation of these protective Duties. ...

It appears from the evidence of Mr Porter, of the Board of Trade, that the total amount of Customs Revenue received in the United Kingdom in the year ending January 1840, was 22,962,610*l*, of which total amount,

17 articles, each producing more than 100,000*l*, produced 94½ per cent, or	£	21,700,630
That 29 articles produced, 3$\frac{9}{10}$ per cent, or		898,661
And that these 46 articles produced 98$\frac{2}{5}$ per cent, or	£	22,599,291
That all other articles, amounting to 144 in number, produced 1$\frac{3}{5}$ per cent, or		363,319
Showing that 190 articles, exclusive of about 80,000*l* collected upon 531 other articles, and excluding 147 articles, upon which an excess of drawback of 5,398*l*, was allowed, produced the total revenue of	£	22,962,610

It will be seen that 17 articles, affording the largest amount of Customs Revenue, are articles of the first necessity and importance to the community: viz. sugar, tea, tobacco, spirits, wine, timber, corn, coffee, butter, currants, tallow, seeds, raisins, cheese, cotton wool, sheep's wool, and silk manufactures; and that the interests of the Public Revenue have been by no means the primary consideration in levying the Import Duties, inasmuch as competing foreign produce is in some instances excluded, and in others checked by high differential duties, levied for the protection of British colonial interests; and in many cases, such differential duties do not answer the object proposed, for it appears, in the case of foreign clayed sugars, where it was obviously intended they should be excluded from the British market, that the monopoly granted to British colonial sugars has so enormously raised the prices in our market, that they have lately come into consumption, though charged with a duty of 63*s* per cwt, while our plantation sugars pay only 24*s*. ...

Your Committee cannot refrain from impressing strongly on the attention of The House that the effect of prohibitory duties, while they are of course wholly unproductive to the revenue, is to impose an indirect tax on the consumer, often equal to the whole difference of price between the British article and the foreign article which the prohibition excludes. This fact has been strongly and emphatically urged on Your Committee by several witnesses; and the enormous extent of taxation so levied cannot fail to awaken the attention of

The House. On articles of food alone, it is averred, according to the testimony laid before the Committee, that the amount taken from the consumer exceeds the amount of all the other taxes which are levied by the Government. And the witnesses concur in the opinion that the sacrifices of the community are not confined to the loss of revenue, but that they are accompanied by injurious effects upon wages and capital; they diminish greatly the productive powers of the country, and limit our active trading relations.

Somewhat similar is the action of high and protective duties. These impose upon the consumer a tax equal to the amount of the duties levied upon the foreign article, whilst it also increases the price of all the competing home-produced articles to the same amount as the duty; but that increased price goes, not to the Treasury, but to the protected manufacturer. It is obvious that high protective duties check importation, and consequently, are unproductive to the revenue; and experience shows, that the profit to the trader, the benefit to the consumer, and the fiscal interests of the country, are all sacrificed when heavy import duties impede the interchange of commodities with other nations.

The inquiries of Your Committee have naturally led them to investigate the effects of the protective system on manufacture and labour. They find on the part of those who are connected with some of the most important of our manufactures, a conviction, and a growing conviction, that the protective system is not, on the whole, beneficial to the protected manufactures themselves. Several witnesses have expressed the utmost willingness to surrender any protection they have from the Tariffs, and disclaim any benefit resulting from that protection; and Your Committee, in investigating the subject as to the amount of duties levied on the plea of protection to British manufactures, have to report that the amount does not exceed half a million sterling; and some of the manufacturers, who are supposed to be most interested in retaining those duties, are quite willing they should be abolished, for the purpose of introducing a more liberal system into our commercial policy. ...

... Your Committee are persuaded that the best service that could be rendered to the industrious classes of the community, would be to extend the field of labour, and of demand for labour, by an extension of our commerce; and that the supplanting the present system of protection and prohibition, by a moderate Tariff, would encourage and multiply most beneficially for the State and for the people our commercial transactions.

Your Committee further recommend, that as speedily as possible the whole system of differential duties and of all restrictions should be reconsidered, and that a change therein be effected in such a manner that existing interests may suffer as little as possible in the transition to a more liberal and equitable state of things. Your Committee is persuaded that the difficulties of modifying the discriminating duties which favour the introduction of British colonial articles would be very much abated if the Colonies were themselves allowed the benefits of free trade with all the world.

Although, owing to the period of the Session at which the inquiry was begun, Your Committee have not been able to embrace all the several branches which come within the scope of their instructions, they have thought themselves warranted in reporting their strong conviction of the necessity of an immediate change in the Import Duties of the kingdom: and should Parliament sanction the views which Your Committee entertain on these most important matters, they are persuaded that by imposts on a small number of those articles which are now most productive, the amount of each impost being carefully considered with a view to the greatest consumption of the article, and thereby the greatest receipt to the Customs, no loss would occur to the revenue, but, on the contrary, a considerable augmentation might be confidently anticipated.

The simplification they recommend would not only vastly facilitate the transactions of commerce, and thereby benefit the revenue, but would at the same time greatly diminish the cost of collection, remove multitudinous sources of complaint and vexation, and give an example to the world at large, which, emanating from a community distinguished above all others for its capital, its enterprize, its intelligence, and the extent of its trading relations, could not but produce the happiest effects; and consolidate the great interests of peace and commerce by associating them intimately and permanently with the prosperity of the whole family of nations.

4 Peel's income tax—the price of commercial reform

(*Parliamentary Debates*, third series, LXI (11 March 1842), cols. 430, 439–40, 450–1)

...We have a deficiency of nearly 5,000,000*l* in two years; is there a prospect of reduced expenditure? Without entering into details, but looking at your extended empire at the demands that are made for the protection of your commerce, and the general state of the world, and calling to mind the intelligence that has lately reached us, can you anticipate, for the year after next, the possibility, consistent with the honour and safety of this country, of greatly reducing the public expenses? I am bound to say I cannot calculate upon that. Is this a casual deficiency for which you have to provide a remedy? Is it a deficiency for the present year on account of extraordinary circumstances? Is it a deficiency for the last two years? Sir, it is not. This deficiency has existed for the last seven or eight years. It is not a casual deficiency. In the year ending the 5th of April, 1838, the deficiency was 1,428,000*l*. In the year ending the 5th April, 1839, the deficiency was 430,000*l*. In 1840 it was 1,457,000*l*. In 1841 the deficiency was 1,851,000*l*; in 1842 I estimate the deficiency will be 2,334,030*l*. The deficiency in these five amounts to 7,502,000*l*; and to that actual deficiency I must add the estimated deficiency for the year ending the 5th of April, 1843, 2,570,000*l*, making an aggregate deficiency in six years of 10,072,000*l*.

*   *   *

...I will now state what is the measure which I propose, under a sense of public duty, and a deep conviction that it is necessary for the public interest; and impressed at the same time, with an equal conviction that the present sacrifices which I call on you to make will be amply compensated ultimately in a pecuniary point of view, and much more than compensated by the effect they will have in maintaining public credit, and the ancient character of this country. Instead of looking to taxation on consumption—instead of reviving the taxes on salt or on sugar—it is my duty to make an earnest appeal to the possessors of property, for the purpose of repairing this mighty evil. I propose, for a time at least, (and I never had occasion to make a proposition with a more thorough conviction of its being one which the public interest of the country required)—I propose, that for a time to be limited, the income of this country should be called on to contribute a certain sum for the purpose of remedying this mighty and growing evil. I propose, that the income of this country should bear a charge not exceeding 7*d* in the pound; which will not amount to 3 per cent, but speaking accurately, 2*l* 18*s* 4*d* per cent; for the purpose of not only supplying the deficiency in the revenue, but of enabling me with confidence and satisfaction to propose great commercial reforms, which will afford a hope of reviving commerce, and such an improvement in the manufacturing interests as will re-act on every other interest in the country; and, by diminishing the prices of the articles of consumption, and the cost of living, will, in a pecuniary point of view, compensate you for your present sacrifices; whilst you will be, at the same time, relieved from the contemplation of a great public evil. (*Interruption, and cries of 'Order!'*) I hope hon. Gentlemen will allow me to make the statement I have yet to lay before the House uninterruptedly. In 1798, when the prospects of this country were gloomy, the Minister had the courage to propose, and the people had the fortitude to adopt, an income-tax of 10 per cent. The income-tax continued to the close of the war in 1802; and in 1803, after the rupture of the peace of Amiens, a duty of 5 per cent was placed upon property. It was raised in 1805 to 6¼ per cent, and in 1806 again to 10 per cent; and so it continued to the end of the war. I propose that the duty to be laid on property shall not exceed 3 per cent, or, as I said before, exactly 2*l* 18*s* 4*d*, being 7*d* in the pound. Under the former tax, all incomes below 60*l* were exempt from taxation, and on incomes between 60*l* and 150*l*, the tax was on a reduced rate. I shall propose, that from the income-tax I now recommend all incomes under 150*l* shall be exempt. Under the former income-tax, the amount at which the occupying tenants were charged, was estimated at three-fourths of the rent. It is admitted, I believe, that to calculate the profits of the tenants at the three-fourths of the rent, was too high an estimate. I propose, therefore, that in respect of the occupying tenant, the occupation of land shall be charged at one-half, instead of three-fourths of the rent.

*　　*　　*

... I calculate on a surplus of 1,800,000*l* after providing for the excess of expenditure on actual votes. Having that surplus, then, Sir, in what way shall we apply it? I propose to apply it, namely, in a manner which I think will be most conducive to the public interest, and most consonant with public feeling and opinion—by making great improvements in the commercial tariff of England, and in addition to these improvements to abate the duties on some great articles of consumption. Sir, I look to the tariff, and find that it comprises not less than 1,200 articles subject to various rates of duty. During the interval which I have been blamed for taking to consider the subject, I can only say, that each individual item in that tariff has been subjected to the most careful consideration of myself and Colleagues. In the case of each article we have endeavoured to determine, as well as we can, the proportion borne by the duty to the average price of the article, for the purpose of ascertaining to what extent it may be desirable to make reductions of the several duties; and the measure which I shall propose will contain a complete review, on general principles, of all the articles of the tariff, with a very great alteration of many of the duties. We have proceeded, Sir, on these principles (observe, that I am speaking of general views; there may be individual articles which should form exceptions, but I wish a general result); first, we desire to remove all prohibition, and the relaxation of duties of a prohibitory character; next, we wish to reduce the duties on raw materials for manufactures to a considerable extent—in some cases the duty we propose being merely nominal, for the purpose more of statistical than revenue objects; in no case, or scarcely any, exceeding, in the case of raw materials, 5 per cent. I speak of course, in a general way. Then we propose that the duties on articles partly manufactured shall be materially reduced, never exceeding 12 per cent. Again, I say, I speak only as to general principles, and without reference to parti- cular cases that may be excepted; while as to duties on articles wholly manufactured we propose that they shall never exceed 20 per cent. These are the general views of the Government as to the maximum duties to be imposed, not referring to certain commodities which I will mention subsequently.

5 A lightly taxed nation, 1869

(Robert D. Baxter, *The taxation of the United Kingdom* (Macmillan, 1869,) pp. 8–9)

Our Income has increased by rapid strides, through the growth of profitable and highly-paid industries, to a gross total of £800,000,000 per annum, of which £325,000,000 is derived from the weekly wages of the Manual Labour Classes.

Property has accumulated in the hands of a comparatively small number of landed proprietors and capitalists, to an estimated total of £6,000,000,000 of land and personalty.

We are burdened with a public Debt, which (including the £50,000,000 Capital of the Terminable Annuities) amounts in the whole to nearly £800,000,000, and involves an annual taxation for interest of £26,600,000, or $3\frac{1}{3}$ per cent on our gross Income.

We maintain, for the national defence and for the garrisons of our Colonies and possessions, an Army, which, in 1867–8, cost £15,400,000; and for the protection of our coasts and commerce, a Navy, costing £11,200,000; amounting together to £26,600,000, or another $3\frac{1}{3}$ per cent on our gross Income.

We carry on the government and internal administration of the country by a Civil Service, a Diplomatic Corps, Courts of Justice, a Postal Service, Educational Grants, and Collectors of the Revenue, at a cost of £16,000,000, or 2 per cent on our gross Income.

We support the destitute Poor, keep up the Police and Highways, pave, light, and sewer our towns and cities, and maintain Harbours, Bridges, and Markets, by a local taxation estimated at £22,500,000, or nearly 3 per cent on our gross Income.

Such are the four great heads of our expenditure, amounting together to £91,500,000, or $11\frac{1}{2}$ per cent on our gross Income.

6 Taxation and social reform, 1909

(*Commons Debates, 1909*, IV (29 April 1909), cols. 474–6, 546; VI (7 June 1909), cols. 22–5)

(i) *The needs of government—Lloyd George*

Now I come to the expenditure side of my balance sheet, and it is to this, after all, that must mainly be ascribed the exceptionally heavy deficit. Were I dealing with a shortage due only to a temporary cause like forestalments I might have resorted to some temporary shift which would have carried me over until next year, when the revenue would resume its normal course. But, unfortunately, I have to reckon not merely with an enormous increase in expenditure this year, but an inevitable expansion of some of the heaviest items in the course of the coming years. To what is the increase of expenditure due? It is very well known that it must be placed to the credit of two items and practically two items alone. One is the Navy and the other is Old Age Pensions. Now, I have one observation which I think I am entitled to make about both, and I think that now I am about to propose heavy increased taxation it is an observation that I am entitled to make on behalf of the Government. The increased expenditure under both these heads was substantially incurred with the unanimous assent of all political parties in this House. There was, it is true, a protest entered on behalf of hon. Members below the Gangway against increased expenditure in the Navy, but as far as the overwhelming majority of Members in this House are concerned the increase has received their sanction and approval. I am entitled to say more. The attitude of the Government towards these two branches of increased expenditure has not been one of rushing a reluctant House of Commons into expense which it disliked, but rather of resisting persistent appeals coming from all quarters of the House for still further increases under both heads.

As to the Navy, we are now in the throes of a great agitation to double and even treble the cost of our Construction Vote this year; and as to old age pensions, the responsibility was cast upon me of piloting that Bill through the Committee, and the one difficulty I experienced was to persuade the House of Commons not to press Amendments which would enormously augment the very heavy bill we were incurring under the original proposals. I had constantly to appeal to the party loyalty of the supporters of the Government to resist Amendments which commended themselves to hon. Members, and which we ourselves should like to have seen carried, in order to confine within something like reasonable limits the Bill which we were encouraged to pass. And these were not Amendments moved by small sections of the House; on the contrary, they were moved from all quarters and almost invariably received the official sanction and support of the Opposition. I say that in order to show that, in the main, the two great items of expenditure which are responsible for this deficit are items which a vast majority of the Members of the House of Commons have not merely sanctioned, but in regard to which they brought a considerable amount of pressure to bear upon the Government to increase.

*   *   *

Final Balance Sheet

I am now in a position to present my final balance sheet for 1909–10.

The Revenue, on the present basis of taxation, being	£148,390,000
And the Expenditure, on the basis of the Estimates already presented to Parliament	164,152,000
The Account, before adjustment, shows, as I have already explained, an anticipated deficit of	£15,762,000

To the Revenue side of the Account must be added:
Under Customs and Excise:

New duty of 3d a gallon on petrol	£340,000
Increase of spirit duties	1,600,000
Increase of tobacco duties	1,900,000
Revision and increase of liquor licence duties	2,600,000
Motor-car licences	260,000

Making a total addition under the heads of Customs and
Excise of £6,700,000

Under the various Inland Revenue duties the new proposals
are estimated to produce:

Estate duties	£2,850,000
Stamps	650,000
Income tax (net)	3,500,000
And the new land taxes	500,000

Or a total from new and increased Inland Revenue duties of £7,500,000

These amounts (namely, £6,700,000 from Customs and
Excise and £7,500,00 from Inland Revenue) added together
give as the total estimated yield of new taxation £14,200,000

(ii) Canons of sound finance—Austen Chamberlain

Let me turn to the other claims made for these canons of sound
finance, for which this Government are the self-appointed guardians.
The first is that you should tax for revenue only. I think that has now
gone by the wall. It is hardly worth while to argue it; in fact, the defence
made is not that these taxes will produce revenue, but that they will
get round the House of Lords on the Licensing Bill, that they will
cheapen the price of land, that they will force land into the market
and develop towns, and will generally facilitate the progress and development of the country. Take another of these canons of sound finance,
namely, that you are not to take more from the taxpayer than you
get into the Exchequer. I do not think the Government will pretend
any longer that they have fulfilled that canon. The Chancellor of the
Exchequer explained to an astonished Committee just before we parted
at Whitsuntide that in order to get £1,600,000 a year from the distillers
of whisky, he was going to endow them with £4,000,000 for themselves,
and, in order to take his new revenue from the brewers of beer, he is
going to put upon the consumers of that article an increased tax
which he estimates at no less than £20,000,000. There never was a case
where so little went into the Exchequer and so much went into private
pockets, and this upon the showing of the finance Minister himself.

And now for the third canon, namely, that, having to raise this
great sum for purposes in which all classes of the people are interested,
the burden should be fairly spread. Does anyone pretend that this
burden is fairly spread? Have you listened to the speeches from Irish
and Scotch Members as to how an industry peculiarly localised in their
countries is affected by a part of this taxation? I have never been able
to understand why, if we grant the premises that this expenditure is

necessary for the common weal and for the benefit of all classes, those who happen to be teetotalers and non-smokers should be exempt from any contribution unless they come up to the Income Tax level. But it is peculiarly unfortunate that you should have chosen for your indirect contribution in the first place a tax which causes, owing to its special incidence, a special sense of hardship in Ireland and the poorer parts of Scotland; and in the second place a tax which is already abnormally high in proportion to the value of the article, namely, tobacco, on which it is imposed, and which has the grievous disadvantage that no one has been able to contrive a system by which the tax should bear a proportion to the value of the articles purchased, and that it falls with wholly disproportionate weight on the cheapest tobaccos, which are smoked by the poorest people.

I may add that the Chancellor of the Exchequer himself anticipates as not improbable that the effect of raising the whisky duty will be to transform some whisky drinkers into beer drinkers, and therefore to deprive distillers of customers for the benefit of the brewers and when I add that his land taxes will fall with repeated blows upon what is after all a minority, though not so small a minority as some hon. Gentlemen opposite think, then I have made out briefly, but I hope plainly, my contention that the Budget fails, judged by this test also. What are we to say of the claim, which is the special boast of the Government, that the Budget causes no disturbance of business? Are they still of that opinion? That I think was the proud boast of the President of the Board of Trade. Ask anyone concerned in any of the taxes which you are imposing, and you will get an answer that should at any rate open your eyes. Ask hotel keepers and restaurant keepers, ask distillers and the agriculturists depending on them—ask anyone connected with land or house property—ask the small-cultivator owner, say in Worcestershire or other parts of the country—ask the owners of garden cities, who have made a most desirable experiment —ask the insurance companies and the friendly societies with their vast investments in mortgages and ground rents—ask the small tradesmen and the thrifty workmen, with whom no form of investment is more popular than that in small house property—ask builders, who are engaged in the development of land round your great cities, and one and all will reply that you have upset every calculation on which they have hitherto proceeded. They will also tell you that they are now obliged to consider if at all they are to carry on in their old forms businesses which your Budget renders impossible. The taxes which the Government impose by this Budget are bad in themselves.

That is not all. The principles of the Government are worse. Your super-tax is to stand at 6*d*. An income over £3,000 is considered by the Government to be a superfluity when the total income exceeds £5,000 a year. That is for the present. The hon. Member for Blackburn says that is very well for a beginning. He says. 'You are apt pupils. We will build on this foundation in future years.' Your undeveloped land duty stands at a halfpenny in the pound. A poor thing, but it is mine own, says the Lord Advocate. Again that will be a beginning. Your reversionary duty is 10 per cent, your increment duty is 20 per cent. Increments are considered as unearned windfalls, and their owners should be happy if we leave them any part of them. If you take only a tenth or a fifth, you say, 'See how moderate we are.' (*Labour cheers*). Again comes that sinister echo from the benches below the Gangway, where the true authors of this Budget sit. What security or confidence have you left to any man who has invested money in the classes of property which come under your ban? Every argument in defence of these taxes—every statement you make—justifies a further raid on the lines upon which the Government have started. This is no temporary disturbance of a trade you are making —I mean of a particular trade—in this Budget, such as occurs when you increase or decrease a duty like the tea duty. You are creating and maintaining a general and permanent spirit of unrest, disturbance and distrust which will have an effect on the fortunes of our country long after you have ceased to be responsible.

7 Back to gold —consequences for national expenditure, 1918
(First interim report of the [Cunliffe] committee on currency and foreign exchanges after the war, *B.P.P.* 1918, VII, pp. 3, 5-6)

1. ...We have had the advantage of consultation with the Bank of England, and have taken oral evidence from various banking and financial experts, representatives of certain Chambers of Commerce and others who have particularly interested themselves in these matters. We have also had written evidence from certain other representatives of commerce and industry. Our conclusions upon the subject dealt with in this Report are unanimous, and we cannot too strongly emphasise our opinion that the application, at the earliest possible date, of the main principles on which they are based is of vital necessity to the financial stability and well-being of the country. Nothing can contribute more to a speedy recovery from the effects of the war, and to the rehabilitation of the foreign exchanges, than the re-establishment of the currency upon a sound basis. Indeed, a sound system of currency will, as is shown in paragraphs 4 and 5, in itself

secure equilibrium in those exchanges, and render unnecessary the continued resort to the emergency expedients to which we have referred. We should add that in our inquiry we have had in view the conditions which are likely to prevail during the ten years immediately following the end of the war, and we think that the whole subject should be again reviewed not later than the end of that period.

The Currency System before the War

2. Under the Bank Charter Act of 1844, apart from the fiduciary issue of the Bank of England and the notes of Scottish and Irish Banks of Issue (which were not actually legal tender), the currency in circulation and in Bank reserves consisted before the war entirely of gold and subsidiary coin or of notes representing gold. Gold was freely coined by the Mint without any charge. There were no restrictions upon the import of gold. Sovereigns were freely given by the Bank in exchange for notes at par value, and there were no obstacles to the export of gold. Apart from the presentation for minting of gold already in use in the arts (which under normal conditions did not take place) there was no means whereby the legal tender currency could be increased except the importation of gold from abroad to form the basis of an increase in the note issue of the Bank of England or to be presented to the Mint for coinage, and no means whereby it could be diminished (apart from the normal demand for the arts, amounting to about £2,000,000 a year, which was only partly taken out of the currency supply) except the export of bullion or sovereigns.

3. Since the passing of the Act of 1844 there has been a great development of the cheque system. The essence of that system is that purchasing power is largely in the form of bank deposits operated upon by cheque, legal tender money being required only for the purpose of the reserves held by the banks against those deposits and for actual public circulation in connection with the payment of wages and retail transactions. The provisions of the Act of 1844 as applied to that system have operated both to correct unfavourable exchanges and to check undue expansions of credit.

4. When the exchanges were favourable, gold flowed freely into this country and an increase of legal tender money accompanied the development of trade. When the balance of trade was unfavourable and the exchanges were adverse, it became profitable to export gold. The would-be exporter bought his gold from the Bank of England and paid for it by a cheque on his account. The Bank obtained the gold from the Issue Department in exchange for notes taken out of

its banking reserve, with the result that its liabilities to depositors and its banking reserve were reduced by an equal amount, and the ratio of reserve to liabilities consequently fell. If the process was repeated sufficiently often to reduce the ratio in a degree considered dangerous, the Bank raised its rate of discount. The raising of the discount rate had the immediate effect of retaining money here which would otherwise have been remitted abroad and of attracting remittances from abroad to take advantage of the higher rate, thus checking the outflow of gold and even reversing the stream.

5. If the adverse condition of the exchanges was due not merely to seasonal fluctuations, but to circumstances tending to create a permanently adverse trade balance, it is obvious that the procedure above described would not have been sufficient. It would have resulted in the creation of a volume of short-dated indebtedness to foreign countries which would have been in the end disastrous to our credit and the position of London as the financial centre of the world. But the raising of the Bank's discount rate and the steps taken to make it effective in the market necessarily led to a general rise of interest rates and a restriction of credit. New enterprises were therefore postponed and the demand for constructional material and other capital goods was lessened. The consequent slackening of employment also diminished the demand for consumable goods. . . .

*   *   *

Restoration of Conditions necessary to the Maintenance of the Gold Standard recommended

15. We shall not attempt now to lay down the precise measures that should be adopted to deal with the situation immediately after the war. These will depend upon a variety of conditions which cannot be foreseen, in particular the general movements of world prices and the currency policy adopted by other countries. But it will be clear that the conditions necessary to the maintenance of an effective gold standard in this country no longer exist, and it is imperative that they should be restored without delay. After the war our gold holdings will no longer be protected by the submarine danger, and it will not be possible indefinitely to continue to support the exchanges with foreign countries by borrowing abroad. Unless the machinery which long experience has shown to be the only effective remedy for an adverse balance of trade and an undue growth of credit is once more brought into play, there will be very grave danger of a credit expansion in this country

34*

and a foreign drain of gold which might jeopardise the convertibility of our note issue and the international trade position of the country. The uncertainty of the monetary situation will handicap our industry, our position as an international financial centre will suffer and our general commercial status in the eyes of the world will be lowered. We are glad to find that there was no difference of opinion among the witnesses who appeared before us as to the vital importance of these matters.

Cessation of Government Borrowings

16. If a sound monetary position is to be re-established and the gold standard to be effectively maintained, it is, in our judgment, essential that Government borrowings should cease at the earliest possible moment after the war. A large part of the credit expansion arises, as we have shown, from the fact that the expenditure of the Government during the war has exceeded the amounts which they have been able to raise by taxation or by loans from the actual saving of the people. They have been obliged, therefore, to obtain money through the creation of credits by the Bank of England and by the Joint Stock Banks, with the result that the growth of purchasing power has exceeded that of purchasable goods and services. As we have already shown, the continuous issue of uncovered currency notes is inevitable in such circumstances. This credit expansion (which is necessarily accompanied by an ever-growing foreign indebtedness) cannot continue after the war without seriously threatening our gold reserves and, indeed, our national solvency.

17. A primary condition of the restoration of a sound credit position is the repayment of a large portion of the enormous amount of Government securities now held by the Banks. It is essential that as soon as possible the State should not only live within its income but should begin to reduce its indebtedness. We accordingly recommend that at the earliest possible moment an adequate sinking fund should be provided out of revenue, so that there may be a regular annual reduction of capital liabilities, more especially those which constitute the floating debt. We should remark that it is of the utmost importance that such repayment of debt should not be offset by fresh borrowings for capital expenditure. We are aware that immediately after the war there will be strong pressure for capital expenditure by the State in many forms for reconstruction purposes. But it is essential to the restoration of an effective gold standard that the money for such expenditure should not be provided by the creation of new credit, and that, in so far as such

expenditure is undertaken at all, it should be undertaken with great caution. The necessity of providing for our indispensable supplies of food and raw materials from abroad and for arrears of repairs to manufacturing plant and the transport system at home will limit the saving available for new capital expenditure for a considerable period. This caution is particularly applicable to far-reaching programmes of housing and other development schemes.

The shortage of real capital must be made good by genuine savings. It cannot be met by the creation of fresh purchasing power in the form of bank advances to the Government or to manufacturers under Government guarantee or otherwise, and any resort to such expedients can only aggravate the evil and retard, possibly for generations, the recovery of the country from the losses sustained during the war.

Use of Bank of England Discount Rate

18. Under an effective gold standard all export demands for gold must be freely met. A further essential condition of the restoration and maintenance of such a standard is therefore that some machinery shall exist to check foreign drains when they threaten to deplete the gold reserves. The recognised machinery for this purpose is the Bank of England discount rate. Whenever before the war the Bank's reserves were being depleted, the rate of discount was raised. This, as we have already explained, by reacting upon the rates for money generally, acted as a check which operated in two ways. On the one hand, raised money rates tended directly to attract gold to this country or to keep here gold that might have left. On the other hand, by lessening the demands for loans for business purposes, they tended to check expenditure and so to lower prices in this country, with the result that imports were discouraged and exports encouraged, and the exchanges thereby turned in our favour. Unless this two-fold check is kept in working order the whole currency system will be imperilled. To maintain the connection between a gold drain and a rise in the rate of discount is essential to the safety of the reserves. When the exchanges are adverse and gold is being drawn away, it is essential that the rate of discount in this country should be raised relatively to the rates ruling in other countries. Whether this will actually be necessary immediately after the war depends on whether prices in this country are then substantially higher than gold prices throughout the world. It seems probable that at present they are on the whole higher, but, if credit expansion elsewhere continues to be rapid, it is possible that this may eventually not be so.

8 National economy, 1922

(*Commons Debates*, *1922*, CLI (1 March 1922), cols. 427–8, 430, 431–2, 448–50)

I think that the Government is entitled to look with great gratification upon the reception which the Report of the Geddes Committee has obtained. When we suggested the setting up of this Committee, we were denounced in all quarters in the most unmeasured terms. The Committee was set up, it was said, merely to provide a shelter behind which we might hide our extravagant heads. A notorious prodigal, it was said, had been deliberately selected as Chairman of the Committee in order to cover up our prodigality. Indeed, I think I recollect a public speech by the right hon. Gentleman the Member for Paisley (Mr Asquith) in which, in a phrase which I may be permitted to describe as somewhat extravagant, he described the Chairman of the Committee as a Landru among Bluebeards. To-day this Committee, which was so uniformly traduced, is hailed by the very organs which vilified it as having performed a service to the country such as has been accomplished by no other Committee within living memory, and the Chairman, who has suffered from the abuse of these journals now for a long period of time, is hailed as the finest example this country has yet produced of the high aptitude of a business man applied to the service of the State. . . .

As Chancellor of the Exchequer I asked the Geddes Committee to suggest reductions in public expenditure to the amount of £100,000,00. It has been said that that was a confession that waste to the extent of £100,000,000 was going on in the Government services. That is a complete travesty of the facts. We have found ourselves in the most acute trade depression which has ever been known in this country. We were confronted with a falling revenue. There are many things which in normal times we should regard as necessary objects of expenditure in a great country like ours which under these conditions we might not be able to afford, and the prime object of the inquiry was to discover, through the means of the Committee, those services in which with the least detriment to the efficiency of the country, we could afford to economise. . . .

I propose to deal, in the first place, with the subject of education. The recommendation is for a reduction of £18,000,000 in the expenditure of the Education Departments. Part of that, really, is not a reduction in expenditure, but a transfer of obligations from taxes to rates. It is not a very large amount, but it is appreciable; it amounts to transferring about £3,000,000 from taxes to the rates. While that would be a reduction so far as the taxpayer is concerned, yet in the capacity of ratepayer I do not think he would be particularly grateful for it. The two items which in the main make up the £18,000,000 recommended by the Geddes Committee are the reduction in teachers' salaries and the exclusion of children from school below the age of six years. I wish to say a word upon both these matters. The Government have very carefully considered the recommendations of the Committee, and are of the view that neither of these proposals can be put into operation.

Mr J. Jones. The red light.

Sir R. Horne. On the question of teachers' salaries I have only one observation to make, and I think the House will find it conclusive. The salaries which are paid to the teachers of this country to-day are, for the most part, or practically entirely, the result of engagements which have been entered into between the local authorities and the teaching staffs. These engagements subsist in the case of London until 1923, and in the case of the country at large until 1925. Accordingly, whatever view we take as to whether these salaries are too high or too low—the Geddes Committee have undoubtedly put on record their opinion that they are too high under modern conditions—and whatever be the correct point of view in regard to that, it is perfectly certain that the local authorities are under engagements with the teachers which cannot be broken without a violation of what is indeed a contract, and so far as the Government is concerned we cannot on our part, and would not, take any action which would have the effect of creating breaches of faith. It is quite certain that any Government which took part in what could fairly be regarded as the breaking of a contract and a breach of faith would set an example in this country which would be attended by serious consequences.

*   *   *

I confidently anticipate that as the result of what we shall be able to do in the coming year, we shall be able to present an account of far greater saving for the year which is to follow it. Generally, so far as the Government are concerned, we mean to use all our efforts to that end. I come back to a question which was asked by my hon. Friend the Member for the St. Rollox Division of Glasgow (Mr G. Murray) —how much of this £64,000,000 will be realised in the next year? We shall have to take off from that figure, just as the Geddes Committee would have required to deduct from theirs, a certain sum which will represent the savings not immediately effective. We cannot dismiss your men from the Army and Navy and dockyards all upon a given day, and, therefore, there must be, however much of the Report we take, a considerable lag in the saving. I estimate—it is only a rough estimate—the deduction which ought to be made from the full figure of the year at £10,000,000. By the savings which we expect to effect through the advice of the Geddes Committee, and the operations of the Departments, we shall be able to reduce the Provisional Estimate put previously before us by £54,000,000 in the next financial year. I have no doubt that some hon. Member may remind me to-day that I asked the Geddes Committee to provide £100,000,000 of savings. I have described to you what that figure was, and how I arrived at it.

Savings Already Effected

Now I should like the House to know what we actually have in anticipation in the way of savings between the Estimates of the present financial year and the Estimates of the next financial year. I am dealing at this moment only with the ordinary Supply services. That is all with which the Geddes Report deals. The Estimates for the present year for ordinary Supply services, including in these Estimates the Supplementary Estimates, amount to £665,000,000. As a result of the reductions, of which I have been speaking, obtained through the counsel of the Geddes Committee, and through the activities of the Departments, I think that the Estimates for the next financial year may be reckoned at a sum of £484,000,000; that is to say, a figure of £181,000,000 less than the Estimates for the present year. I venture to put to the House that that is a great achievement. It is not achieved yet, as my right hon. Friend says, but I am assuming that we are able to carry out the proposals which I have been putting before you.

A great deal of talk with regard to economy by the critics of the Government has been on the footing that we have done nothing up to now, and that only as a despairing effort, when we were drifting on the rocks, did we appoint the Geddes Committee. I hope hon. Members will not repeat that kind of fallacy in their constituencies again. I mean to enlighten them as to what the actual facts are. These are the facts:

In 1918–19, which contained eight months of war and four months of peace, our Gross Supply services took £2,965,000,000.

In 1919–20, we had reduced that sum to £1,690,000,000—a reduction of £1,275,000,000.

In the year 1920–21, we had reduced the figure for these services to £1,015,000,000, making a reduction for that year of £675,000,000, and,

In the current year, my estimate is that these services will cost us £825,000,000, or a reduction of £190,000,000.

I have already disclosed to the House that I anticipate from the ordinary Supply services next year I shall get a further reduction of £181,000,000 compared with the present year.

That is a record incomparable in the world at the present time. There is no other country which has made efforts in the least like it, or which has achieved nearly so much. There are many criticisms that we are still keeping War Departments going, where limpets cling, where they do nothing but go to sleep, and it seemed to be assumed that it was possible for us to get rid of all the War Departments as soon as the Armistice came. I wonder if the House realises what has been taking place in connection with these so-called War Departments. They were created to carry on business which there was no Department to conduct. We took control of enormous amounts of material in this country. We were preparing for a continuance of the War, and we had in hand, at the time the War ceased, enormous stocks. The Disposal Board since the Armistice has sold £600,000,000 worth of material. Was that to be done without a staff? Let anybody who has ever been connected with an ordinary liquidation of even a comparatively small company consider the time which elapses before you can put it into liquidation, and the number of people you have to employ if you are to save the assets of the company. Were we to leave all that property derelict? If so, the country would have been a loser to the extent of £600,000,000. The Food Ministry has sold since the Armistice £550,000,000 worth. The Wheat Commission has sold £370,000,000 worth. The Sugar Commission has sold £160,000,000 worth of sugar, and the Shipping Ministry £91,000,000 worth of shipping.

9 The fall of free trade, 1932

(Import Duties Act, 1932—22 Geo. V, c. 8)

An Act to provide for the imposition of a general ad valorem duty of customs and of additional duties on any goods chargeable with the duty aforesaid, for the imposition of duties on goods produced or manufactured in a foreign country which discriminates in the matter of importation as against goods produced or manufactured in the United Kingdom, in certain other parts of His Majesty's dominions, in protectorates or in mandated territories, and for purposes connected with the matters aforesaid.

(29th February 1932)

Most Gracious Sovereign,

We Your Majesty's most dutiful and loyal subjects, the Commons of the United Kingdom in Parliament assembled, with a view to the restricting in the national interest of the importation of goods into the United Kingdom, to the providing of a remedy in cases where a foreign country discriminates in the matter of importation as against goods produced or manufactured in the United Kingdom, in certain other parts of Your Majesty's dominions or in territories under Your Majesty's protection or in respect of which a mandate is being exercised by Your Majesty's Government of the United Kingdom, and to the making of an addition to the public revenue, have freely and voluntarily resolved to give and grant unto Your Majesty the duties for which provision is hereinafter contained; and do therefore most humbly beseech Your Majesty that it may be enacted, and be it enacted, by the King's most Excellent Majesty, by and with the advice and consent of the Lords Spiritual and Temporal, and Commons, in this present Parliament assembled, and by the authority of the same, as follows:

PART I

General ad valorem Duty and Additional Duties

1—(1) As from the first day of March, nineteen hundred and thirty-two, there shall, subject to the provisions of this Act, be charged on all goods imported into the United Kingdom, other than goods exempted as hereinafter provided from the provisions of this section, a duty of customs equal to ten per cent of the value of the goods.

(2) The following goods shall be exempted from the provisions of this section—

(*a*) goods for the time being chargeable with a duty of customs by or under any enactment other than this Act, but not including (subject to the provisions of this Act) any composite goods in the case of which duty is chargeable under any such enactment as aforesaid because some (but not all) of their components are articles so chargeable;

(*b*) goods of any class or description specified in the First Schedule to this Act or added to that Schedule by an order made under the next following subsection.

(3) The Treasury, after receiving a recommendation from the Committee to be constituted under the following provisions of this Act that goods of any class or description ought to be exempted from the provisions of this section, and after consultation with the appro-

priate department, may by order direct that goods of all or any of the classes or descriptions specified in the recommendation shall be added to the First Schedule to this Act: ...

2—(1) For the purpose of giving advice and assistance in connection with the discharge by the Treasury of their functions under this Act, there shall be constituted a committee, to be called 'the Import Duties Advisory Committee', consisting of a chairman and not less than two or more than five other members to be appointed by the Treasury.

(2) The members of the Committee shall hold office for a period of three years and shall be eligible for re-appointment from time to time on the expiration of their term of office.

If a member becomes, in the opinion of the Treasury, unfit to continue in office or incapable of performing his duties under this Act, the Treasury shall forthwith declare his office to be vacant and shall notify the fact in such manner as they think fit, and thereupon the office shall become vacant.

(3) The Committee shall, as soon as may be after the commencement of this Act, take into consideration the provisions of this Act, and shall, from time to time, take into consideration any representations which may be made to them with respect to matters in which, under the provisions of this Act, action may be taken on a recommendation by the Committee, and may make recommendations to the Treasury with respect to the matters aforesaid. ...

3—(1) Where it appears to the Committee that an additional duty of customs ought to be charged in respect of goods of any class or description which are chargeable with the general ad valorem duty and which, in their opinion, are either articles of luxury or articles of a kind which are being produced or are likely within a reasonable time to be produced in the United Kingdom in quantities which are substantial in relation to United Kingdom consumption, the Commitee may recommend to the Treasury that an additional duty ought to be charged on goods of that class or description at such rate as is specified in the recommendation.

(2) In deciding what recommendation, if any, to make for the purposes of this section, the Committee shall have regard to the advisability in the national interest of restricting imports into the United Kingdom and the interests generally of trade and industry in the United Kingdom, including those of trades and industries which are consumers of goods as well as those of trades and industries which are producers of goods.

FIRST SCHEDULE

Goods Exempted from the General Ad Valorem Duty

Gold and silver bullion and coin; platinum in grain, ingot bar, or powder.

Wheat in grain.

Maize in grain.

Meat, that is to say, beef, veal, mutton, lamb, pork, bacon, ham and edible offals, but not including extracts and essences of meat or meat preserved in any airtight container.

Live quadruped animals.

Fish of British taking, including shell fish.

Whale oil and whale products shown to the satisfaction of the Commissioners to have been produced or manufactured in floating factories which are British concerns.

Tea.

Cotton (raw) (including unmanufactured cotton waste and unbleached cotton linters).

Flax and true hemp (cannabis sativa), not further dressed after scutching or decorticating; tow of flax and true hemp (cannabis sativa).

Cotton seed, rape seed and linseed.

Wool and animal hair (raw), whether cleaned, scoured or carbonised or not; rags of wool not pulled; wool noils; and wool waste not pulled or garnetted.

Hides and skins (including fur skins, but not including goat skins), raw, dried, salted or pickled, but not further treated.

Newspapers, periodicals, printed books and printed music.

Newsprint, that is to say, paper in rolls containing not less than 70 per cent of mechanical wood pulp and of a weight of not less than 20 lbs or more than 25 lbs to the ream of 480 sheets of double crown, measuring 30 inches by 20 inches.

Wood pulp and esparto.

Rubber (raw) including crepe; rubber latex; gutta-percha (raw).

Metallic ores, concentrates and residues; scrap metals and wastes fit only for the recovery of metal.

Iron pyrites, including cupreous pyrites.

Copper unwrought, whether refined or not, in ingots, bars, blocks, slabs, cakes, and rods.

Wooden pit-props.

Sulphur.

Mineral phosphates of lime.

Potassium carbonate, chloride and sulphate; kainite and other mineral potassium fertiliser salts.

Cinchona bark.

Coal, coke, and manufactured fuel of which coal or coke is the chief constituent.

Unset precious and semi-precious stones and pearls.

Radium compounds and ores.

Scientific films, that is to say, cinematograph films exempted under the provisions of section eight of the Finance Act, 1928, from the customs duty imposed by section three of the Finance Act, 1925.

Flint, unground.

Soya beans.

Cork, raw and granulated, cork shavings and waste.

Ramie, not dressed. [A fibre used as a substitute for cotton.]

10 A policy for full employment, 1944

(Employment policy, *B.P.P.*, 1943–44, VIII, p. 3)

The Government accept as one of their primary aims and responsibilities the maintenance of a high and stable level of employment after the war. This Paper outlines the policy which they propose to follow in pursuit of that aim.

A country will not suffer from mass unemployment so long as the total demand for its goods and services is maintained at a high level. But in this country we are obliged to consider external no less than internal demand. The Government are therefore seeking to create, through collaboration between the nations, conditions of international trade which will make it possible for all countries to pursue policies of full employment to their mutual advantage.

11 'Stop-go': Modern policy and its methods

(*Commons Debates, 1966–67*, 732 (20 July 1966), cols. 627–32, 635–8)

Economic Measures

(*The Prime Minister Mr Harold Wilson*) With permission, Mr Speaker, I wish to make a statement.

Sterling has been under pressure for the past two and a half weeks. After improvement in the early weeks of May we were blown off course by the seven-week seamen's strike and when the bill for that strike was presented in terms of the gold and convertible currency figures in June the foreign exchange market reacted adversely. But there were deeper and more fundamental causes. Many have been at home and of these I shall speak in a moment. Several have been overseas. For several weeks past there has been an increasing pressure on liquidity in the world's financial centres. Action taken by the United States' authorities to strengthen the American balance of payments has led to an acute shortage of dollars and Euro-dollars in world trade and this has led to a progressive rise in interest rates in most financial centres and to the selling of sterling to replenish dollar balances. Last Thursday, action was taken by the Bank of England to raise its discount rate and to double its call on the clearing banks for special deposits. On that day I informed the House that I would shortly be announcing further measures to deal not only with the short-run pressure on sterling, but also with the underlying economic situation.

Action is needed for the purpose of making a direct impact on our payments balance, and particularly on certain parts of our overseas expenditure which, in recent years, has been growing rapidly. Action is needed equally to deal with the problem of internal demand, public and private, and to redeploy resources, both manpower and capacity, according to national priorities, and check inflation.

Exports until the seamen's strike have been rising. By value, in the first five months of this year they were 9 per cent higher than in the same period last year. By volume, the increase over the same period last year was 6 per cent, a rate of increase higher than that laid down in the National Plan. But abundant market opportunities abroad for British products—which are competitive enough in terms of quality, performance and price—are being lost owing to the shortage of labour. Order books are too long, and delivery dates excessively protracted. Hours of work have been reduced and incomes have been rising faster than productivity.

What is needed is a shake-out which will release the nation's manpower, skilled and unskilled, and lead to a more purposive use of labour for the sake of increasing exports and giving effect to other national priorities. This redeployment can be achieved only by cuts in the present inflated level of demand, both in the private and public sectors. Not until we can get this redeployment through an attack on the problem of demand can we confidently expect growth in industrial production which is needed to realise our economic and social policies.

I will begin with the measures needed to restrain private demand at home. The economy is carrying too heavy a burden of production financed by hire purchase which means that too high a proportion of today's production is being paid for by a mortgage on tomorrow's earnings.

My right hon. Friend the President of the Board of Trade has today made Orders, which will come into effect at midnight tonight, tightening up still further the regulations governing hire purchase.

The down payment on cars, motor cycles and caravans is raised to 40 per cent and the repayment period shortened to 24 months. The down payment on furniture is raised to 20 per cent and the repayment period shortened to 24 months. The down payment on domestic appliances is raised to $33\frac{1}{3}$ per cent; the repayment period remains at 24 months. There is no change in the present regulations for cookers and water heaters. Corresponding changes are made in the regulations governing rental payments. It is estimated that this will cut hire-purchase borrowing by £160 million.

This of itself is not enough. The Government have, therefore, decided to activate the regulator created under Section 9 of the Finance Act of 1961, renewed in successive Finance Acts and given greater flexibility in Section 8 of the Finance Act, 1964.

The Treasury has today made an Order the effect of which is to put a surcharge of 10 per cent on the duties on beer, wines and spirits; on hydrocarbon oils, petrol substitutes and power methylated spirits; and on Purchase Tax.

I wish to make it clear that in the case of Purchase Tax the increase is the equivalent of 10 per cent of the existing rates. Thus, for goods now chargeable at 10 per cent, the new effective charge will be 11 per cent: for goods chargeable at 15 per cent it will be $16\frac{1}{2}$ per cent; and for goods chargeable at 25 per cent it will be $27\frac{1}{2}$ per cent.

The surcharge will take effect from midnight tonight. Its effect will be further to increase the revenue at the rate of about £150 million in a full year. This is a net figure after allowing for the effect of the hire-purchase proposals and for the additional export rebate which will become payable following the increases in oil duty and Purchase Tax.

The increase in the duty on petrol and derv will, following the precedent set last year, be refunded to bus operators. The necessary administrative arrangements will be made as soon as possible and Parliamentary authority sought in the ordinary way. In addition, a further £20 million will be taken out of the economy as a result of changes announced today by my right hon. Friend the Postmaster-General in certain postal and telecommunications tariffs; parcels, registration and overseas rates will be increased from 3rd October. The telecommunications changes affecting certain call charges will involve no net increase in Post Office revenue, but will be designed to rationalise charges on a basis more closely related to costs. These will take effect from 1st January next. In addition, my right hon. Friend will be requiring a year's rental in advance for new telephones instead of a quarter's rental as at present. This will apply to orders accepted from tomorrow. Details will be published in the Official Report and are now available in the Vote Office.

In the field of direct taxation the Government propose that a one-year surcharge of 10 per cent be imposed on Surtax. This will be levied on Surtax liabilities for 1965–66 for payment on 1st September, 1967. The necessary legislation will be introduced in next year's Finance Bill. The extra yield is estimated at £26 million.

These measures on private current expenditure will be reinforced by action to restrain private sector building outside the housing and industrial fields and outside the development areas. The Government have decided to intensify the control on less essential building work and thus to reinforce the priority accorded to building programmes in the fields of housing, schools, hospitals—and new factories.

When the Building Control Bill, now before Parliament, receives the Royal Assent, the Minister of Public Building and Works will make an Order reducing the cost limit above which a project is subject to control from £100,000 to £50,000. The Order will require an affirmative Resolution in both Houses. As before this control will not apply in the development areas as now defined. With a cost limit of £100,000 the control would affect about 500 projects worth £180 million in a year. The lowering of the limit to £50,000 will extend control to cover a total of about 1,000 projects worth £220 million in a year.

This will give the Government more power to adjust the volume of privately-sponsored construction work as the economic situation develops. Hitherto, approval has been withheld from less than 10 per cent of the projects about which the Minister has been consulted, but it will be necessary to defer a larger proportion of privately-sponsored work in future. The lowering of the limit will give the Minister scope to concentrate the postponement control on less urgent smaller schemes instead of having to rely on deferring some of the larger projects which are more in the public interest. This measure will be supplemented by a tighter control on office building.

My right hon. Friend the President of the Board of Trade has made an Order, coming into force at midnight, extending control of office building to the whole of Britain south of a line from the Wash to the borders of Hampshire and Dorset, by including within the control the whole of the East Midlands, West Midlands and South-East Region. Projects for buildings of more than 3,000 sq. ft. of office space which were not the subject of an application for planning permission at the time the Order comes into operation will require an Office Development Permit. In addition to tightening the control on building, this measure will reinforce those already taken by the Government for the prevention of undue congestion in these parts of the country and will supplement the strict policy which is being applied to the issue of industrial development certificates.

Now I turn to public investment programmes. The investment programmes in the public sector have been reviewed and the Government are introducing a number of deferment measures which will reduce demands on resources in this field by £150 million in 1967–68, though these steps will also lead to significant reductions in demand in the current year. While they will involve forgoing for the present a number of desirable projects, they will be concentrated on those activities which are not vital to our production capacity and for the development of the economy as a whole.

Housing, schools, hospitals, Government-financed factories built in development areas, including advance factories, will not be affected. On investment by central and local government, we are making cuts amounting to £55 million in 1967–68. This will mean cutting back projects designed to contribute to local amenities, but which, in present circumstances, must be postponed without any set back to our major projects. These will cover such items as swimming baths and new local government offices. The Covent Garden Market project will similarly be deferred.

The programme for investment in nationalised industries has also been carefully scrutinised to ensure that essential industrial investment within the public sector shall go on. Nevertheless, the Government are arranging, in consultation with the chairmen of these industries, for a reduction in the total demand on resources made by public industry investment to be reduced by £95 million in 1967–68. This is in addition to programmes which have fallen behind schedule owing to slippage in construction, or in the delivery by contractors of the necessary plant and machinery.

The measures I have so far announced, by reducing the level of demand within the domestic economy, will make a vital contribution to our balance of payments by freeing resources, particularly labour, for work on exports and essential investment. But this, of itself, is not enough. More direct action on the balance of payments is required.

In accordance with the policy foreshadowed in the Defence White Paper, we have been urgently reviewing how far we can make a major saving in overseas Government expenditure without altering the basic lines of external policy on which the Defence Review was founded. We have also reviewed the level of military and economic aid which we can afford next year. The Government have decided on firm programmes which will reduce our overseas Government expenditure, military and civil, by at least £100 million. ...

To sum up the measures I have so far outlined. I estimate that they will reduce demand on the domestic economy by more than £500 million. This is in addition to the earlier budgetary measures by this Government, reducing the pressure of demand in the private sector by over £700 million. They are in addition to the monetary policy which has been in force and which was reinforced by the three further measures announced last week. They are in addition, also, to the impact of this year's Finance Bill which is yet to have its effect on the economy, which is due to reduce demand and which will extract a further £300 million from the economy, with all which this means in terms of imports and of redeployment of labour towards exports and other essential industries.

In addition to the measures designed to reduce the domestic pressure, the economies in overseas expenditure, public and private, will, as I have said, make a direct saving of £150 million. But the House will recognise that the whole operation stands or falls on the extent to which we can keep our costs and prices under control. In recent years money incomes have been increasing at a rate far faster than could be justified by increasing production; in 1965, we paid ourselves increases in money incomes of about £1,800 million compared with the previous year. About £1,300 million of this represented increases in wages and salaries. Over the same period we earned only £600 million by way of increased production. These trends are continuing.

The Declaration of Intent of 16th December, 1964, was a great landmark when, for the first time in our history, employer, trade union and Government signed a compact designed to restrain the growth of incomes to a norm within the national capacity to pay. Yet ever since that time wage increases have outrun the figure allowed for, and pre-empted the amount available for such increases for a considerable period ahead. The time has come to call a halt.

The Government are now calling for a six-month standstill on wages, salaries and other types of income, followed by a further six months of severe restraint, and for a similar standstill on prices.

Where a definite commitment already exists to increase pay or reduce hours, its implementation should be deferred for six months. New commitments should not be implemented during the rest of 1966 and in the following six months only if the grounds for exceptional treatment are particularly compelling. In this way it is intended to secure virtual stability in incomes for a period of six months followed by a limited growth of incomes in accordance with national priorities during the first six months of 1967. Thereafter, it will be essential to secure that the growth of incomes is resumed in an orderly manner in step with national output.

The same principles apply to other types of money income. Companies, for example, must hold down their dividends during the 12-month period. The Government similarly call for a 12-month standstill on prices of all goods and services, except to the limited extent that increases are necessitated by increases in the cost of imported materials, by seasonal factors or by the action of the Government, for example through increased taxation.

It is not our intention to introduce elaborate statutory controls over incomes and prices. This is a situation in which the Government look with confidence to everyone concerned with these matters to act in accordance with the public interest. Many individual salaries and other forms of remuneration are fixed outside the normal process of collective bargaining. Here, too, the same canons of restraint must apply. This applies also to emoluments of directors and high executives: companies will be required to publish details of these fees and salaries with comparable figures for the previous year.

Within the main field of collective bargaining we shall rely in the first instance on voluntary action. Nevertheless, in order to ensure that the selfish do not benefit at the expense of those who cooperate, it is our intention to strengthen the provisions of the Prices and Incomes Bill, to speed its passage through Parliament and to redefine the role of the National Board for Prices and Incomes. Meanwhile the Government will not hesitate to act within the powers they enjoy, or may further seek, to deal with any actions involving increases outside and beyond this policy.

The Government will be consulting the T.U.C., C.B.I. and other interested organisations on the detailed application of the standstill within the next few days and a White Paper will be issued in the near future.

The House will not under-rate the deep significance of what I have just announced, its implications for industry and the degree of co-operation and restraint which will be required on the part of those affected by it. But, equally, the House, and, I believe, the country, will recognise the urgency of these measures, if we are to get our economy into balance and to keep our costs under control.

35*

I should not feel justified in making this demand on industry, if I did not feel that we had done everything in our power—

(*Hon Members*) Resign.

(*Mr Speaker*) Order. If the right hon. Member for Hexham (*Mr Rippon*) wishes to intervene, he must do so in a proper manner.

(*The Prime Minister*) I should not feel justified in making this demand on industry, if I did not feel that we had done everything in our power to secure social justice for the first time in the broader fiscal and social policies of the Government. For no Government have the right to ask for restraint, still less for an effective standstill, unless they have done everything a Government can do to create a climate of social justice, which alone can justify such a policy.

Inevitably, today I have dealt with measures which, taken by themselves, involve restriction and restraint. But the House will realise that their whole purpose is to provide industry with the opportunity to achieve a major increase in productivity by streamlining production and labour utilisation. They must be seen against a background of policies designed to speed the application of scientific methods and techniques—already well-known to progressive managers—to increase efficiency in private industry and in the public sector.

Industry by industry, the Economic Development Committees are tackling the practical problems of raising efficiency and spreading knowledge of how performance can be improved in individual companies. Industry by industry—shipbuilding, printing, the docks, rail transport—the Government are engaged in urgent discussions designed to increase productivity and to eliminate overmanning and restrictive practices. We have sought to proceed by voluntary agreement. Where this is not forthcoming, other action must be taken. The Government have indicated to all concerned their determination that the freight liner train services shall go ahead on the basis of open terminals.

The problems with which I have been dealing are problems that have beset Britain's economy virtually since the end of the war. The unsung achievements of keen executives and of hard-working, responsible trade unionists, of inventive scientists and creative designers are all too often overshadowed by attitudes of selfishness and indifference, of indolence and indiscipline on both sides of industry.

For while the Government can and must do all in their power to create conditions to lay down the rules within which the economy must operate, the determination and resolve which today's measures demonstrate must be matched by effort and endeavour on the part of the whole British people.

12 Britain and Europe: A new start?

(*i*) *First approaches to the Common Market, 1962*
(The United Kingdom and the European Economic Community: report by the Lord Privy Seal [Mr Edward Heath] on the meeting with Ministers of Member States of the European Economic Community at Brussels from August 1–5, 1962, *B.P.P.*, 1961–62 XXXVI, pp. 3–5. Over a period of ten years Britain made three applications to join the European Economic Community, 'The Common Market'. The negotiations that followed the third application were successfully concluded in 1971, and Britain joined the Community on 1 January 1973.)

The British Government and the British people have been through a searching debate during the last few years on the subject of their relations with Europe. The result of the debate has been our present application. It was a decision arrived at, not on any narrow or short-term grounds, but as a result of a thorough assessment over a considerable period of the needs of our own country, of Europe and of the Free World as a whole. We recognise it as a great decision, a turning point in our history, and we take it in all seriousness. In saying that we wish to join the E.E.C., we mean that we desire to become full, wholehearted and active members of the European Community in its widest sense and to go forward with you in the building of a new Europe.

Perhaps you will allow me to underline some of the considerations which have determined our course of action. In the first place, ever since the end of the war, we in Britain have had a strong desire to play a full part in the development of European institutions. We, no less than any other European people, were moved by the enthusiasms which gave birth to the Brussels Treaty, the Council of Europe, the O.E.E.C., the Western European Union and the North Atlantic Treaty. These organisations, based on the general principle of co-operation between sovereign states, played an important role in developing amongst us all the practice of working together. They gave us that knowledge of one another's institutions, practices and modes of thought, which is the necessary foundation for common action. Many are the the tables round which we have all sat—round which our officials and experts have sat—during the last 15 years, creating bit by bit the habit of international co-operation and joint action on which our present friendships and understandings are based.

Then there came a point when you decided to move a stride ahead towards a more organic type of unity; and my country, though understanding this move, did not then feel able to take part in it. It is true to say, however, that it was never agreeable to us to find that we were no longer running with the stream toward European unity. There were reasons for it and we knew them; but we did not feel comfortable to be outside. Nor, I believe, did you feel entirely comfortable to see us outside. One of our main purposes today is to discover afresh the inspiration and the stimulus of working together in a new effort of political and economic construction.

˙The second consideration has been the increasing realisation that, in a world where political and economic power is becoming concentrated to such a great extent, a larger European unity has become essential. Faced with the threats which we can all see, Europe must unite or perish. The United Kingdom, being part of Europe, must not stand aside. You may say that we have been slow to see the logic of this. But all who are familiar with our history will understand that the decision was not an easy one. We had to weigh it long and carefully.

In particular, we had to think very deeply about the effect on the Commonwealth of so important a development in United Kingdom policy. I hope you will agree with me that the Commonwealth makes an essential contribution to the strength and stability of the world, and that sound economic foundations and prospects of development go hand in hand with this. We believe that it is in the interests of all of us round this table that nothing should be done which would be likely to damage the essential interests of its Member Countries. Some people in the United Kingdom have been inclined to wonder whether membership of the Community could in fact be reconciled with membership of the Commonwealth. The task of reconciliation is complex, but we are confident that solutions can be found to Commonwealth problems fully compatible with the substance and the spirit of the Treaty of Rome.

The third factor determining our decision has been the remarkable success of your Community and the strides which you have made towards unity in both political and economic fields. This has been in many ways an object lesson. You have shown what can be done in a Community comprising a group of countries with a will to work closely together. Our wish is to take part with you in this bold and imaginative venture; to unite our efforts with yours; and to join in promoting, through the E.E.C., the fullest possible measure of European unity. ...

Her Majesty's Government are ready to subscribe fully to the aims which you have set yourselves. In particular, we accept without qualification the objectives laid down in Articles 2 and 3 of the Treaty of Rome, including the elimination of internal tariffs, a common customs tariff, a common commercial policy, and a common agricultural policy.

(ii) The European Communities Act 1972 (1972 chapter 68)

An Act to make provision in connection with the enlargement of the European Communities to include the United Kingdom ....

In this Act and, except in so far as the context otherwise requires, in any other Act (including any Act of the Parliament of Northern Ireland)—

'the Communities' means the European Economic Community, the European Coal and Steel Community and the European Atomic Energy Community;

'the Treaties' or 'the Community Treaties' means, subject to subsection (3) below, the pre-accession treaties, that is to say, those described in Part I of Schedule 1 to this Act, taken with—

(a) the treaty relating to the accession of the United Kingdom to the European Economic Community and to the European Atomic Energy Community, signed at Brussels on the 22nd January 1972 .... .

All such rights, powers, liabilities, obligations and restrictions from time to time created or arising by or under the Treaties, ... shall be recognised and available in law, and be enforced, allowed and followed accordingly; and the expression 'enforceable Community right' and similar expressions shall be read as referring to one to which this subsection applies. ...

... Her Majesty may by Order in Council, and any designated Minister or department may by regulations, make provision—

(a) for the purpose of implementing any Community obligation of the United Kingdom, or enabling any such obligation to be implemented, or of enabling any rights enjoyed or to be enjoyed by the United Kingdom under or by virtue of the Treaties to be exercised .... .

Index